Eddy,

Welcome on board at

Masayori

Protein Engineering

Applications in Science,
Medicine, and Industry

Protein Engineering

Applications in Science, Medicine, and Industry

Edited by

Masayori Inouye

Raghupathy Sarma

Department of Biochemistry
State University of New York at Stony Brook
Stony Brook, New York

1986

ACADEMIC PRESS, INC.

Harcourt Brace Jovanovich, Publishers

Orlando San Diego New York Austin
Boston London Sydney Tokyo Toronto

ACADEMIC PRESS, INC.
Orlando, Florida 32887

United Kingdom Edition published by
ACADEMIC PRESS INC. (LONDON) LTD.
24–28 Oval Road, London NW1 7DX

Library of Congress Cataloging in Publication Data

Protein engineering.

Includes index.
1. Proteins—Synthesis. 2. Genetic engineering.
3. Proteins—Structure-activity relationships.
I. Inouye, Masayori. II. Sarma, Raghupathy.
[DNLM: 1. Amino Acid Sequence. 2. DNA, Recombinant.
3. Genetic Intervention. 4. Protein Conformation.
QU 55 P96625]
QP551.P6959 1986 660'.63 86-8068
ISBN 0–12–372485–6 (alk. paper)

PRINTED IN THE UNITED STATES OF AMERICA

86 87 88 89 9 8 7 6 5 4 3 2 1

Contents

8 Design and Total Chemical Synthesis of a Gene for Bovine Rhodopsin

Daniel D. Oprian, Michael Nassal, Luca Ferretti, Sadashiva S. Karnik, and H. Gobind Khorana

9 Surface-Simulation Synthesis and Its Applications in Protein Molecular Recognition

M. Zouhair Atassi

II Mutant Analysis

10 Functional Analysis of the Signal Peptide for Protein Secretion with Use of Oligonucleotide-Directed Site-Specific Mutagenesis

Susan Lehnhardt, Sumiko Inouye, and Masayori Inouye

11 Physical Properties of Genetically Defined Synthetic Signal Sequences Suggest Initial Steps in Protein Export

Lila M. Gierasch, Martha S. Briggs, and Donald G. Cornell

12 Studies on the Mechanism of Membrane Fusion

Mary-Jane Gething, Robert W. Doms, Judy White, and Ari Helenius

13 Expression and Site-Specific Mutagenesis of an Integral Membrane Protein, Bacterio-Opsin

Sadashiva S. Karnik, Mark S. Braiman, Betty H. Chao, Robert Dunn, Marie A. Gilles-Gonzalez, Neil R. Hackett, Ernest Jay, K. Anne Kronis, Michael Nassal, and H. Gobind Khorana

IV Applications

21 Enzymatic Reactions in Organic Media

Alexander M. Klibanov

22 Antibody Targeting of Toxin Polypeptides

A. E. Frankel, M. J. Bjorn, and J. L. Winkelhake

23 Production of Novel Antibiotics by Gene Cloning and Protein Engineering

Richard H. Baltz, Jeffrey T. Fayerman, Thomas D. Ingolia, and R. Nagaraja Rao

24 Genetic Transformation of Plants

Ernest G. Jaworski, Robert T. Fraley, Stephen G. Rogers, Robert B. Horsch, Roger N. Beachy, and Nam-Hai Chua

Preface

One of the most exciting developments in molecular biology has been the ability to design and construct new proteins or enzymes with novel functions. Thanks to recombinant DNA technology it is now possible to modify the amino acid sequence of a protein at will, including substitution, deletion, or insertion of amino acid residues, or to fuse two or more proteins to form hybrid proteins, or even to totally synthesize genes for new proteins.

Thus the crucial question is no longer how to construct new protein molecules, but how to design proteins with desired functions. We believe that it is most timely and appropriate to put together various approaches and strategies used by scientists in the field of protein engineering. We have been extremely fortunate in being able to present chapters written by the leaders in this field. We believe that this book will provide new directions for those who are already working on this subject and will serve as a useful guide for those who are about to start research in protein engineering.

Several of the contributing authors participated in the Fourth Stony Brook Symposium on Molecular Biology held in May 1985 on Protein Engineering: Applications in Basic Science, Industry, and Medicine.

We wish to thank Janet Koenig for her invaluable assistance throughout this project.

Stony Brook
July 1986

Masayori Inouye
Raghupathy Sarma

Protein Engineering

Applications in Science, Medicine, and Industry

I

Structure and Design

1

Classical Protein Chemistry in a World of Slicing and Splicing

CHRISTIAN ANFINSEN

Department of Biology
The Johns Hopkins University
Baltimore, Maryland 21218

The rather wistful title of this short lecture gives the impression that the world of protein chemistry is hanging by its teeth to a small corner of biology. The explosive development of DNA recombinant technology has made it possible, in principle at least, to carry out large-scale production of a number of medically and industrially important proteins and peptides. At the same time, new or improved methods in protein chemistry have appeared which should, increasingly, permit the efficient isolation, structure determination, and synthesis of large and small polypeptides. Indeed, the constant advances in solid-phase peptide synthesis and in the stitching together by enzymatic means of peptide fragments makes it quite possible that the insulins and growth hormones and interferons of the future may very well be produced by old-fashioned organic chemical methods brought up to date.

The one point that needs to be made strongly has to do with spatial conformation. We all know that linear structures, especially in larger molecules such as proteins and polynucleotides, have little meaning in biological processes. Function, in the life of the cell, must be described in terms of three-dimensional structure and, for that reason, protein molecules will always be with us. The most intelligent oncogene or messenger RNA is gibberish without translation into polypeptide chain language. At the same time, the protein chemist must agree that DNA recombinant technology has been an enormously valuable catalyst in the unearthing of both normal and pathological biological processes. From the standpoint

PROTEIN ENGINEERING
Applications in Science,
Medicine, and Industry

of sequencing, there appears to be nothing quite as good as the determination of the sequence of the polynucleotide that codes for a particular protein rather than the traditional and laborious procedure of polypeptide digestion, fragment separation, and sequence determination on smaller pieces.

Let me say a few words about the current trends in protein chemistry which should make so-called protein engineering of greater and greater value. The standard methods of protein isolation have been greatly facilitated over the past few years by the development of high-performance liquid chromatography (HPLC) and affinity chromatography. The latter method generally uses columns to which have been bound either substrate analogs or specific antibodies. In many cases, affinity chromatography can lead directly from the crude source of a protein to homogeneous material in a single step. At this point, we may capitalize on still another exciting development in modern protein chemistry—microsequencing, as carried out strikingly, for example, by Hunkapillar and Hood at Cal Tech. Their improvements on the original method of Edman have led to the commercial availability of equipment that even novices can use with considerable success and, most important, on very small samples of material, frequently less than 50 picomoles. Having determined the amino acid sequence of a protein and the nature of any non-amino acid components attached to side chains, we can then turn the protein over to our crystallographic colleagues who employ the sophisticated equipment that can efficiently feed diffraction data into intelligent computers that produce three-dimensional representations with outstanding speed.

Although it is now pretty much old-hat, I would like to spend a few moments discussing the relation between the primary structures of proteins and their three-dimensional structures. I think that most of us are, by this time, convinced that three-dimensional structure is completely determined by the primary sequence and that the folding process proceeds spontaneously. Let me show a few illustrations to support this fairly obvious conclusion. Figure 1 shows a two-dimensional version of the ribonuclease molecule with its four disulfide bonds. Following total reduction of these SS bridges, one can remove reagents for reduction and allow the chain to fold up and form SS bonds from the eight SH groups. Random pairing would yield 105 isomers of RNase. Reactivation does indeed take place, but in an uncatalyzed solution being exposed to air the length of time required for total folding and full activity can be as long as 24 hr. We discovered that this sluggishness was due in large part to incorrect pairing of SH groups to form what might be called scrambled molecules, but that there existed in the tissues of the body an enzyme (which we named disulfide interchange enzyme) that could correct the

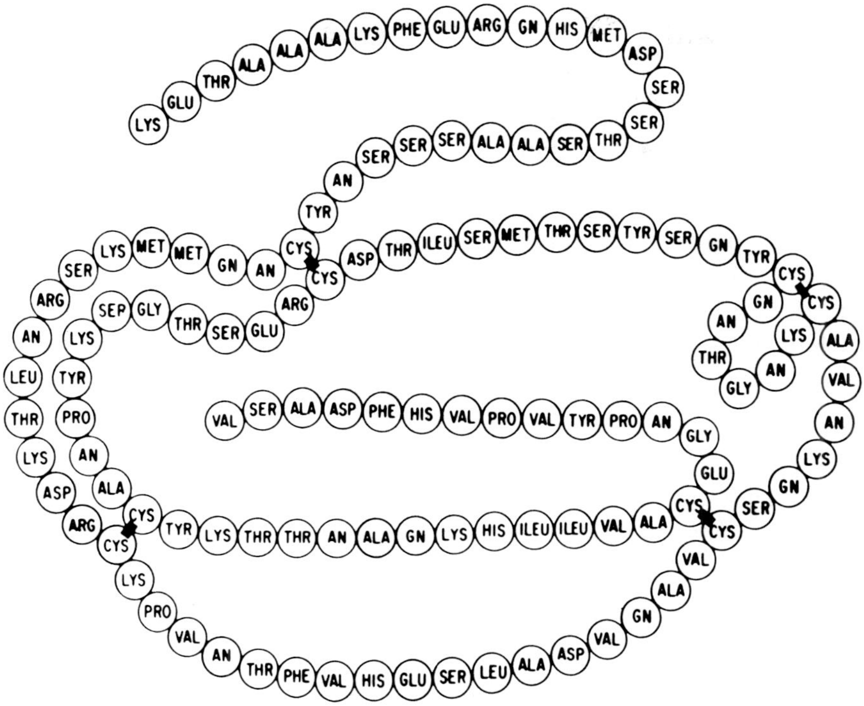

Fig. 1. Amino acid sequence of bovine pancreatic ribonuclease.

wrong SS bridges and, by reshuffling, yield the native form with the correct bridges (1). This enzyme, incidentally, is located in largest quantities in the endoplasmic reticulum of cell types associated with active synthesis and secretion of proteins and large polypeptides that contain a good deal of cysteine—such tissues as thyroid, liver, and pancreas. Using this enzyme, which is not too difficult to purify, it is possible to show that the refolding of a solution of reduced ribonuclease to yield the correct structure with full activity was completed in about 2 min, which, we now know, is about the length of time required for ribonuclease to be synthesized *in vivo*.

So much for the conversion of one-dimensional to three-dimensional structure in proteins. The process as I have described it for ribonuclease is characteristic, as far as we know, of most proteins that have been studied except "derivative" proteins such as insulin and chymotrypsin, which are multichained and which do not fold into the proper SS-linked structures but can only fold correctly when in the form of the single chains, proinsulin and chymotrypsinogen.

I would like now to say a few words about some aspects of protein chemistry that might be referred to as "protein engineering." I will use examples that are familiar to me but are, I believe, representative of many other situations. Most of our work along these lines was done with the staphlyococcal enzyme staphylococcal nuclease (2) which is a single chain and has no SS bridges and is, consequently, a very labile and easily denatured and refoldable substance with a half-time of refolding of about 250 msec. The chain sequence is shown in Fig. 2. Let me point out two areas of the molecule containing the sequence lysyl-lysine at residues 5–6 and residues 48–49. This molecule is quite motile in solution, but when calcium and deoxythymidine diphosphate are added to the solution, these two ligands attach firmly to the protein and greatly decrease its motility. Even more important, they make the molecule unsusceptible to trypsin attack except at the two lysyl-lysine positions. Figure 3 shows what occurs when nuclease, in the presence of the two stabilizing ligands, is exposed to trypsin (3). The molecule loses its amino terminus by a cleavage between positions 5 and 6 and is also cleaved either between 49 and 50 or between 48 and 49. The two pieces produced have no particular structure in solution and are individually inactive. However, when the 6-to-49 fragment is mixed with the 50-to-149 fragment, activity is regenerated to the level of about 10% of the native activity, and the two-fragment complex (named nuclease T), crystallographically isomorphous with native nuclease, is resistant to trypsin as long as the ligands are present.

For several years, experiments were carried out on the synthesis of one or both of these large fragments. The main effort was with the fragment 6 to 49 rather than the rather large fragment 50 to 149 (which also happens to contain a tryptophan residue, making life more difficult for the peptide chemist). A large number of analogs of the nuclease T molecule were prepared by standard or slightly modified solid-phase synthesis (4). It could be determined, for example, that the calcium atom which is essential for activity is complexed with four particular carboxyl groups. These are essential since synthetic replacement with an amide or with some neutral amino acid side chain destroys the calcium binding and the enzyme activity.

When it was clear that a peptide the size of 6 to 49 could be made in a relatively routine way in quite pure form, it was realized that the connection of this synthetic peptide to the natural 50–149 fragment would reconstitute a nearly complete nuclease molecule and might serve as the basis of some protein engineering involving deliberate replacement or deletion of residues in the synthetic portion.

It was shown several years ago by Homandberg and Laskowski (5,6) that peptides that had been cleaved with trypsin could be rejoined by

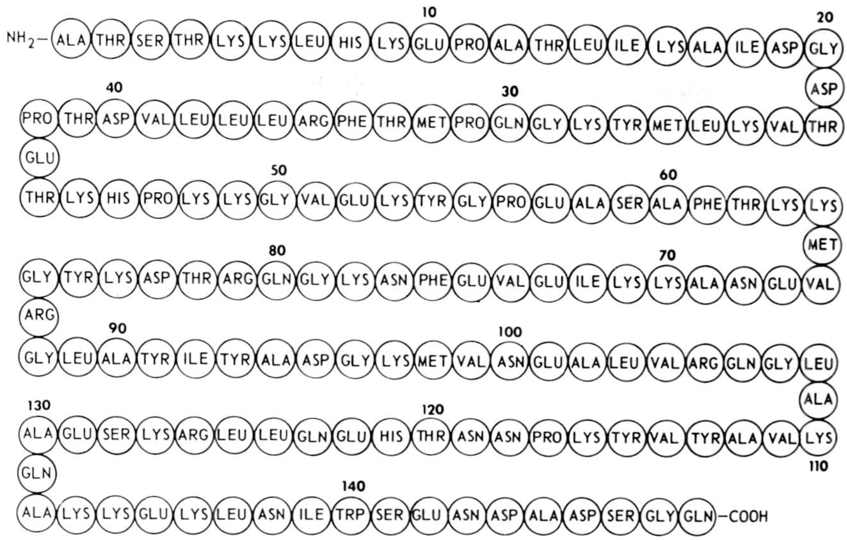

Fig. 2. Amino acid sequence of staphylococcal nuclease.

incubating the two fragments with trypsin in the presence of high levels of nonaqueous solvents such as glycerol. The presence of 90% glycerol altered the equilibrium constant in such a way that rejoining of the amino terminal fragment to the carboxyl terminal fragment was favored, and the original polypeptide could be regenerated. This relatively simple method was applied by Akira Komoriya, Homandberg, and Chaiken to the case of nuclease T. They found that incubation of the fragment of nuclease from residue 6 to 49 with the rest of the nuclease molecule between lysine 50 and the C-terminal 149 with trypsin in 90% glycerol did indeed yield a reconstituted chain. However, this reconstituted "protein" was inactive. The reason for this was that lysine 49 was lost during the trypsin reconstitution step. (I should emphasize that these incubations must be carried out in the presence of the ligands calcium and thymidine diphosphate so that any formed material will be protected from indiscriminate trypsin

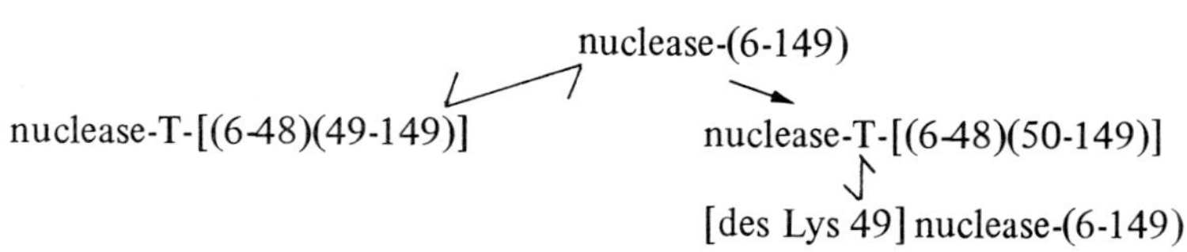

Fig. 3. Products of cleavage of staphylococcal nuclease by trypsin in the presence of calcium ions and 3′,5′-deoxythymidine diphosphate.

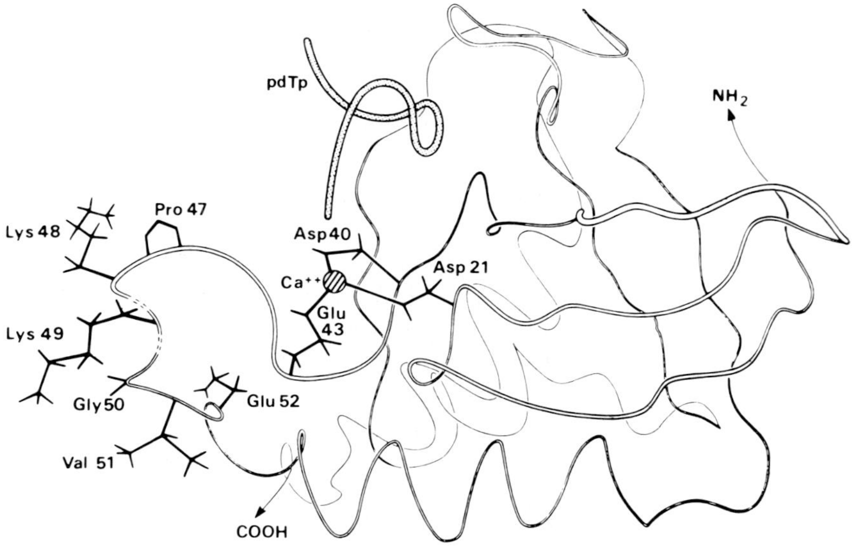

Fig. 4. Schematic representation of the backbone of staphylococcal nuclease emphasizing the "loop" of the left of the figure.

cleavage.) Thus, a model of the nuclease chain could be formed but was inactive because of the deletion of lysine 49(7). However, by synthesis of the 6–49 sequence with Lys 48–Lys 49 replaced by Gly 48–Lys 49, loss of Lys 49 was prevented and fully active nuclease analog could be formed by tryptic attachment to native (50–149) (Fig. 4). It would appear that maintenance of a correct *length* in the loop that contains the (48–50) sequence is critical for formation of a competent active site but that the changed group on Lys 48 is not (8).

These synthetic experiments are quite clear-cut and could potentially give us a great deal of information about the function and three-dimensional features of this and other proteins. However, synthesis of long peptides is clearly tedious and may not be as useful in the engineering of other proteins with less favorable sequence and conformation. It is for this reason that I find myself enormously intrigued by the kind of work now going on in David Shortle's laboratory (see Shortle, Chapter 14, this volume). He and his colleagues are able to introduce point mutations leading to the modification of more than 90 different residues in the chain of 149 residues in staphylococcal nuclease. Such site-specific mutagenesis will certainly become a major tool in protein chemistry.

I would like, finally, to spend a short time in the consideration of the potential for synthesizing totally synthetic vaccines. This kind of work is now popular in quite a few laboratories and promises to be an important feature in the management of infectious diseases. My own exposure to this area of peptide chemistry began with some work in the Weizmann Institute of Science in which I collaborated with Ruth Arnon and Michael Sela. These studies were concerned with the immunogenicity of a small portion of the lysozyme molecule—a small disulfide-stabilized loop on the exterior of the protein (Fig. 5) (9). When this loop was prepared syntheti-

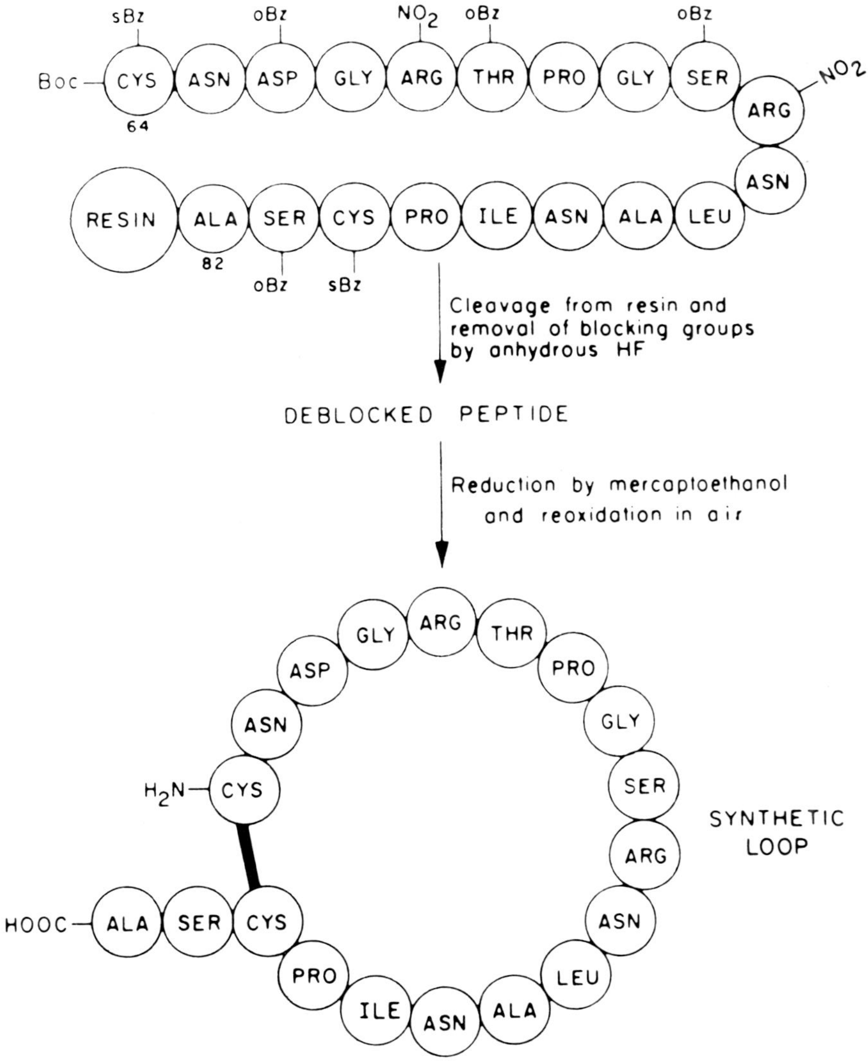

Fig. 5. Synthesis of a loop of amino acids as the structure occurs in egg white lysozyme.

cally and attached to a nonimmunogenic carrier, injection into animals produced antibodies which indeed recognized the parent enzyme molecule. One might say that this material constituted a synthetic vaccine against the lysozyme molecule. As I mentioned, this kind of work is now going on in a number of laboratories—I might mention the Scripps Clinic in La Jolla, the Weizmann Institute in Rehovot, Dr. Atassi's laboratory at the Texas Medical Center, and studies at the National Institutes of Health.

A basic question that one must ask in this kind of work is how small a piece of a protein, forming an antigenic site, can fold up into the conformation that it assumes when it is part of the native protein, before its excision by some enzymatic means. It certainly is clear that a bit of peptide carved out of the center of a protein would be useless since it is not normally exposed to the antibody-forming machinery of the organism. Some years ago, Dr. David Sachs, Anne Eastlake, Alan Schecter, and I

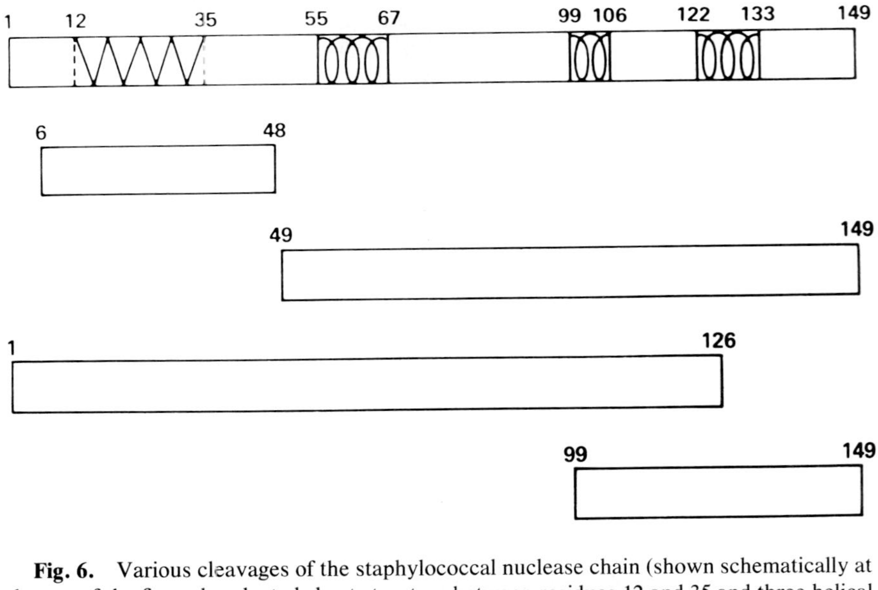

Fig. 6. Various cleavages of the staphylococcal nuclease chain (shown schematically at the top of the figure by pleated sheet structure between residues 12 and 35 and three helical structures at other points in the molecule). The second block represents the fragment produced by trypsin digestion of staphylococcal nuclease in the presence of tightly binding inhibitory ligands. The next column is the second half of the molecule produced by trypsin digestion. The fragment from residue 1 to 126 is produced by trypsin digestion of a trifluoroacetyl derivative of the staphylococcal nuclease molecule. The final peptide at the bottom of the figure is produced by cyanogen bromide cleavage of the molecule, residue 98 being a methionine residue.

(10) began to attack this problem, using, once again, the staphylococcal nuclease polypeptide chain as a working material. The chain of staphylococcal nuclease can be cleaved into a number of convenient fragments by a combination of tryptic digestion and cyanide bromide cleavage (Fig. 6). I will concentrate here on one particular portion of the chain, namely the part consisting of residue 99 to residue 149, a product of CNBr cleavage. We found it necessary to first fractionate antiserum prepared against native nuclease by affinity chromatographic methods (Fig. 7). The multi-

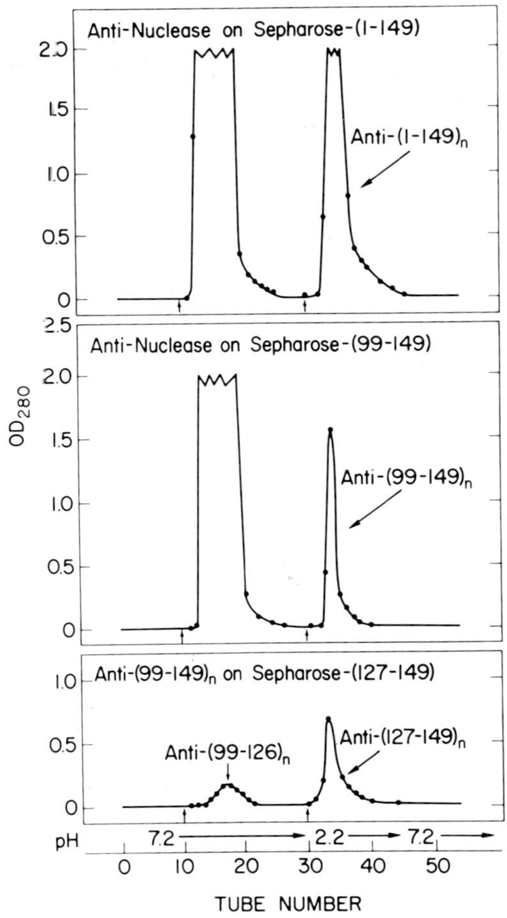

Fig. 7. Fractionation of antinuclease antibodies by successive absorption of a multicomponent mixture of antibodies against native nuclease by successive passage through three columns to yield anti-$(99–126)_n$.

component mixture of antibodies in the total serum of an inoculated test animal was first passed through a Sepharose–nuclease affinity column and thus a large amount of nonimmunogenic material could be washed away. The part that stuck was then passed through a second column, to which the peptide (99–149) had been attached. This peptide, presumably in an equilibrium between "native" and "denatured" forms, bound the fraction of the antibody which recognized the (99–149) portion. The antibodies retained by the column, following elution, were obviously polydeterminant because they precipitated when added to native nuclease and thus contained more than two antigenic determinants. This multivalent anti-(99–149) fraction was then subsequently passed through a third column in which the fragment (127 to 149) had been attached. This column bound antibodies recognizing the terminal portion of the nuclease chain, allowing the antibody specific for (99 to 126) to pass through. This final antibody fraction against a relatively limited portion (Fig. 8) of the chain—(I should emphasize, in its native form)—inactivated nuclease but was nonprecipitating. That is to say, a population of antibodies had been prepared that was monodeterminant, equivalent to a monoclonal preparation. This anti-(99–126) material was, as I mentioned, inhibitory but nonprecipitating. By adding incremental competing amounts of the peptide (99–149), release of inhibition by the antibody was observed, and the freed native nuclease could be assayed by activity measurements. By an analysis of

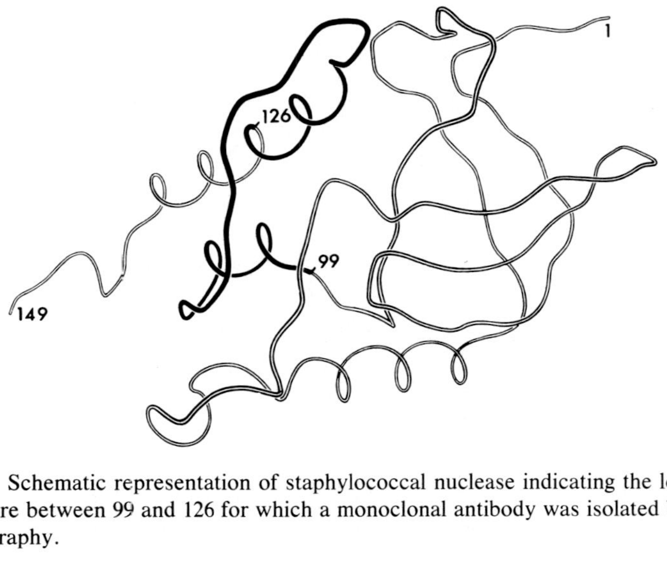

Fig. 8. Schematic representation of staphylococcal nuclease indicating the location of the structure between 99 and 126 for which a monoclonal antibody was isolated by affinity chromatography.

the stoichiometry between antibody, nuclease, and competing fragment, one could calculate that the random (99–149) peptide was able to assume a native conformation about 0.02% of the time.

In closing, I must state my own strong feelings about the likelihood that this kind of approach—that is, the use of synthetic protein fragments to study the biology of various substances such as growth hormones, cell receptors, and oncogene products—will be of increasing value in the burgeoning effort and success of the current collaboration between genetic manipulation and protein chemistry.

REFERENCES

1. Givol, D., DeLorenzo, F., Goldberger, F. F., and Anfinsen, C. B. Disulfide interchange and the three-dimensional structure of proteins. *Proc. Natl. Acad. Sci. U.S.A.* **53,** 676–684 (1965).
2. Taniuchi, H., Anfinsen, C. B., and Sodja, A. The amino acid sequence of an extracellular nuclease of *Staphylococcus aureus,* V8. III. Complete amino acid sequence. *J. Biol. Chem.* **242,** 4752–4758 (1967).
3. Taniuchi, H., Anfinsen, C. B., and Sodja, A. Nuclease-T: an active derivative of staphylococcal nuclease composed of two noncovalently bonded peptide fragments. *Proc. Natl. Acad. Sci. U S.A.* **58,** 1235–1242 (1967).
4. Chaiken, I. M., and Anfinsen, C. B. A solid phase synthetic study of the active site region of staphylococcal nuclease-T *J. Biol. Chem.* **246,** 2285–2290 (1971).
5. Homandberg, G. A., and Laskowski, M., Jr. *Biochemistry* **18,** 586–592 (1979).
6. Homandberg, G. A., Mattis, J. A., and Laskowski, M., Jr. *Biochemistry* **17,** 5220–5227 (1978).
7. Homandberg, G. A., and Chaiken, I. M. Trypsin-catalyzed conversion of staphylococcal nuclease-T fragment complexes to covalent forms. *J. Biol. Chem.* **255,** 4903–4904 (1980).
8. Komoriya, A., Homandberg, G. A., and Chaiken, I. M. Enzyme-catalyzed formation of semisynthetic staphylococcal nuclease using a new synthetic fragment, 48-glycine synthetic-(6–49). *Int. J. Pept. Protein Res.* **16,** 433–439 (1980).
9. Arnon, R., Maron, E., Sela, M., and Anfinsen, C. B. Antibodies reactive with native lysozyme elicited by a completely synthetic antigen. *Proc. Natl. Acad. Sci. U.S.A.* **68,** 1450–1455 (1971).
10. Sachs, D. H., Schechter, A., Eastlake, A., and Anfinsen, C. B. Antibodies to a distinct antigenic determinant of staphylococcal nuclease. *J. Immunol.* **109,** 1300–1310 (1972).

2

Protein Sequence Data Banks: The Continuing Search for Related Structures

R. F. DOOLITTLE

Department of Chemistry
University of California, San Diego
La Jolla, California 92093

I. INTRODUCTION

The number of new protein sequences being reported is reaching staggering proportions, as many as 4000 now having been published. As might be expected, the number of reports in which a new sequence resembles one already known is also growing. Even allowing for a certain amount of redundancy for the same sequence from different species, a substantial fraction of the world's existing proteins have now been cataloged, and the task of reconstructing their evolutionary history can be begun in earnest. Many of these new sequences are the result of DNA sequencing, and often the proteins themselves are known only through their inferred amino acid sequences. That being the case, one of the most revealing aspects of a new sequence may lie in its relationship to one from some other better characterized protein. In any event, it is essential that every new sequence be searched against an appropriate data bank to find out whether it resembles an already known sequence.

What is required is access to a sequence bank, suitable programs for searching the data, and a basis for deciding what constitutes a significant similarity. Of the three, the last is the most elusive. At this point researchers worldwide have access to several well-maintained data banks. Most of these also offer excellent utility programs for searching sequences, mak-

PROTEIN ENGINEERING
Applications in Science,
Medicine, and Industry

ing alignments, and providing certain statistical evaluations. Not all are in a position to offer final judgment about marginal relationships, however; or, if they are, they naturally choose to err on the conservative side. Casual users, on the other hand, often attribute more significance to similar sequences than is deserved. There are several pitfalls that can catch the unwary, including overzealous "gapping," the promiscuous "boxing" of arbitrarily chosen "similar" amino acids, and the comparison of short segments where statistical proofs are difficult. In this chapter I confront some of these problems and offer some guidelines for assessing significance, particularly when an occasion demands the comparison of short segments. The arithmetic of comparisons is such that it is relatively easy for short sequences to be similar merely by chance. Conversely, the longer a similarity extends in a sequence, the more likely that it is not due to chance alone (Fig. 1). All the same, there are pressing reasons for comparing short sequences. Sometimes a short, incomplete sequence is all that is available. In other cases, short segments of a gene may have been exchanged in the course of evolution, as has recently been observed in the case of shuffled exons in numerous modern proteins (for a review, see Doolittle, 1985). Finally, in the case of ancient proteins particularly, brief outcroppings of similarity may be all that endure, the remainder of sequence resemblance having been eroded by numerous amino acid replacements over the eons.

In those situations where one is forced to deal with relatively short sequences, proofs of relationship often must come from experiments that demonstrate commonality of function. For example, melanoma tumor antigen was isolated by a Seattle group several years ago, but in such small amounts that a sequence determination with a microsequencer was only able to proceed through 13 cycles from the amino terminus, and even then it was only possible to identify 10 of the 13 residues with any assurance. Nonetheless, the fragmentary sequence was searched against a data base, and only one protein sequence was retrieved: human transferrin. The match amounted to only 6 identities among the 10 certain residues, and, although tantalizing, the similarity could easily have come about by chance, even considering that the matched regions both occurred at the amino terminus. The relationship was confirmed, however, when a sample of melanoma tumor antigen was found to bind radioactive iron with an avidity characteristic of transferrin (Brown *et al.*, 1982). Since then the protein has been cloned and the findings completely verified (Rose *et al.*, 1986). The point to be made is that searching even a very short sequence may be very rewarding, but some form of experimental verification is desirable before constructing too much of a story.

On the other hand, when sequences are sufficiently long, even low-level

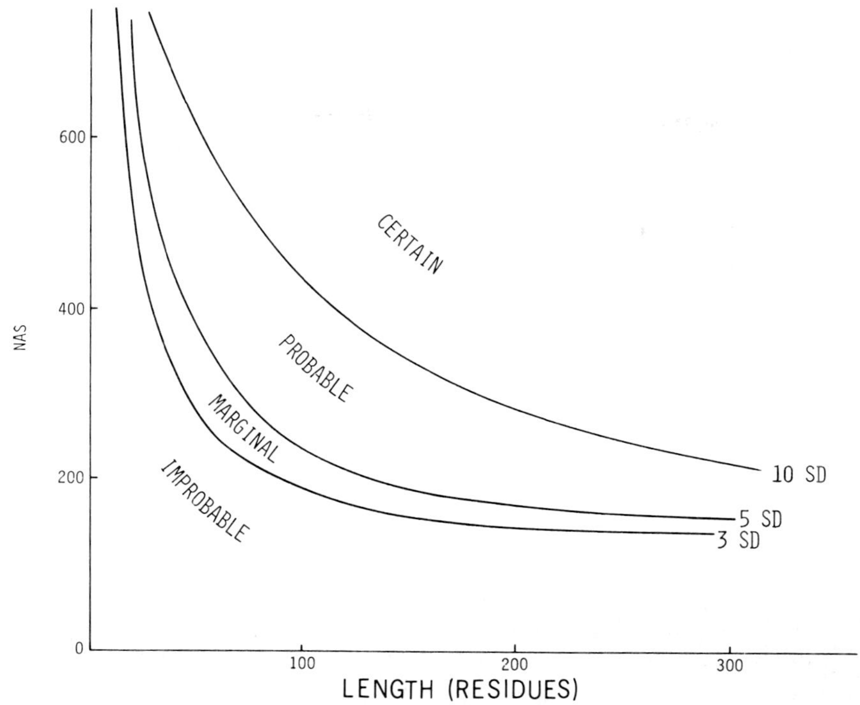

Fig. 1. Relationships between sequence length (abscissa) and significance of alignment scores (ordinate). NAS, Normalized alignment score; sequence length is the average of the two sequences compared. The NAS is calculated by multiplying the number of matches by 10, subtracting 25 for every gap, and normalizing the result to 100 residues; as such it amounts to an adjusted percent identity × 10. The significance in standard deviations is taken from a compilation of a large number of real sequences that have been subjected to randomization testing (Doolittle, 1981; Feng *et al.,* 1985).

resemblances can be detected and shown to be significant. For example, a computer search of rat angiotensinogen suggested that it might be related to certain protease inhibitors found in vertebrate blood plasma (Doolittle, 1983). The identity for these sequences was less than 20%, even after suitable gapping, but because the proteins involved were more than 400 residues in length, the relationship was statistically very significant. Still, it was reassuring when further experimental studies revealed that the intron distributions in these distantly related proteins were virtually identical (Tanaka *et al.,* 1984), removing any doubt that the similarity was a statistical fluke. The fact that exon shuffling had not been involved in the evolution of these proteins also helped to make the interpretation straightforward.

II. COMPUTER SEARCHING METHODS

Different search routines are employed depending on the circumstances. For sequences as short as the melanoma tumor antigen, a simple sliding segment search was employed in which the condition to be met was merely that some number of residues *x* had to be identical in a moving window *y* residues long. For the routine searching of longer sequences it is desirable to use a scheme that employs a weighted scale that takes into account similar amino acids. The scale most often employed is the Mutation Data Matrix developed by Dayhoff (1978) from a compilation of amino acid replacements in closely related proteins. Occasionally, also, a search is conducted for a specific sequence, such as an active site or a consensus sequence derived from parts of proteins presumed to function in the same way, and in these cases a simple identity program is used.

III. NUCLEOTIDE BINDING SEQUENCES

In 1974 Michael Rossmann, on the basis of X-ray crystallographic results from several laboratories, reported that a number of dehydrogenase enzymes share a similar folding pattern in the regions of the proteins that bind nucleotides (Rossmann *et al.*, 1974). The same structural motif was observed in flavodoxin and, less strikingly, in a kinase (Schulz and Schirmer, 1974). Rossmann's observation was of considerable interest and led to much meaningful conjecture about protein evolution. His bold proposal was that the nucleotide-binding fold was invented once and used repeatedly in the concoction of a variety of proteins.

Most of these "folds" are composed of segments that are only 40–50 amino acids long, however, and the sequence resemblances between them are slight. As such, there was initially little hope that the observation could be extended to other proteins on the basis of amino acid sequence data alone, or even without benefit of a crystal structure. Recently, however, a number of reports have appeared which claim to have identified a set of nucleotide-binding segments on the basis of amino acid sequence resemblances. The latter provide a good exercise for the problem of comparing short sequences and warrant our close scrutiny. If, as seems reasonable, most contemporary proteins have descended from a very small number of starter sequences, then resemblances of this sort could furnish a trail back to the most ancient proteins.

The roots of this extended sequence-based study can be traced to Walker *et al.* (1982), who, having determined the sequences of the ATP synthases from *Escherichia coli* and bovine mitochondria, found them to have been highly conserved during the course of evolution. At the same time they scrutinized the sequences of several other proteins that interact with ATP and noted a number of similarities with the ATP synthases. Their original list included—in addition to the α and β subunits of *E. coli* ATP synthase and the β subunit of the bovine mitochondrial enzyme—myosin, adenylate kinase, and the Rec A protein of *E. coli* (Fig. 2). Two sets of segments were compared, each of the order of 30 residues in length. Shortly thereafter, Gay and Walker (1983) reported an even more intriguing relationship: one of the alleged ATP-binding segments from the ATP synthases also resembled a section of the potentially oncogenic protein known as p21 ras, a protein reported to bind GTP. Moreover, the ras segment involved a critical glycine at position 12, replacement of which was thought to render the protein oncogenic. The resemblance amounted to only eight identities over the course of a 35-residue segment, however, and by any ordinary statistical criterion based on the sequences alone was hardly overwhelming. Still, the functional connection of nucleotide binding was enticing and the importance so great that the matter was obviously worth pursuing.

And pursued it was. Comparable levels of similarity were soon found for the ras protein and the elongation factors of the protein biosynthetic machinery (Halliday, 1984). The elongation factors are well known to bind GTP, and they are also known to be related to initiation factors (Sacerdot *et al.,* 1984) that also take part in protein biosynthesis. The list was subsequently expanded to include DNA helicase II from *E. coli* and two prokaryotic DNA-binding proteins (Finch and Emmerson, 1984). In still another case, a computer search of a membrane protein of unknown function from *E. coli* showed it to be an obvious relative of elongation factors. Following in the best tradition of computer searching as a guide, the authors promptly performed the experiment to find out whether the protein bound GTP; it did (March and Inouye, 1985). Resemblances were also found between the elongation factor proteins, ras, and the transducin known as the photoreceptor G protein (Leberman and Egner, 1984; Lochrie *et al.,* 1985) and among some prokaryotic transport proteins (Hobson *et al.,* 1984; Higgins *et al.,* 1985). Viral thymidine kinases were implicated, first by Finch and Emmerson (1984) and then by Gentry (1985).

The network was extending—perhaps overextending. How credible are these findings? Do the similarities imply some ancient common ancestry? Has convergence played a role in shaping these similar sequences? Indeed, how similar are these segments? Can any of the similarity be due to chance? Are any of these segments related to the mononucleotide fold first described by Rossmann?

To examine these questions I took the putative consensus sequence

protein (source)	sequence			*		G*
ATPase, alpha (E. coli)	Q R E L I I	G	D R Q T	G K T	A L A I D A	180
ATPase, beta (E. coli)	G K V G L F	G	G A G V	G K T	V N M M E L	155
ATPase, beta (bovine)	G K I G L F	G	G A G V	G K T	V F I M E L	161
Rho protein (E. coli)	Q R G L I V	A	P P K A	G K T	M L L Q N I	184
Elongation factor G (E. coli)	R N I G I S	A	H I D A	G K T	T E R I L F	21
Elongation factor Tu (E. coli)	V N V G T I	G	H V D H	G K T	T L T A A I	23
Initiation factor 2B (E. coli)	P V V T I M	G	H V D H	G K T	S L L D Y I	246
Lep A protein (E. coli)	R N F S I I	A	H I D H	G K S	T L S D R I	19
Elongation factor (yeast)	I N V V V I	G	H V D S	G K S	T T T G H L	19
Rho protein, Ras-like (aplysia)	K K L V I V	G	D G A C	G K T	C L L I V F	12
Ras protein (yeast)	Y K I V V V	G	G G G V	G K S	A L T I Q F	22
Ras (drosophila)	Y K L V V V	G	P G G V	G K S	A L T I Q L	15
Ras, p21 (human)	Y K L V V V	G	A V G V	G K S	A L T I Q N	15
Photoreceptor, alpha (bovine)	V K L L L L	G	A G E S	G K S	T I V K Q M	45
Oligopeptide transport (Salmonella)	E T L G I V	G	E S G S	G K S	Q S R L R L	59
Maltose transport (E. coli)	E F V V F V	G	P S G C	G K S	T L L R M I	41
Histidine transport P (Salmonella)	D V I S I I	G	S S G S	G K S	T F L R C I	14
Rec A (helicase) (E. coli)	R I V E I Y	G	P E S S	G K T	T L T L Q V	87
Helicase II (E. coli)	S N L L V L	A	G A G S	G K T	R V L V H R	35
DNA-Binding protein (E. coli)	D L I I V A	A	R P S M	G K T	T F A M N L	235
Thymidine kinase (Herpes simplex)	L R V Y I D	G	P H G M	G K T	T T T Q L L	61
Myosin (nematode)	Q S M L I T	G	E S G A	G K T	E N T K K V	181

Fig. 2. Compilation of 22 peptide segments reported to be involved in nucleotide binding (see text for literature citations). Whether or not these particular segments are involved, the parent proteins in each case are known to bind nucleotides. The right-hand column denoted G* identifies the position in the native sequence of the glycine that is part of the characteristic Gly-Lys-Ser or Gly-Lys-Thr tripeptide.

from a compilation of the reportedly similar sequences (Fig. 2) and searched the 1985 version of the NEWAT data bank (Doolittle, 1981) for similarities. The aim was to see if other candidates for nucleotide binding exist. Such a search also serves as an index of reliability of the consensus sequence. If too many proteins are found that are known *not* to bind nucleotides, then one's confidence in the consensus sequence must be lowered.

The consensus sequence focused on the tripeptides G-K-T and G-K-S (Gly-Lys-Thr and Gly-Lys-Ser) that are found in all the alleged nucleotide-binding segments noted in Fig. 2. The other constraint imposed was that the amino acid 5 residues toward the amino terminus had to be G or A (Gly or Ala). At the time of the search the data base contained about 300,000 residues, and the known frequency of occurrence for the amino acids involved allowed the number of sequences that ought to be generated in a random collection to be calculated. Thus, it was anticipated that there should be about 25 segments meeting the consensus condition; actually, more than 50 emerged (in a few cases, including the Rec A protein

protein (source)	sequence					G*
				*		
Large T antigen (polyoma)	R N I L F R	G	P V N S	G K T	G L A A A L	581
Large T antigen (SV40)	R Y W L F K	G	P I D S	G K T	T L A A A L	431
DNA-bound protein (parvovirus)	N T V L F H	G	P A S T	G K S	I I A Q A I	404
Adenovirus 5, IVA2	V I G V I Y	G	P T G C	G K S	Q L L R N L	171
Subunit 3 (hepatitis virus)	V V C Y L Y	G	K R G G	G K S	L T S I A L	155
Protein 8 (encephalitis virus)	V V I V L R	G	D A G Q	G K S	L S S Q V I	124
Protein X (polio virus)	V C L L V H	G	S P G T	G K S	V A T N L I	134
Protein 7 (rhino virus)	V C V L I H	G	T P G S	G K S	L T T S I V	135
Histidine transport Q (Salmonella)	L E D V V K	A	T Q L A	G K S	T W E P F Y	184
Ribose transport (E. coli)	A L G A L R	A	L Q T A	G K S	D V M V V G	230
Alkaline phosphatase (E.coli)	F A E T A T	A	G E W Q	G K T	L R E Q A Q	222
beta-Galactosidase (E. coli)	L I T T A H	A	W Q H Q	G K T	L F I S R K	844
p-Aminobenzoate synthetase (E. coli)	M A Q A F G	A	K V M H	G K T	S P I T H N	102
Rec F endonuclease (E. coli)	G F N F L V	G	A N G S	G K T	S V L E A I	105
Reaction center light chain (R. cap.)	A L V L S A	A	N P V K	G K T	M R T P D H	174
Nitrogenase reductase (cyanobacter)	R Q I A F Y	G	K G G I	G K S	T T S Q N T	18
Phycocyanin c, alpha (cyanobacter)	T S G P N Y	A	A D A R	G K S	K C A R D I	80
Enterotoxin (S. aureus)	S S K D N V	G	K V T G	G K T	C M Y G G I	107
Flavodoxin (azotobacter)	F L P K I E	G	L D F S	G K T	V A L F G L	88
alpha-Sarcin cytotoxin (aspergillus)	E S N S H H	A	P L S D	G K T	G S S Y P H	42
Cytochrome c (brandling worm)	G G I P A	G	D V E K	G K T	I F K Q R C	11
Rhodopsin (drosophila)	C P C C V F	G	K V V D	G K S	S D A Q S Q	355
Myosin (nematode)(2nd segment)	A K E G G G	G	G K K K	G K S	G S F M T V	648
Erb A oncogene (avian virus)	L G V V S D	A	I F D L	G K S	L S A F N L	344
Hemagglutinin (influenza virus)	F I A P N R	A	S F L R	G K S	M G I Q S D	236
Factor VIII (human)	K F I L L F	A	V F D E	G K S	W H S E T K	224
Complement C3 beta (mouse)	G V R P S N	A	D A L V	G K S	L Y V S V T	298
Lactate dehydrogenase H (pig)	Q V G M A C	A	I S I L	G K S	L T D E L A	41
Chymotrypsinogen B (bovine)	P A G M L C	A	T T G W	G K T	K Y N A L K	142
Pancreatic lipase (pig)	P Q M G H Y	A	D R F P	G K T	N G V S Q Y	316

Fig. 3. Compilation of 30 additional segments identified by searching the NEWAT data base with the putative consensus sequence for nucleotide binding. Most of these are not likely to be nucleotide binders, even though a number of them have sequences that closely resembly those in Fig. 2. Several, however, are known to interact with nucleotides or oligonucleotides.

and an invertebrate myosin, sequences met the consensus condition more than once). Ordinarily, one would expect the excess to be largely a reflection of redundancy from related sequences. In this regard, the retrieved sequences included the 22 sequences expected on the basis of published reports (Fig. 2) and an additional 30 not previously considered (Fig. 3).

Are any of the newly identified sequences nucleotide binders? Closer examination of the segments from the authentic nucleotide binders (Fig. 2) reveals that they have more in common than the bare-bones consensus sequence used in the search. For example, most of them have sets of nonpolar residues flanking the central region. It is important, then, that the new candidates be examined with regard to this property also. Casual inspection makes it appear that a good number of them qualify.

How may of the "new" sequences are likely *not* to bind nucleotides? Several must fall into this category. Chymotrypsinogen B, for example, ought to be excludable if only because its close relative chymotrypsinogen A does not meet the conditions and was not retrieved. A similar argument can be made against some others on the list, including the solitary cytochrome *c*.

But some of the others on the list are tantalizing, including several that are known to interact with nucleotides or nucleic acids. Still, the segments in Figs. 2 and 3 are only 20 residues long. Even if one resorts to a resemblance matrix for weighting similar but nonidentical amino acids, it is difficult to validate the relationships. Accordingly, I subjected the full sequences of all the proteins listed in Figs. 2 and 3 to a cross-search in an effort to find further similarities among them. The primary objective was to distinguish those sequences in which the resemblance was limited to the putative nucleotide-binding segment from those in which there were deep-seated resemblances elsewhere in the protein. For this purpose, a search routine was employed that compared all segments of each of the proteins and scored resemblances with the Mutation Data Matrix (Dayhoff, 1978). The stringency could be adjusted and the best matches printed regardless of significance. Those pairs of proteins that survived this secondary searching—which is to say there was evidence for matching segments elsewhere in their sequences—were then subjected to standard alignment and evaluation procedures in order that a degree of confidence could be expressed about the similarity in standard deviations (see, for example, Feng *et al.*, 1985).

The cross-search immediately allowed many of the sequences listed in Figs. 2 and 3 to be grouped into families. Assignment to a family implies that a protein shares a common ancestor with the other members of that family; it is fundamentally different from merely having a similar domain as part of an otherwise different structure. Thus, the "elongation factor family" included the *E. coli* EF-G and Tu proteins, the common ancestry of which had been recognized long ago (Laursen and Duffy, 1978), as well as the more recently characterized *E. coli* initiation factor (Sacerdot *et al.*, 1984) and the elongation factor from yeast (Schirmaier and Phillippsen, 1984; Cottrelle *et al.*, 1985). This group also contains the lep A gene product, which, as noted above, is now known to bind GTP (March and Inouye, 1985). The three prokaryotic transport proteins listed in Fig. 2 also constituted a readily identifiable cluster, as did the four ras proteins (from yeast, aplysia, drosophila, and human). The ATP synthases from *E. coli* and bovine mitochondria were clearly homologous, and, interestingly, so was the *E. coli* rho terminator protein. It is known that this protein has ATPase activity (Pinkham and Platt, 1983), but it was surpris-

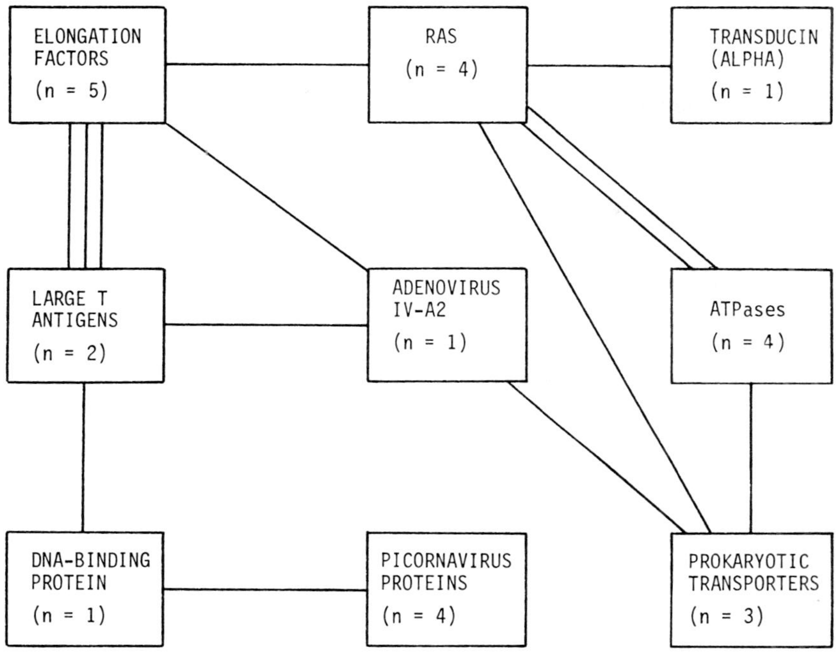

Fig. 4. Network of nine families involving 25 sequences that contain the consensus sequence thought to be indicative of nucleotide binding. Each connecting line represents the best match of a sequence to one outside its own family grouping. The *E. coli* rho terminator protein has been included in the ATPase family.

ing all the same to find that in this case, unlike many of the other resemblances reported, the sequence similarity extends throughout the full lengths of the proteins. A full statistical analysis of the rho protein and either the α or β subunits of the *E. coli* ATP synthase left no doubt that these proteins have all descended from a common ancestor, the significance amounting to more than +7.5 standard deviations.

Among the new entries, one obvious family consisted of four picornavirus sequences, and another was composed of the two large T antigens. All in all, the search scores found within each of these six "families" were very significant. The scores that emerged from the rest of the cross-search were less compelling. Conspicuously absent, for example, were significant connections between the ATP synthases and myosin. All the same, a best match was found for every sequence to one not in its own family (as designated above) and a network of connections developed for those that approached significance (Fig. 4). Of all these, the most unexpected resemblances that emerged were between the elongation fac-

```
Initiation factor       ..G A A A E P R A P V V T I M G H V D H G K T S L L D Y I R S T K V A S G E A G I T..
Elongation factor       ..R T T P I A R Y R N I G I S A H I D A G K T T T T E R I L F Y T G V N H K I G E V..
Elongation factor       ..K F E R T K P H V N V G T I G H V D H G K T T L T A A I T T V L A K T Y G G A A R..
Lep A gene product              M K N I R N F S I I A H I D H G K S T L S D R I I Q I C G G L S D R E M E..
Elongation factor           M G K E K S H I N V V V I G H V D S G K S T T T G H L I Y K C G G I D K R T I E..
                            | |   | |     | | |       |   | | | | | | | |   | | | |     | | |   |
Large T (SV40)          ..M V Y N I P K K R Y W L F K G P I D S G K T T L A A A L L E L C G G K A L N V N L..
Large T (polyoma)       ..L T E N V P K R R N I L F R G P V N S G K T G L A A A L I S L L G G K S L N I N C..
                          |           | |       | |   | |       | |       |   |           |
DNA-Bound protein       ..L N R Q G G K R N T V L F H G P A S T G K S I I A Q A I A Q A V G N V G C Y N A A..
    (parvovirus)
```

Fig. 5. Alignment of elongation factor group with the large T antigens of SV40 and polyoma and a DNA-binding protein from a parvovirus. The vertical lines denote matches between the elongation factors and large T antigens, on the one hand, and between the latter and the parvovirus, on the other (see also Seif, 1984).

tor family and the large T antigens from SV40 and polyoma. It is interesting to note that these papilloma virus proteins also exhibit ATPase activity (Tjian *et al.,* 1979; Griffin *et al.,* 1979). In addition, a segment from a parvovirus DNA-binding protein fell close to this group (Fig. 5).

Although the connections between many of the sequences and the other families were not statistically significant, this is not to say that they are not related, only that a case cannot be made on the basis of amino acid sequences alone. It must be emphasized that in those cases where a relationship did achieve statistical significance, the resemblance extends throughout the proteins involved. As a case in point, a thorough analysis of the transducin α subunit and the ras proteins indicates a level of confidence of +3.8 standard deviations for the similarity being due to something other than chance when the full sequences are compared.

A search was also conducted of two kinases that had been implicated in the nucleotide-binding consensus study of Walker *et al.* (1982), adenylate kinase and fructokinase, as well as creatine kinase, the sequence of which has appeared only recently (Putney *et al.,* 1984). None of these three strictly obeys the consensus sequence, at least one of the four "invariant residues" being different in each case. Interestingly, only the creatine kinase sequence exhibited any significant resemblance to entries in Fig. 2. Like the large T antigens, it has a segment that bears a close resemblance to the elongation factors.

Finally, we might inquire whether the putative nucleotide-binding segment analyzed in this exercise has any connection to the mononucleotide fold identified by Rossmann *et al.* (1974). In fact, a computer graphics study based on the putative GTP-binding site for the ras protein attempted to relate its structure to the Rossmann fold (Wierenga and Hol, 1983). More recently, an authentic crystal structure of a portion of the elongation factor Tu has been published, and if a resemblance between the two nucleotide-binding folds was found, it did not warrant comment (Jurnak, 1985). It is interesting to note, nonetheless, that one of the segments retrieved by the consensus sequence search is from the H chain of lactate dehydrogenase (Fig. 3), and it does indeed correspond to a part of Rossmann's mononucleotide fold. In most cases, however, if there is a connection, the sequence similarity appears to have been obliterated.

IV. CONCLUDING REMARKS

What may we conclude from these observations? First, it is clear that a consensus sequence of the form used in our initial search exercise does not necessarily imply a nucleotide-binding site. Of the more than 50 sequences retrieved in the search, only about half appear to be from proteins exhibiting any interaction with nucleotides or polynucleotides. Conversely, there are many other known proteins that *do* bind nucleotides or interact with nucleic acids that were not retrieved by the search.

Given this uncertainty, it would seem imprudent to presume that the presence of the consensus sequence in a nucleotide-binding protein automatically implies that the segment containing the consensus sequence is actually involved in the binding. Certainly, further proof of the connection is required, either by the demonstration of further similarities in sequence to the point where common ancestry can be expressed with some confidence, or by direct experiment. In some cases it may even require a complete crystal structure determination to prove the point.

In the meantime, full sequence comparisons confirmed the existence of several presumed families and validated a few other relationships, including that of the transducin α subunit to the ras family, and also an unexpectedly strong relationship between the rho terminator protein and the ATP synthases. It must be underscored that in these situations the evidence favors common ancestry for the entire protein and not just for the nucleotide-binding portion. Put another way, the only connections that are readily confirmed are those that involve duplication of all or most of a gene. If binding sites ever existed as prebiotic exchangeable units, not enough information is likely to remain to prove the point with sequence data, mostly because the segments involved are so short.

It is, of course, impossible to prove that two sequences are not related. No matter how long they are, they may merely have changed beyond the point of recognition, or at least beyond the point of statistical verification. In some of these cases there can be hope that connections will be made by way of common relatives (Fig. 6). Surely in some of these cases, also, they may have changed so much that the relationship is no longer of interest.

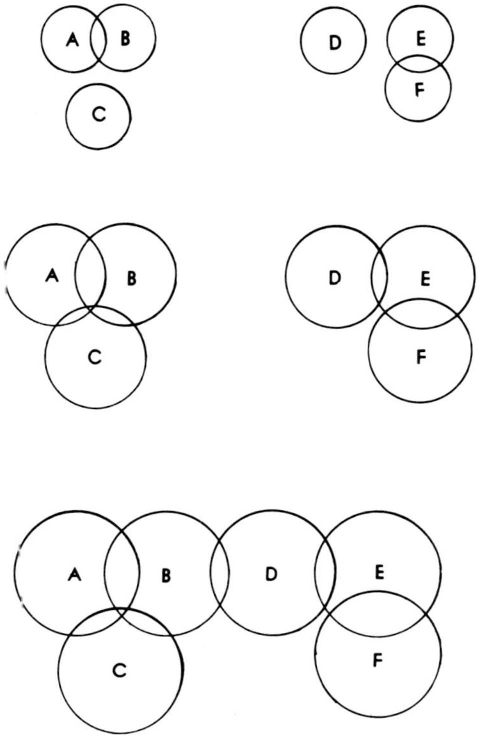

Fig. 6. Ven diagram showing how the evolutionary network may come to include distant relatives as more sequences are reported. Thus, although the similarity between two distantly related sequences may be blurred beyond recognition, the resemblances of each to a third may reveal the relationship. (From Doolittle, 1981.)

ACKNOWLEDGMENTS

I thank Karen Anderson for assistance in the preparation of this manuscript and Mark Johnson and Da-Fei Feng for help in the area of computer programming. I am also grateful to Dr. S. Rohde for sending me the sequence of the parvovirus DNA-binding protein in advance of its publication.

REFERENCES

Brown, J. P., Hewick, R. M., Hellström, J., Hellström, K. E., Doolittle, R. F., and Dreyer, W. J. (1982). *Nature (London)* **296,** 171–173.

Cottrelle, P., Thiele, D., Price, V. L., Memet, S., Micouin, J.-Y., Marek, C., Buhler, J.-M., Sentence, A., and Fromageot, P. (1985). *J. Biol. Chem.* **260,** 3090–3096.

Dayhoff, M. O., ed. (1978). "Atlas of Protein Sequence and Structure," Vol. 5, Suppl. 3. Natl. Biomed. Res. Found., Washington, D.C.
Doolittle, R. F. (1981). *Science* **214,** 149–159.
Doolittle, R. F. (1983). *Science* **222,** 417–419.
Doolittle, R. F. (1985). *Trends Biochem. Sci.* **10,** 233–237.
Feng, D.-F., Johnson, M. S., and Doolittle, R. F. (1985). *J. Mol. Evol.* **21,** 112–125.
Finch, P. W., and Emmerson, P. T. (1984). *Nucleic Acids Res.* **12,** 5789–5799.
Gay, N. J., and Walker, J. E. (1983). *Nature (London)* **301,** 262–264.
Gentry, G. A. (1985). *Proc. Natl. Acad. Sci. U.S.A.* **82,** 6815–6819.
Griffin, J. D., Spangler, G., and Livingston, D. M. (1979). *Cold Spring Harbor Symp. Quant. Biol.* **44,** 113–122.
Halliday, K. R. (1984). *J. Cyclic Nucleotide Protein Phosphorylation Res.* **9,** 435–448.
Higgins, C. F., Hiles, I. D., Whalley, K., and Jamieson, D. J. (1985). *EMBO J.* **4,** 1033–1040.
Hobson, A. C., Weatherwax, R., and Ferro-Luzzi Ames, G. (1984). *Proc. Natl. Acad. Sci. U.S.A.* **81,** 7333–7337.
Jurnak, F. (1985). *Science* **230,** 32–36.
Laursen, R. A., and Duffy, L. (1978). *FEBS Lett.* **92,** 200–202.
Leberman, R., and Egner, V. (1984). *EMBO J.* **3,** 309–341.
Lochrie, M. A., Hurley, J. B., and Simon, M. I. (1985). *Science* **228,** 96–99.
March, P. E., and Inouye, M. (1985). *Proc. Natl. Acad. Sci. U.S.A.* **82,** 7500–7504.
Pinkham, J. I., and Platt, T. (1983). *Nucleic Acids Res.* **11,** 3531–3545.
Putney, S., Herlihy, W., Royal, N., Pang, H., Aposhian, H. V., Pickering, L., Belagaje, R., Biemann, K., Page, D., Kuby, S., and Schimmel, P. (1984). *J. Biol. Chem.* **259,** 14317–14320.
Rose, T. M., Plowman, G. D., Teplow, D. B., Dreyer, W. J., Hellström, K. E., and Brown, J. P. (1986). *Proc. Natl. Acad. Sci. U.S.A.* **83,** 1261–1265.
Rossmann, M. G., Muras, D., and Olsen, K. W. (1974). *Nature (London)* **250,** 194–199.
Sacerdot, C., Dessen, P., Hershey, J. W. B., Plumbridge, J. A., and Grunberg-Manago, M. (1984). *Proc. Natl. Acad. Sci. U.S.A.* **81,** 7787–7791.
Schirmaier, F., and Phillippsen, P. (1984). *EMBO J.* **3,** 3311–3315.
Schulz, G. E., and Schirmer, R. H. (1974). *Nature (London)* **250,** 142–144.
Seif, I. (1984). *Virology* **138,** 347–352.
Tanaka, T., Ohkubo, H., and Nakanishi, S. (1984). *J. Biol. Chem.* **259,** 8063–8065.
Tjian, R., Robbins, A., and Clark, R. (1979). *Cold Spring Harbor Symp. Quant. Biol.* **44,** 103–111.
Walker, J. E., Saraste, M., Runswicke, M. J., and Gay, N. J. (1982). *EMBO J.* **1,** 945–951.
Wierenga, R. K., and Hol, W. G. M. (1983). *Nature (London)* **302,** 842–845.

3

The Analysis of Homologous Tertiary Structures and the Design of Novel Proteins

D. J. BARLOW, T. L. BLUNDELL, M. S. EDWARDS, B. L. SIBANDA, M. J. E. STERNBERG, W. R. TAYLOR, AND J. M. THORNTON

Laboratory of Molecular Biology
Department of Crystallography
Birkbeck College, University of London
London WC1E 7HX United Kingdom

I. INTRODUCTION

As *ab initio* simulation of the folding pathway of a protein is beyond our technical capability, the prediction of protein structure or the effects of site-specific mutations is best achieved by using the wealth of data already available from the high-resolution X-ray analyses of proteins. For site-specific replacements, insertions, or deletions we can use information derived from analyses of structural differences in homologous sequences and conformations derived from supersecondary motifs such as β hairpins, $\alpha\beta$ loops, and β bulges which have been defined by X-ray analysis. When the three-dimensional structure has not been defined by X-ray analysis, the tertiary structure of a homologous protein may be used as a basis for modeling with similar techniques. Even when there is no significant sequence homology, the tertiary structure may still be modeled approximately if a pattern or residue types of a "fingerprint" can be identified which is characteristic of a known structure. These models may then allow the correct structure to be reached by using molecular dynamics or, hopefully, energy minimization. Such computer-aided modeling tech-

PROTEIN ENGINEERING
Applications in Science,
Medicine, and Industry

niques will be most powerful when they are linked to a data base of protein sequences and structural motifs. These knowledge-based techniques provide a rational approach to the design of new molecules into which new features—binding sites, catalytic groups, stabilizing cross-links, etc.—can be engineered.

II. ENGINEERING AMINO ACID REPLACEMENTS, INSERTIONS, AND DELETIONS

Many amino acid replacements and most insertions and deletions occur in the loops and turns between secondary structure elements close to the surface of the protein. Greer (1981) has suggested that loops of a certain length in a family of proteins are likely to have the same conformation, and that this may be useful in modeling where the structures of several homologous proteins have been defined by X-ray analysis. Although this may not be strictly true, it is clear that the conformations of loops are much less variable than has previously been assumed.

The work of Sibanda and Thornton (1985) shows that loops between two antiparallel β strands (β hairpins) have well-defined conformations that frequently recur in protein structures. They analyzed 107 β hairpins occurring in 39 proteins and defined loop residues (L_1, L_2, etc.) as those not participating in the antiparallel β hairpin hydrogen-bonded ladder.

Two-residue loops comprised 29 of the 107 hairpins. Although in general 60% of turns are of type I + type III and 15% are type II, Sibanda and Thornton (1985) show that for β hairpins 25 were type I′ or type II′ and only 4 were type I. It appears that although type I′ and II′ turns may be less favorable due to steric hindrance, they give a twist which is compatible with that of the β-hairpin ladder. Thus when modeling two-residue turns in β hairpins, we should first consider type I′ or II′, a conclusion which will be a surprise to most protein modelers.

Three-residue turns are less common than those with four or five residues. Six of the 13 four-residue loops contain a type I turn with residues L_1, L_2, and L_3 in the α-helical region of the ϕ, ψ plot and residue L_4 (glycine or asparagine) in the +/+ quadrant (Fig. 1). In a similar way, 8 of the 19 five-residue loops contain a type I turn, with a hydrogen bond between the carbonyl of residue L_1 and the amide of residue L_4, and a G_1-type β-bulge. This structure has a glycine, asparagine, or aspartate at L_4.

Each of the types of loops has associated restrictions on the sequence both in the loop and on each side of the loop. Type I′ turns prefer L_1 as Asn, Asp, or Gly and L_2 as Gly; type II′ prefer L_1 as Gly and L_2 as Ser or Thr; and four- and five-residue loops often have the preferences noted

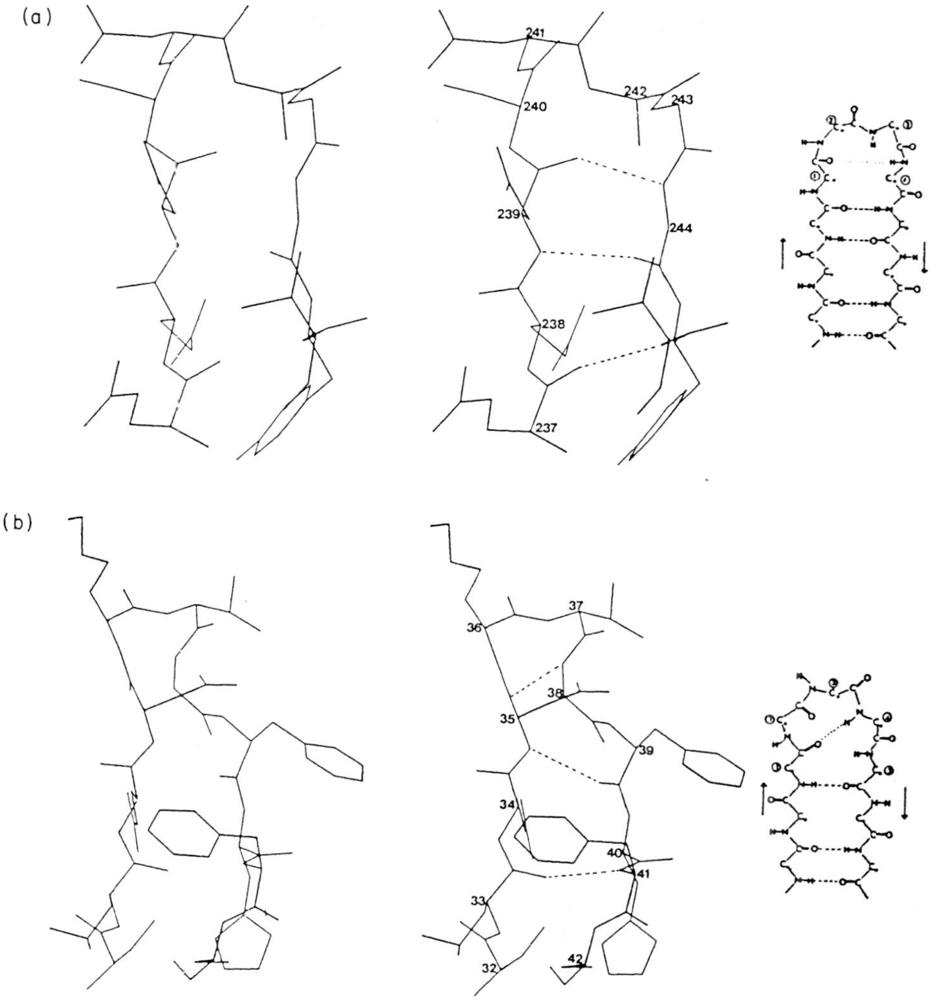

Fig. 1. Stereo diagrams of representative structures for four- and five-residue hairpin loop families. (a) Four-residue hairpin loop with a type I β turn; penicillopepsin, residues 237–246. The amide NH groups of residues L_2, L_3, and L_4 point inward toward the center of the turn and to the carbonyl oxygen of the residue immediately before the loop (-B_1). The side chain of this residue may help to neutralize the cluster of NH groups. A schematic diagram indicating the hydrogen bonding of this family is shown at the right. (b) Five-residue hairpin loops with a type I β turn and a G_1 β-bulge; γ-chymotrypsin A, residues 32–42. This structure is reminiscent of the four-residue family in that the nitrogens of residues L_3, L_4, and L_5 all point inward toward the carbonyl of residue L_1, and an aspartyl oxygen is often observed nearby. A schematic hydrogen-bonding diagram of this family is shown at the right.

(a)

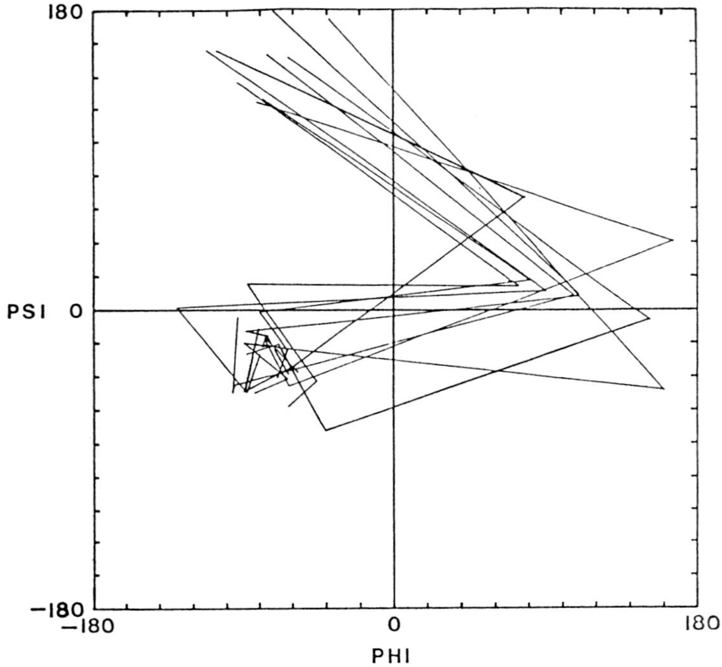

Fig. 2. Conformations of loops between adjacent α helices and β strands. (a) Characteristic ϕ, ψ angles of the loop residues; (b) an example from triose-phosphate isomerase.

above for L_4 as glycine. These observations often allow the loop conformations to be identified from a combination of length and sequence.

Similar features occur for other structures, including β arches (i.e., loops between different β sheets), $\alpha\beta$ loops, and $\alpha\alpha$ loops. For example, M. Edwards, M. J. E. Sternberg, and J. M. Thornton (unpublished results, 1985) have found that loops joining α helices to β strands often have a type I turn followed by a glycine with a positive value for ϕ. An example is give in Fig. 2.

For loops between two helices the following pattern often occurs: hydrophobic—hydrophobic—X—residue favoring β conformation—Pro—X—hydrophobic—hydrophobic—. This gives rise to helices related by 50–90° (Figure 3) (D. J. Barlow and J. M. Thornton, unpublished results, 1985). Thus specific sequence and structural patterns are particularly evident for short loops of all classes. Glycine is often involved and these

(b)

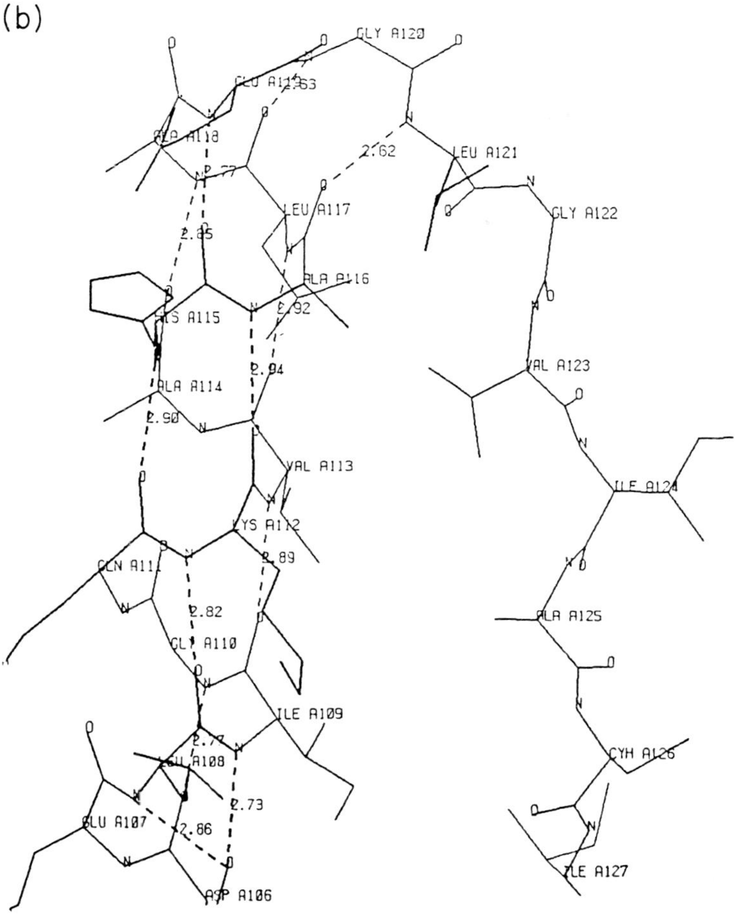

Fig. 2. *(Continued)*

conformations will be most straightforwardly used in modeling. For longer loops the conformations are more variable.

These observations can be used in modeling insertions and deletions. If an insertion or deletion is to be made in a loop region, we begin by characterizing the residues before and after the loop to be modified in terms of their main-chain torsion angles and interatomic distances between C^{α} atoms. Similar information will be stored in a relational data base for all precisely defined loops of proteins whose three-dimensional structures are defined by X-ray analysis. We then search for all loops of the correct length, matching the torsion angles and distance matrices of

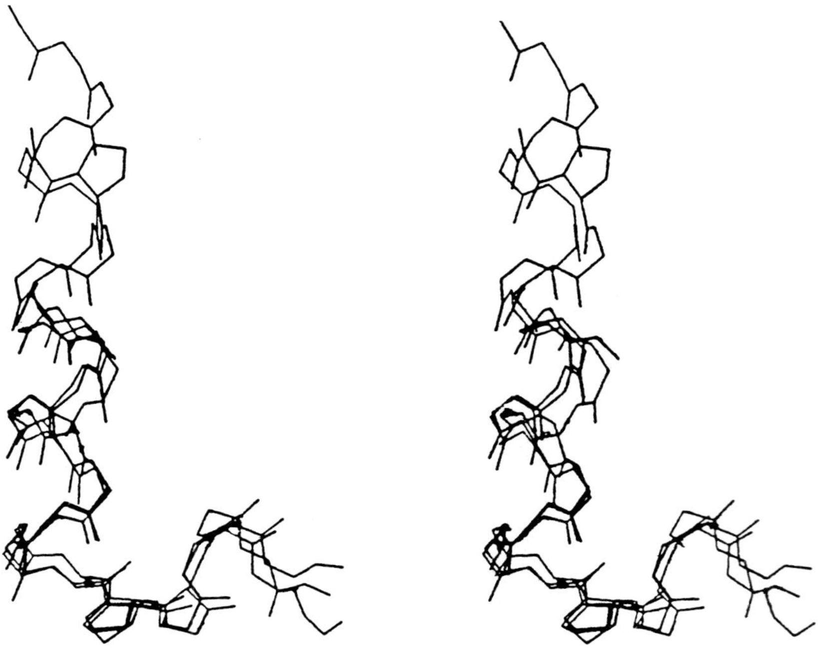

Fig. 3. A family of loops joining consecutive α helices, in which the key residue is a proline. The three structures illustrated (involving residues 20–41 in myoglobin, residues 19–36 in erythrocruorin, and residues 60–75 in cytochrome *c*) are superimposed with an rms error (calculated for the C^{α} atoms of the proline, the three residues following it, and the four residues preceding it) of ~0.7 Å.

the secondary structure strands and the sequence patterns defined by the analysis of protein structure. The distance matrices are particularly important when α helices are involved. We select the best matches and weight by their frequency of occurrence to obtain a rank order.

We must then test the loops against the rest of the known wild-type structure. We achieve this by docking the loop so that it occupies approximately the volume of that in the wild type and does not overlap with the rest of the structure, using logical operations on volumes with the program BILBO of Honegger and Blundell (1984). We join the chains by using the Hermans and McQueen (1974) algorithm to regularize the geometry. The conformations of the side chains are guided by those in the data base of structures and the whole system is energy-minimized, constraining movements of the protein that are more than 10 Å distant from the loop residues. Other local minima can be explored by using restrained molecular dynamics. The optimal conformations are chosen not only from energy criteria but also in terms of the accessibilities of hydrophobic and hydrophilic groups, ion pairs, etc.

III. MODELING HOMOLOGOUS PROTEINS

Thus far we have assumed that the structure of the wild-type protein has been defined by X-ray analysis at high resolution. In many cases this will not be the case, and we need to model the protein before we can start to consider the conformational effects which result from mutagenesis. If the structure of a homologous protein is available, this may provide a basis for the generation of such a model.

The first model of this kind reported was for α-lactalbumin on the basis of lysozyme (Browne *et al.*, 1969). We have used the method over the past 15 years to model many proteins, including members of the insulin family (Bedarkar *et al.*, 1977; Blundell *et al.*, 1978, 1983a), the aspartic proteinases (Blundell *et al.*, 1983b; Sibanda *et al.*, 1984), lens β/γ-crystallins (Wistow *et al.*, 1981; Inana *et al.*, 1983), and histocompatibility antigens (Travers *et al.*, 1984). Although the general approach is the same, the quality of the model will be very much dependent on the degree of divergence of the homologous proteins.

The three-dimensional structures of several insulins of differing sequences have been defined by X-ray analyses. Although crystal packing, ionic strength, and the presence of specific ions can cause large conformational changes, it has been shown that a closely similar conformer can exist for primitive vertebrate insulins and those of mammals (Dodson *et al.*, 1979). This is because the hydrophobic core is conserved and so the secondary structural elements can retain an identical scaffold. For the insulin-like growth factors (IGF) the residues of the hydrophobic core are also identical and the pattern of glycines in the main chains at A1, B8, B20, and B23 is retained with the exception of B20 in rat IGF2. We are thus confident that our models of the insulin-like growth factors are reliable indications of their true structures (Blundell *et al.*, 1978, 1983a). The main ambiguities are in the C- and D-peptide regions, which represent small loops on the surface and can be predicted by methods described in the previous section.

What happens when there are differences in the residues of the hydrophobic core between two homologous proteins? Lesk and Chothia (1980) have shown for α-helical proteins (globins) that the secondary structural elements tend to have different relative positions although the nature of the interactions (ridge to groove) is largely retained. There are, of course, complementary changes which tend to lead to a rough retention of the volume. For example, in modeling relaxin on the homologous insulin structure we noted that, although most residues of the core were varied, they were retained as hydrophobic with complementary changes such as A2 Ile, A16 Leu in porcine insulin compared with A2 Leu, A16 Ile in porcine relaxin, and B6 Leu, B14 Ala in porcine insulin compared with B6 Ala, B14 Leu in porcine relaxin (Bedarkar *et al.*, 1977). Similar complementarity also occurs on the surface and allows a rough model to be constructed by assuming a rigid scaffold of secondary structural elements. We then allowed the molecule to relax into a minimum by using energy minimization; the minimum will only be the global minimum if the initial model was sufficiently correct. For the model of protein S based on γ-crystallin (Wistow *et al.*, 1985) this has led to a contraction of volume (S. Najmudin, A. M. Hemmings, and T. L. Blundell, unpublished results, 1985) which is greater than that for γ-crystallin as there is a decrease in core volume. However, it is unlikely that energy minimization techniques will effectively model the reorientation of secondary structural elements noted by Lesk and Chothia (1980). It will be necessary to attempt systematic searches of packing of rigid secondary structure elements to increase the radius of convergence, although molecular dynamics calculations or the searching of energy minima along low-frequency normal modes appear to offer attractive alternative approaches.

For structures which have an evolving hydrophobic core we have made systematic searches of the packing relationships of secondary structures in members of the family already defined by X-ray analysis. For example, topologically equivalent C^{α} positions for the Greek key motifs of γII-crystallin defined at 1.6 Å resolution (Blundell *et al.*, 1981; Wistow *et al.*, 1983; Summers *et al.*, 1984) show root-mean-square (rms) differences of 0.69 to 1.48 Å for homologies ranging from 50% down to 20%. In modeling protein S we were careful to use those motifs which showed closest sequence homology; thus we used γII1, γII2 to model both S_2, S_1, and S_4, S_3 (Wistow *et al.*, 1985).

For the aspartic proteinases we have compared the known three-dimensional structures of porcine pepsin, penicillopepsin, endothiapepsin, and rhizopuspepsin (Sibanda *et al.*, 1984). For penicillopepsin and endothiapepsin, sequences are defined by chemical methods and the structures are refined to $R < 17\%$ by using restrained least squares at 1.8 and 2.1 Å resolution, respectively. These have 55% homology and have 207 pairs of C^{α} atoms with an rms difference of 0.48 Å. Closer examination of the structures with computer graphics (see Fig. 4) shows that the active site regions are very closely similar in conformation, as they are in sequence. Most differences occur in loops at the periphery and in the hydrophobic cores toward the edges of the bilobal enzymes.

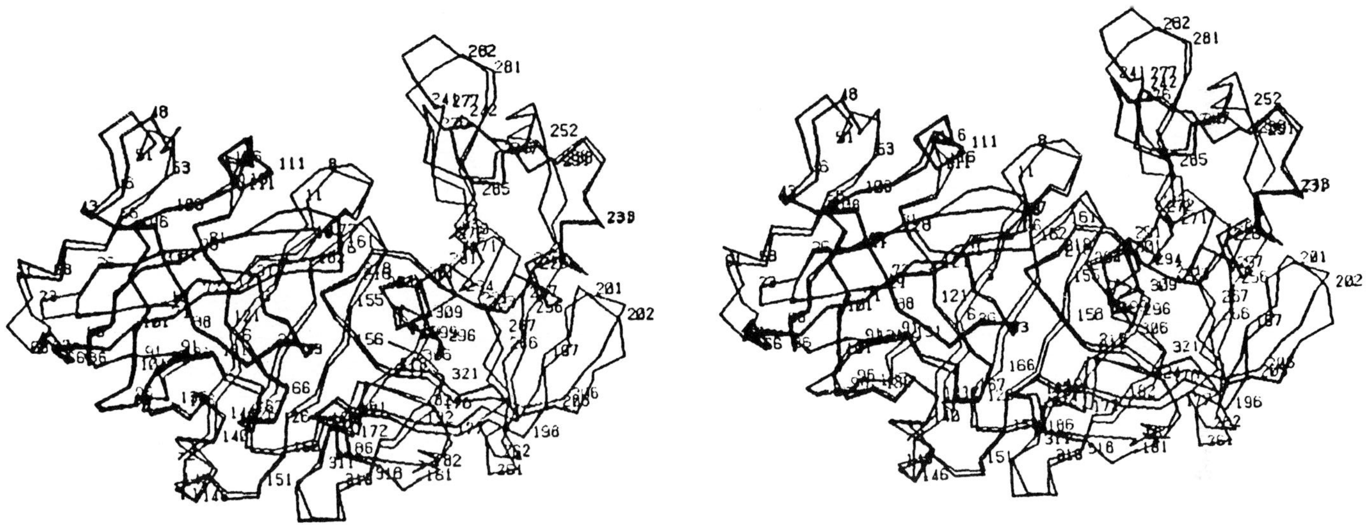

Fig. 4. A least-squares fit of endothiapepsin and penicillopepsin indicating the similar main-chain arrangements in the central core regions and in the active site cleft.

The analyses of rhizopus pepsin and porcine pepsin with respect to endothiapepsin and penicillopepsin showed similar active site regions and arrangements of the central β strands but even larger differences in the peripheral strands and loops consistent with the lower degree of sequence identity. Such information is of great value in the modeling of other aspartic proteinases such as mouse and human renins (Blundell *et al.*, 1983b; Sibanda *et al.*, 1984) and in the assessment of probable errors in the models.

In general, we opt for a structure which is well-defined chemically and by X-ray analysis. However, we remember that those regions (active site and cleft) which are conserved in evolution between other aspartic proteinases are most likely to be correctly modeled in human renin.

Perhaps the most critical stage in modeling by homology is the alignment of the sequences. The alignment is best obtained by noting the relationship between the sequences and the known three-dimensional structures and listing residues at topologically equivalent positions. This will often reveal the equivalence of residues which do not give the highest sequence homology when based on a simple residue identity count. It will also identify residues which are invariant and positions which are conservatively varied for structural reasons within certain categories of residues. Alignment of the new sequence then proceeds by first identifying the critical residues which are invariant or nearly invariant. These often turn out to be glycines (either with conformations with ϕ positive or where no side chain is possible); cis-prolines; buried disulfides such as that in the immunoglobulins; or buried polar residues such as Thr 33 and Thr 216 in aspartic proteinases. We then check patterns of hydrophobics required by the tertiary structure. A program has been devised to carry this out automatically (Taylor, 1986). If these criteria are met, it is likely that those of Novotny *et al.* (1984) concerning accessibility of hydrophobics and hydrophilics will also be satisfied. Although our criteria are more subjective at first sight, they will probably distinguish between local modeling differences to which the overall criteria will be relatively insensitive. In summary, before any model building is started we must understand the structural data already available for the homologous proteins.

The elaboration of the homologous structure once the main relationships between secondary structural elements are achieved can proceed in the same manner described above for point mutations, insertions, and deletions.

IV. CONCLUSIONS FOR PROTEIN ENGINEERING

We believe that an intelligent use of the data base of known conformations of not only secondary structures but also supersecondary structural motifs and tertiary interactions provides the best starting model for prediction, whether it be amino acid replacements, insertions, or deletions or the design of completely novel structures. A rough model derived by such procedures can then be refined by the developing techniques of molecular mechanics (see, for example, Weiner *et al.,* 1984; Wodak *et al.,* 1984; Shih *et al.,* 1985; and the references therein).

REFERENCES

Bedarkar, S., Turnell, W. G., Blundell, T. L., and Schwabe, C. (1977). *Nature* (*London*) **270,** 449–451.

Blundell, T. L., Bedarkar, S., Rinderknecht, E., and Humbel, R. E. (1978). *Proc. Natl. Acad. Sci. U.S.A.* **75,** 180–184.

Blundell, T. L. Lindley, P., Miller, L., Moss, D., Slingsby, C., Tickle, I. J., Turnell, W. G., and Wistow, G. (1981). *Nature* (*London*) **289,** 771-777.

Blundell, T. L., Bedarkar, S., and Humbel, R. E. (1983a). *Fed. Proc., Fed. Am. Soc. Exp. Biol.* **42,** 2592–2597.

Blundell, T. L., Sibanda, B. L., and Pearl, L. H. (1983b). *Nature* (*London*) **304,** 273–275.

Browne, W. J., North, A. C. T., Phillips, D. C., Brew, K., Vanamann, T. C., and Hill, R. L. (1969). *J. Mol. Biol.* **42,** 65–86.

Dodson, E. J., Dodson, C. G., Hodgkin, D. C., and Reynolds, C. D. (1979). *Can. Biochem.* **57,** 469–479.

Greer, J. (1981). *J. Mol. Biol.* **153,** 1027–1042.

Hermans, J., and McQueen, J. E. (1974). *Acta Crystallogr., Sect. A* **A30,** 730–738.

Honegger, A. M. and Blundell, T. L. (1984). *In* "Insulin-like Growth Factors/Somatomedius: Basic Chemistry, Biology and Clinical Importance" (E. M. Spencer, ed.) pp. 9–113. Gruyter, Berlin.

Inana, G., Piatigorsky, J., Norman, B., Slingsby, C., and Blundell, T. L. (1983). *Nature* (*London*) **302,** 310–315.

Lesk, A. M., and Chothia, C. (1980). *J. Mol. Biol.* **136,** 225–270.

Novotny, J., Bruccoleri, R., and Karplus, M. (1984). *J. Mol. Biol.* **177,** 787–818.

Shih, H-L., Brady, J., and Karplus, M. (1985). *Proc. Natl. Acad. Sci. U.S.A.* **82,** 1697–1700.

Sibanda, B. L., and Thornton, J. M. (1985). *Nature* (*London*) **316,** 170–174.

Sibanda, B. L., Blundell, T. L., Hobart, P. M., Fogliano, M., Bindra, J. S., Dominy, B. W., and Chirgwin, J. M. (1984). *Febs Lett.* **174,** 102–111.

Summers, L., Wistow, G., Narebor, M., Moss, D.S., Lindley, P. F., Slingsby, C., Blundell, T. L., Bartunik, H., and Bartels, K. (1984). *Pept. Protein Rev.* **3,** 147–168.

Taylor, W. R. (1986). *J. Mol. Biol.* **188,** 233–258.

Travers, P., Blundell, T. L., Sternberg, M. J. E., and Bodmer, W. F. (1984). *Nature* (*London*) **310,** 235–238.

Weiner, S. J., Kollman, P. A., Case, D. A., Singh, V. C., Ghio, C., Alagone, G., Proteta, S., and Weiner, P. (1984). *J. Am. Chem. Soc.* **106,** 765–784.

Wistow, G., Slingsby, C., Blundell, T. L., Driessen, H., de Jong, W., and Bloemendal, H. (1981). *FEBS Lett.* **133,** 9–16.

Wistow, G., Turnell, W. G., Summers, L., Slingsby, C., Moss, D. S., Miller, L., Lindley, P. F., and Blundell, T. L. (1983). *J. Mol. Biol.* **170,** 175–202.

Wistow, G., Summers, L., and Blundell, T. L. (1985). *Nature* (*London*) **315,** 771–773.

Wodak, S. J., Alard, P., Delhaise, P., and Renneboog–Squilbin, C. (1984). *J. Mol. Biol.* **181,** 317–322.

4

Structural Implications for Macromolecular Recognition and Redesign

ELIZABETH D. GETZOFF,*
ROBERT A. HALLEWELL,*,†
AND JOHN A. TAINER*

** Department of Molecular Biology*
Scripps Clinic and Research Foundation
La Jolla, California 92037
and
† Chiron Corporation
Emeryville, California 94608

I. INTRODUCTION

A. Importance of Structural Information

A full understanding of macromolecular interactions requires a detailed knowledge of their structural basis at a level that approaches atomic resolution. In the broadest sense, such interactions include the intramolecular folding and assembly of single molecules, which depend on forces similar to those acting between molecules. By combining new genetic techniques based on cloning with improved X-ray crystallographic methods, the redesign of proteins to satisfy specific preconceived requirements and to test theoretical models of the folding, stability, and mechanism can now be attempted. Accordingly, protein crystallography of site-directed mutants has produced new information concerning the relationship between macromolecular structure and function (see, for example, Craik *et al.,* 1985). The challenge is to apply structural information to rational

PROTEIN ENGINEERING
Applications in Science,
Medicine, and Industry

protein design through genetic techniques. If we completely understand the molecular recognition processes involved in protein folding, interaction, and assembly, then we should be able to choose specific changes that will influence these processes in predictable ways. At present, our limited understanding results in some unpredicted results, which can be used to improve the design of future experiments.

1. The Need for Rational Design

Genetic techniques can now create almost any single site mutation. Yet, even if specific mutation sites have been chosen, a quick calculation shows the enormous task involved in examining the effects of putting all 20 naturally occurring amino acids into a limited set of sites. Desired mutations might also require concerted changes in different areas of the amino acid sequence, further increasing the problem. Thus, although almost any possible amino acid sequence can be made, the technical limitations on examining the results require that we have some specific ideas of what to make. These specific ideas must ultimately depend on a knowledge of the three-dimensional structure. In the following sections, we consider the nature of the information in protein structures and how best to interpret structural information for the purposes of understanding and controlling macromolecular recognition processes. Such processes include the formation of complex macromolecular machinery such as ribosomes and the multienzyme complexes of metabolic pathways, the assembly of viruses, and the recognition and specific binding of antibodies to antigens and hormones to receptors. Determination of the specific structural arrangement of the amino acid residues in a protein provides substantial insight into their functions at the level of detailed atomic geometry and, in addition, yields information about the important covalent and noncovalent forces that stabilize structural conformation and control activity. Now that X-ray crystallographic studies have provided the three-dimensional structures of over 200 proteins and examples of both DNA and RNA (Bernstein *et al.*, 1977), the properties controlling such macromolecular recognition, interaction, and assembly processes can begin to be examined and understood at a molecular level.

2. Cu, Zn Superoxide Dismutase: Opportunities for Analysis and Design

Although our analyses are designed for general application to many systems, our current focus is the enzyme Cu, Zn superoxide dismutase (SOD) (Tainer *et al.*, 1982; Getzoff, 1982), which dismutes the toxic superoxide radical (O_2^-) to oxygen and hydrogen peroxide. Although the structure of SOD at 2 Å resolution is generally representative of other protein structures known at atomic resolution, a number of its features offer important opportunities for analysis. SOD is unusually stable. Bovine SOD has a conformational melting temperature of 83°C and is apparently among the most stable of the currently characterized globular proteins, including those from extreme thermophiles (Stellwagen and Wilgus, 1978). The dimeric enzyme remains active in 8 *M* urea (Malinowski and Fridovich, 1979) and is not dissociated by 4% sodium dodecyl sulfate (Forman and Fridovich, 1973). Based on our knowledge of the structural interactions in the bovine enzyme, improved stability of the human enzyme should be obtainable by site-specific mutagenesis of the human gene, which has now been cloned and expressed (Hallewell *et al.,* 1985). The mechanism and active site of SOD are also of interest. The remarkably high catalytic rate, due to precollision guidance of the substrate by local electrostatic forces (Getzoff *et al.,* 1983), makes the enzyme a simple model system in which to study the role of electrostatic forces in molecular recognition. The Cu and Zn ions at the active site are joined by a bridging histidine ligand. Both of these metal ions can be selectively removed and replaced by other metal ions to obtain different derivatives useful for studying numerous aspects of metalloprotein chemistry (Valentine and Pantoliano, 1981).

B. Use of Computer Graphics Models

A high-resolution X-ray crystal structure gives detailed information defining the positions of all atoms in three-dimensional space. However, it is impossible to understand and analyze a protein structure directly from these atomic coordinates. Instead models are needed to express the numerical results, and the appropriate model depends on the nature of the question being addressed. Computer graphics offers the unique opportunity to display models of macromolecules and their interactions based on virtually any mathematically describable parameters. By using the computer graphics language GRAMPS (O'Donnell and Olson, 1981) and the molecular modeling program GRANNY (Connolly and Olson, 1985), both arbitrary objects and molecules can be displayed simultaneously to better examine structural features. One major power of computer graphics molecular modeling and analysis is pattern discovery: the identification and classification of characteristic patterns for a variety of protein interactions.

Computer graphics analysis of crystallographically determined protein structures can provide many insights into macromolecular interactions, thereby allowing the design of specific and useful experiments. Diverse molecular properties including electrostatic forces, solvent structure, shape, surface area, hydrophobicity, and patterns of mobility can be examined and integrated into models for protein structural and functional interactions (Tainer *et al.*, 1985c). We have focused on three stages of macromolecular interactions by (1) examining the role of electrostatic forces in promoting efficient macromolecular interaction through the recognition and orientation of macromolecules before collision, (2) exploring the control of specificity by the degree of chemical, electrostatic, and shape complementarity of macromolecular interfaces, and (3) studying how molecular flexibility may promote transition states and improve complementarity by allowing induced fit. Studies of this type represent an opportunity to contribute to the background knowledge necessary for major advances in molecular biology and medicine resulting from bioengineering.

II. DISSECTION OF A PROTEIN STRUCTURE

A. Stereochemical Fold

Protein structures can be classified according to the internal arrangement of their domains, which are contiguous portions of the polypeptide chain that fold into compact, semi-independent units. These domains generally fall in a limited size range. A major achievement resulting from the comparative analysis of three-dimensional protein structures has been the discovery that all known structural domains can be grouped in a small number of categories. Richardson's comprehensive protein structure taxonomy gives four major categories of folding domains: (1) antiparallel α, (2) parallel α/β, (3) antiparallel β, and (4) small disulfide or metal-rich (Richardson, 1981).

The two identical single domain subunits of SOD belong to the antiparallel β category. As illustrated by the α-carbon backbone shown in Fig. 1, each 16,000-dalton subunit (151 amino acid residues) is composed primarily of eight antiparallel β strands, which form a flattened cylinder, plus three external loops. The SOD β-barrel structure consists of four very regular extended β strands on one side (the back in Fig. 1) and four β strands in more twisted or irregular conformations on the other side (the front in Fig. 1). The two sides of the β barrel might also be interpreted as two β sheets forming a sandwich. The loops contain many tight turns and are primarily as well defined and highly ordered as the β strands. The single disulfide bridge forms a left-handed spiral conformation and ties one loop to the β barrel. The active site of the enzyme, which contains one Cu and one Zn ion, is located on the irregular or twisted side of the β

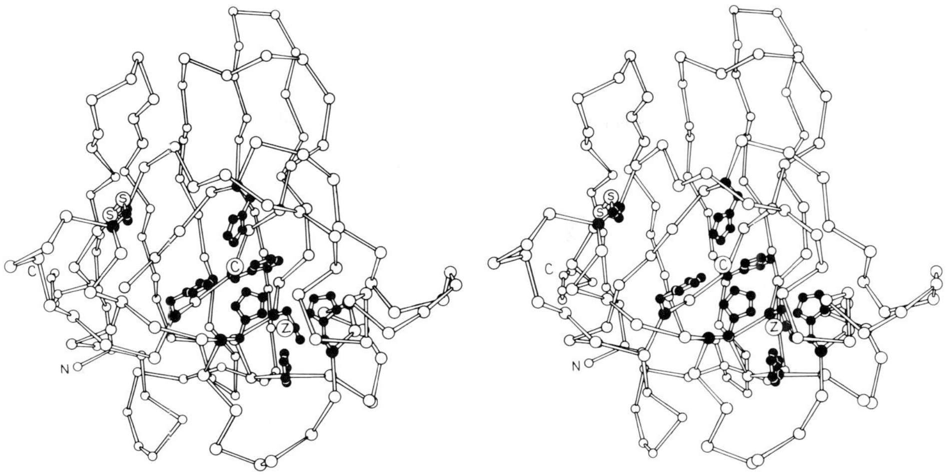

Fig. 1. Stereo view of the superoxide dismutase (SOD) subunit showing the active site side chains. The α-carbon backbone of the subunit is viewed down the axial direction of the Cu from the solvent. Solid spheres represent the α-carbon and side-chain atoms of the Cu ligand residues (His 44, 46, 61, and 118), the Zn ligand residues (Asp 81 and His 61, 69, and 78), and the nonsulfur atoms in the disulfide bond (Cys 55 and 144). The disulfide bond (circled S, upper left) covalently joins the largest loop subdomain to the β barrel. The side chain of His 61 bridges between the Cu (circled C) and the Zn (circled Z) ions. Residue 1 (labeled N) is at the lower left with residue 151 (labeled C) and above it. (Adapted from Tainer *et al.*, 1982.)

barrel between two large loops. Within the major category of antiparallel β, the overall topology of the SOD subunit classifies the enzyme as a Greek key β barrel, a very common folding motif that is also found in immunoglobulin domains, plant lectins, and capsid subunits of spherical viruses (Richardson, 1981).

B. Subdomains

The behavior and stability of a folding domain depend on the interaction of intrasubunit structural elements or subdomains. The subdomains can be defined as regions having more interactions within themselves than between themselves and the rest of the protein. Figure 2 shows a two-dimensional schematic of the subdomains forming the overall structural fold of SOD. Beta strands are identified by a number beginning at the N terminus and proceeding clockwise around the top of the β barrel as viewed in Fig. 1 and also by a letter assigned in order of amino acid sequence. The three long loops of nonrepetitive secondary structure are

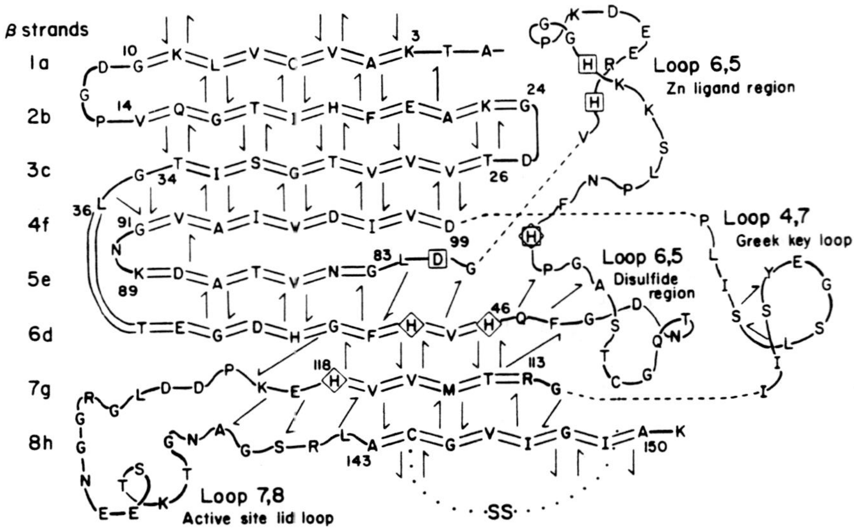

Fig. 2. Diagram showing the SOD β barrel and loops spread out flat and shown from the outside, with the amino acid sequence given in one-letter code. The initial Ala is acetylated. The eight β strands are shown as zigzag pairs of parallel lines with main-chain hydrogen bonds drawn as arrows from hydrogen donor to hydrogen acceptor. Hydrogen bonds joining β strands 1a and 8h are shown in duplicate pointing outward from these strands. Some main-chain hydrogen bonds outside the β sheet are shown. Residues in the valleys of β strand zigzags have side chains pointing toward the interior of the barrel, while those on the peaks have side chains extending outward from the barrel. The β-strand residues are numbered at the beginning and end of each strand and β strands are labeled at the left. The three loops (distinguished by heavy single lines) are named for the β strands at the beginning and end of each loop and by descriptive names. The dashed lines between three pairs of adjacent residues (Val 79–Gly 80, Asp 99–Pro 100, and Ile 111–Gly 112) identify places where the polypeptide chain was artificially stretched to make this schematic drawing. A dotted line shows the disulfide bond (SS) between Cys 55 and Cys 144. The Cu ligands are outlined with diamonds and the Zn ligands with squares. (Adapted from Tainer *et al.*, 1982.)

named for the β strands that they join (Fig. 2). Our analysis of the subunit structure (Getzoff, 1982) is based on a total of six subdomains: β strands 1–4 (the regular side of the β sandwich); β strands 5–8 (the irregular side of the β sandwich); Greek key loop 4, 7; Zn ligand region of loop 6, 5; disulfide region of loop 6, 5; and active site lid loop 7, 8.

C. Strategy and Methods of Approach

The complex interactions present in the folded enzyme structure were analyzed using computer search and calculation programs and interactive color computer graphics models (Getzoff, 1982). Interactive computer graphics allowed the accurate display of accessible and buried molecular surfaces and geometry and easy manipulation of complex three-dimensional data. Color coding of various types of information and grouping of related objects for display facilitated the application of human pattern recognition to the data (Getzoff *et al.*, 1986b; Olson *et al.*, 1984).

Points defining molecular surfaces were determined by using Connolly's program MS (Connolly, 1983a,b). The program places solvent probe spheres around the van der Waals surface of each atom in the protein to calculate points forming the solvent-accessible molecular surface described by Richards (1977). It has been adapted for the calculation of two types of intramolecular surfaces, which can be called buried and packing surfaces. The buried surface is the molecular surface made inaccessible to the solvent by the interaction of two pieces of the protein. The packing surface, calculated with a probe sphere of 0.5 Å radius, includes the molecular surface of each independent atom that overlaps or is in contact with the molecular surfaces of any other atoms. These packing surfaces indicate the most tightly packed regions within each subdomain. Because packing surfaces are much more sensitive to small changes in the structure than are buried or accessible surfaces calculated with a water-sized probe, they permit the study of internal contacts in well-determined protein structures at a level not previously possible.

These different types of surfaces were used to examine the packing within and between subdomains and domains of SOD including the β barrel, the two β sheets forming the β-barrel sides, the individual loops, and the dimer contact. Each surface was characterized by the composition of contributing residues, the relative hydrophobicity versus hydrophilicity, and the relative involvement of side-chain versus main-chain interactions. The qualitative and quantitative information resulting from the calculation of buried and packing molecular surfaces provides objective, definable criteria for assessing the interactions between individual atoms, residues, and structural fragments. Shape complementarity was assessed by the volumes of gaps and interpenetrations between pairs of surfaces buried in interfaces as well as by the twist and degree of interdigitation of the interacting surfaces. The determination of side-chain interactions between individual residues complements the surface area calculations by identifying the important contact pairs. Together, the data from surface area and contact pair computations provide a concise summary of the structural interactions between pairs of subdomains or domains. The major patterns found in these data have theoretical implications for protein folding and stability and provide an overview of the important interactions in SOD.

D. Intramolecular Interactions and Protein Stability

The stability of a domain depends on the interactions of its underlying structural elements. By definition, the structural integrity within a subdomain is greater than that between a subdomain and the other parts of the protein. Thus, the energy required to unfold a protein can be expected to be a reflection of that necessary to break down intrasubdomain interactions, and the characterization of these interactions as outlined below is a first step toward design of more stable proteins.

1. SOD β Barrel

The calculation of separate surfaces for each β strand revealed the packing between strands in cross sections perpendicular to the long axis of the β barrel. Most of the contacts are made between adjacent strands around the barrel. Two particular contacts (Ile 18–Phe 43; Ile 33–Val 117) are noteworthy because they tie together the two opposite sides of the β barrel with orthogonal cross bracing. These cross-barrel interactions stabilize the β-barrel fold and also anchor the active site through the β barrel to the more regular side of the structure.

Color coding of the exposed molecular surface of the β barrel by atom type revealed a predominance of hydrophobic (carbon and carbon-bound hydrogen) atoms (67%) and also illustrated the patterns of main-chain hydrogen bonding underlying the alternation of hydrophobic and hydrophilic side chains. The external surface characteristics of the β barrel (62% hydrophobic, 70% side chain) resemble those of the whole molecule (61% hydrophobic, 69% side chain).

As a result of the twist of the individual strands, the two β-barrel sides have an overall twist of about 25° for β 1–4 and about 50 ° for β 5–8. Due to the differential twist of these two β sheets, the more highly twisted β 5–8 is wrapped around the four straighter β strands. The continuous surface buried between the two sides of the β barrel is quite bumpy, conflicting with the description of prealbumin and the immunoglobin packing by Chothia and Janin (1981), but matching that of Cohen *et al.* (1981) for these two structures. In SOD, the interdigitation of the two internal β-barrel surfaces resulting from interlocking side chains is a major feature of the barrel packing.

The β-strand molecular surfaces buried in the formation of the two 4-stranded β sheets are 61% hydrophobic and 54% side chain by area. In contrast, the surface area buried between the two β sheets is 78% hydrophobic and 65% due to side chains. The hydrophilic and main-chain components result primarily from interactions between the edges of the two sheets, whereas the buried surface area internal to the β barrel is almost entirely composed of hydrophobic side-chain atoms. The edge strands of the two β sheets participate in fewer hydrogen bonds and have poorer packing than the other strands. Within the β barrel only one-third of the surface area of the individual β strands is buried between the two sheets, whereas the remaining two-thirds are buried in the formation of the sheets. Similar values have been found for concanavalin A, prealbumin, and several immunoglobins. This weak link between the two sheets is considerably strengthened by the Greek key fold, which covalently locks together the middle strands of the two sheets while simultaneously capping the ends of the β barrel.

2. *SOD Loops*

The loop subdomains are similar to the β barrel in terms of packing and stereochemical complementarity, but have the important distinction of being largely determined by the interaction of sequence-local residues. Loops are therefore especially promising candidates for redesign. Detailed analysis of the SOD loops (Getzoff, 1982) led to the following overview of loop packing.

The packing interactions of the Greek key loop 4, 7 indicate its role in closing off one end of the β barrel. Loop 4, 7 is the least self-contained of the identified subdomains; of the loops, it has the least average internal packing surface per residue within the loop and the most average buried surface per residue against the β barrel. The internal and external packing interactions of loop 6, 5 indicate that this longest loop can be properly divided into two separate structural fragments: the disulfide loop (Gln 47–Pro 60) and the Zn ligand loop (His 61–Leu 82). The disulfide loop does not interact strongly with the other structural elements within the subunit, but does play a significant role in intersubunit interactions. The Zn ligand loop is the most tightly packed of the loops and also has the largest side-chain contributions to the internal packing. The active site lid loop 7, 8 has the most solvent-exposed area per residue and also the most hydrophilic buried and packing surface areas. The largest contributions to its surface area buried by the β barrel come from residues with important roles in the active site.

To illustrate some of the detailed data contributing to the structural analysis of loops, we focus on the active site lid loop 7, 8. This loop (Fig. 3), consisting of residues Glu 119 to Leu 142, extends outward from β strands 7 and 8 to form one side of the active site channel. Loop 7, 8 has the highest average solvent exposure of any structural element (50 Å^2 per residue). In accord with this exposed position, both the packing surface within loop 7, 8 and the surface buried by its interaction with the β barrel

Fig. 3. Stereo view of the loop 7, 8 (Glu 119–Leu 142) subdomain. The two extended antiparallel sides of this loop are spanned by a β bulge to form a lid for the active site. The open turn joining the sides is stabilized by two pseudo tight turns in which side-chain conformation mimics that found for the main chain in a true tight turn. At the lower left, initial and final tight turns are interdigitated with a six-residue α helix. The orientation is rotated to the left about 45° and downward about 45° from the view of the subunit in Fig. 1. Glu 119 (top) is labeled by number. As with other subdomains formed by nonrepetitive secondary structural elements, the conformation of this loop subdomain is stabilized by side-chain packing and hydrogen-bonding interactions. The packing interactions between this subdomain and the β barrel are summarized in Table I. (Adapted from Tainer *et al.*, 1982.)

(Table I) have a larger percentage of hydrophilic surface area than those of the other loops.

Of the 35 residue pair side-chain contacts of loop 7, 8, 16 occur within the loop, 14 connect to β-barrel residues, and 5 join with loop 6, 5. Five of these loop to β-barrel contacts are contributed by Asp 122 and Arg 141, which together account for 35% of the loop surface buried by the loop 7, 8/β-barrel interface (Table I). Both of these residues appear to have important roles in the active site (see Sections III,C and III,D) in addition to their importance to the integrity of the tertiary interactions between the

TABLE I

Surface Buried between the β Barrel and Loop 7,8[a]

Residue		Surface area (Å^2)				
No.	Type	Hydrophobic	Hydrophilic	Side chain	Main chain	Total
β Strands 1–8						
10	Gly	8	—	—	8	8
11	Asp	—	2	—	2	2
12	Gly	4	8	—	13	13
13	Pro	3	—	3	—	3
14	Val	26	3	26	3	29
36	Leu	10	—	9	1	10
37	Thr	5	—	5	—	5
40	Asp	7	11	4	14	18
41	His	20	9	20	9	29
42	Gly	9	6	—	15	15
44	His	14	10	24	—	24
46	His	14	7	22	—	22
83	Gly	15	3	—	19	19
84	Asn	4	31	34	—	34
116	Val	22	—	22	—	22
117	Val	7	8	7	8	15
118	His	51	23	50	24	74
143	Ala	3	5	—	8	8
144	Cys	6	—	6	—	6
Total		227	128	232	124	355
Loop 7,8						
119	Glu	5	11	—	16	16
120	Lys	7	8	—	15	15
121	Pro	24	—	17	7	24
122	Asp	13	27	28	12	40
124	Leu	18	—	18	—	18
135	Thr	10	6	7	9	16
138	Ala	31	—	27	3	31
139	Gly	—	12	—	12	12
140	Ser	4	—	—	4	4
141	Arg	46	14	50	9	60
142	Leu	50	17	46	21	67
Total		207	95	194	108	302

[a] Surface buried to a 1.4 Å radius probe. (From Getzoff, 1982.)

loop and β barrel. Nine residues located at both the beginning and the end of the loop (Glu 119–Asp 122 and Ala 138–Leu 142) make up 89% of its molecular surface buried by the β barrel.

3. Determinants for Subdomain Conformation

The intrasubunit structural fragments originally identified on the basis of main-chain conformation and hydrogen-bonding pattern (Tainer *et al.*, 1982) have considerable validity in terms of the observed contact and packing interactions. The majority of the packing surface occurs within rather than between these fragments (44% within the β barrel, 34% within the individual loops, and only 21% among all these structural elements). Three of the segments identified as separate loops (the disulfide loop, the Zn ligand loop, and loop 7, 8) have a significant amount of local structural integrity. However, loop 4, 7 (with an average of 16 Å^2 per residue of packing surface within the loop) is less an individual structural element than a connection between two nonadjacent β strands that acts to close off one end of the barrel. At the other end of the β barrel, the short Greek key connection between β strands 3c and 6d similarly caps the barrel, primarily with interactions from a single residue (Leu 36). The Zn ligand loop (which shares a common pattern of four tight turns with loop 4, 7) is more tightly packed, with an average of 25 Å^2 per residue of packing surface within the loop. Despite their local structural integrity and almost complete independence of each other, the loops pack closely against the β barrel to bury only slightly less surface area than is buried within each β sheet.

Overall, the surface buried between the loops and the β barrel has about the same 2 : 1 ratio of hydrophobic to hydrophilic surface as that buried among the eight β strands or between the two sides of the β barrel. In the packing surfaces, however, the loops appear to have a greater hydrophilic composition than the β barrel. Packing within the β barrel also is distinguished from that in the loops by the organized pattern of a hydrophilic shell with a hydrophobic core. Among the individual loops, there is a significant difference in the relative amount of local interactions versus interactions with other intrasubunit structural elements. Loops with a large portion of internal interactions (such as the Zn loop) have as much potential to act as folding nuclei as the repetitive secondary structure of β sheet or α helix, although the structure of others (such as the Greek key loop) may be almost totally dependent on their interactions with other structural fragments. Since the nature of the packing interactions in loops has not been characterized for other proteins, this analysis of the packing within the loops and between the individual loops and the β barrel in SOD could be profitably extended to determine whether the observed patterns are representative of other known protein structures, and to explore the role of the loops in protein folding and stability. At present, the possible role of subdomains as nucleation sites for protein folding should be a consideration in design; amino acid changes that affect interactions within subdomains may be more likely to disturb the folding process than those affecting interactions between subdomains.

4. SOD Dimer Interface

The molecular surfaces buried on each subunit by the dimer contact are continuous and complementary in both shape and chemical composition: the buried surfaces are interdigitated without any gaps the size of a water molecule, and the areas of contact match in terms of hydrophobic and hydrophilic atom types. Twenty-two residues contribute to the average of 519 $Å^2$ of molecular surface buried on each subunit by the contact, which represents about 8% of the total external molecular surface. When viewed along the surfaces perpendicular to the twofold symmetry axis of the dimer, the two closely fitted surfaces in the contact have a right-handed twist of about 60° (from the edge of the active site face to the edge of the β-barrel face of the subunit). This twist results in a larger interacting buried surface area on the two globular subunits than would occur by the interaction of other comparable parts of the subunit surface. The twisted surfaces are also effectively wrapped around each other to provide an overall specificity of fit with regard to possible subunit rotations and translations.

Although the contact is primarily hydrophobic, a significant portion (30%) of the buried surface area is hydrophilic. About 36% of the buried contact area is formed by main-chain atoms. Both the hydrophilic and main-chain contributions are located in patches scattered throughout the surface. Ten β-strand residues and 12 loop residues are buried by the dimeric interaction. In addition to these residues in direct contact, there is a surrounding region in which a number of hydrophilic residues may interact through bridging water molecules, such that the effective contact area is increased. Three residues from each subunit form four main-chain hydrogen bonds between the two subunits, but there are no intersubunit hydrogen bonds involving side-chain atoms. All of these hydrogen bonds link a β-strand residue and a loop residue, and together they form a hydrophilic patch near the center of the contact (Getzoff *et al.*, 1986a).

In sum, the strong dimeric interaction in SOD apparently results from both chemical and shape complementarity of the molecular surfaces buried from solvent. The comparable strength of the tertiary and quaternary interactions in this enzyme is consistent with the similar sizes of the molecular surface buried by the dimeric interaction (1038 $Å^2$) and that buried between the two sides of the β barrel (1327 $Å^2$). The structural determinants for the dimer contact result from the nonbonded interactions of sequence-invariant and main-chain atoms that form a stripe of buried surface area (located roughly perpendicular to the dimeric twofold axis) which accounts for 67% of the entire contact. The structural position, low temperature factors (see Section III), and sequence conservation that characterize this stripe suggest that these residues are important determinants for the dimeric interaction and stability.

III. INTERMOLECULAR INTERACTIONS

A. Complementarity at Molecular Interfaces

Both macromolecular recognition and stability require complementarity provided by specific residues at the molecular interface. This complementarity is a function of shape, stereochemistry, chemical interactions such as hydrogen bonds or hydrophobic forces, electrostatic properties, and perhaps other forces. The characteristics of this complementarity help determine the specificity and stability of the interaction and can thus tune an interface for a particular function.

By analyzing the details of complementarity in different molecular interfaces we should gain an understanding of the features associated with given biological functions. Three useful sources of information on intermolecular interactions are: (1) modeling of complexes based on the separated individual structures of the interacting molecules, (2) examination of multimeric proteins and complexes for which only the structure of the complex is known, and (3) comparison of solved structures both for individual molecules and for the complex. Most information currently comes from sources 1 and 2, since examples of 3 are limited (i.e., trypsin–trypsin inhibitor and carboxypeptidase–inhibitor complexes). However, current work on the structures of antibody–antigen complexes should eventually lead to more information of this type.

Three distinct types of SOD molecular interfaces are relevant to complementarity in intermolecular interaction: the interface between the substrate and its binding pocket (Tainer *et al.*, 1983), the intramolecular interface between the two sides of the core β-barrel structure of each subunit (Getzoff, 1982), and the extremely stable interface between the two subunits of the active dimer (Getzoff *et al.*, 1986a; Getzoff, 1982). The substrate superoxide fits snugly in a depression of the molecular surface located at the bottom of a deep active site channel. The molecular surfaces interpenetrate due to the electrostatic complementarity between the negatively charged substrate and both the positively charged copper ion and the guanidinium side chain of Arg 141 of the protein. The strong dimeric interaction and intramolecular packing were characterized in atomic detail by the examination of molecular surfaces buried from solvent by each interface. In each case, these buried surfaces were continuous and showed shape complementarity.

B. Shape Complementarity and Induced Fit

Shape and chemical complementarity are the established features of known complexes (Getzoff *et al.*, 1986a); however, a growing body of information suggests that induced fit may also play a crucial role in protein interactions. In protein folding, a domain may pass through intermediate structural forms before attaining the final secondary and tertiary structural motif (Goldenberg *et al.*, 1983; Richardson, 1981). Evidence for this can be found in protein active site regions. In SOD the active site binds two metal ions, and several of the metal-liganding residues are part of nonrepetitive β-bulge secondary structure (Richardson *et al.*, 1978), that permits a side-chain flip of 180° from the normal β alternation. Since SOD folds in the absence of metal ions, the implication is that the energetically preferred β alternation is changed to accommodate the induced fit associated with the binding of the active site metal ions (Tainer *et al.*, 1982).

The conformations of macromolecules in a complex are unlikely to be entirely identical to the uncomplexed conformations of the individual molecules. Local relaxation or induced fit of the molecular conformation at an interface would allow correction of imperfections in the initial complementarity. Such an ability to accommodate imperfections would tend to decrease the specificity of the interaction while increasing the stability; the desirability of this feature would depend on the function of the molecular interaction. To analyze flexibility in intermolecular interactions, we have examined additional empirically determined parameters from refined high-resolution X-ray crystal structures. These parameters, termed the atomic temperature factors, describe the relative mobility of different atoms.

Temperature factors can provide important data about the lower-frequency concerted motions of groups of atoms, indicating the relative conformational flexibility of different regions (Holbrook and Kim, 1984; Swaminathan *et al.*, 1982): high temperature factors indicate shallow potential wells, allowing access to multiple conformations at biological temperatures. By examining averaged main-chain temperature factors color-coded onto the molecular surface, we have begun to define roles for mobility and induced fit in different types of intermolecular interactions. For binding of small rigid molecules to proteins, flexibility is not required: the very rapid binding of the superoxide substrate (Tainer *et al.*, 1983) and the binding of heavy metal ions used for X-ray crystallographic derivatives (Getzoff and Tainer, 1986) occur in relatively well-ordered preexisting sites on the molecular surface of SOD. However, the molecular surface regions of trypsin where trypsin inhibitor binds and carboxypeptidase where the potato inhibitor binds contain both ordered and highly flexible parts of the polypeptide chain, suggesting how these interfaces accommodate both recognition and induced fit.

Local flexibility also appears to be important in antibody–antigen interactions (Tainer *et al.*, 1984; Getzoff *et al.*, 1985; Westhof *et al.*, 1984; Moore and Williams, 1980). Our studies of anti-peptide antibodies raised against both mobile and well-ordered sites in the protein myohemerythrin suggest that anti-peptide antibodies against contiguous, highly mobile sites react more strongly with the native protein than do anti-peptide antibodies against well-ordered regions (Tainer *et al.*, 1984). We (Tainer *et al.*, 1985b) and others (Westhof *et al.*, 1984) have found this high correlation between mobility and antigenicity for other proteins as well, by combining previously observed immunological data with high-resolution crystallographic results. Figure 4 illustrates this correlation for insulin. Since mobility is influenced by amino acid type (Karplus and Schultz, 1985) as well as by location in the structure (Thornton *et al.*, 1985), the flexibility of residues at an interface could potentially be changed by site-specific mutation.

C. Precollision Guidance and Interaction Efficiency

Electrostatic forces have been implicated in a variety of biologically important intermolecular interactions, but characterization of their roles in sufficient detail to allow the design of specific experiments has been difficult. Examination of the electrostatic potential mapped onto surfaces buried in macromolecular complexes indicates that electrostatic complementarity is often a feature of stable complexes (Getzoff *et al.*, 1986a), but the poorly defined dielectric constant in regions surrounding macromolecules complicates the calculation for uncomplexed molecules. To define the possible role of electrostatic forces in orienting molecules prior to their actual collision, we have developed a method for both the analysis and color-coded computer graphics display of the three-dimensional electrostatic vector field surrounding a macromolecule (Getzoff *et al.*, 1983; Getzoff *et al.*, 1986b). The electrostatic field direction has the advantage of being relatively insensitive to the dielectric model. We calculate both the electrostatic potential and field from partial charges assigned to each of the atoms of the protein (Weiner and Kollman, 1981; Weiner *et al.*, 1984), thus enabling the contributions of specific residues to be identified and tested experimentally. In another approach, dynamics simulations, which use simple geometric models based on the charge distribution in the

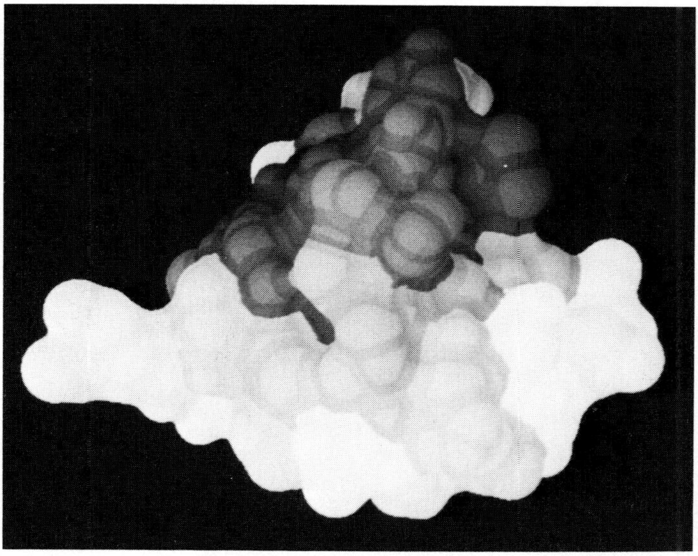

Fig. 4. Pattern of molecular surface mobility for insulin. The average main-chain temperature factors determined by crystallographic refinement have been mapped onto the molecular surface as a brightness scale with the most highly mobile areas being the brightest. The more mobile areas correlate with the insulin antigenic determinants (Tainer *et al.*, 1985b). Surface mobility and shape are highly correlated such that the more protruding areas, which are most available for interaction with other molecules, have the most local flexibility. The apparent high order of the upper portion of the insulin molecule is actually an artifact of the close packing between adjacent molecules in the crystal lattice. Thus, accurate information on mobility from X-ray crystallographic refinement requires that the crystal contacts as well as other factors be taken into account (Sheriff *et al.*, 1985). The solid external molecular surface was calculated with the programs AMS (Connolly, 1983a,b) and RAMS (Connolly, 1985).

atomic structure, are being developed to predict the degree of rate enhancement by electrostatic forces (Allison *et al.*, 1985).

By calculation, display, and examination of the electrostatic potential and electrostatic fields around SOD (Getzoff *et al.*, 1983; Getzoff *et al.*, 1986b) we have identified electrostatic forces involved in both long-range (Koppenol, 1982) and short-range recognition and orientation of the negatively charged superoxide substrate. Color coding the solvent-accessible molecular surface for the entire subunit according to electrostatic potential (Fig. 5) shows a unique feature of the molecule: the large positive area

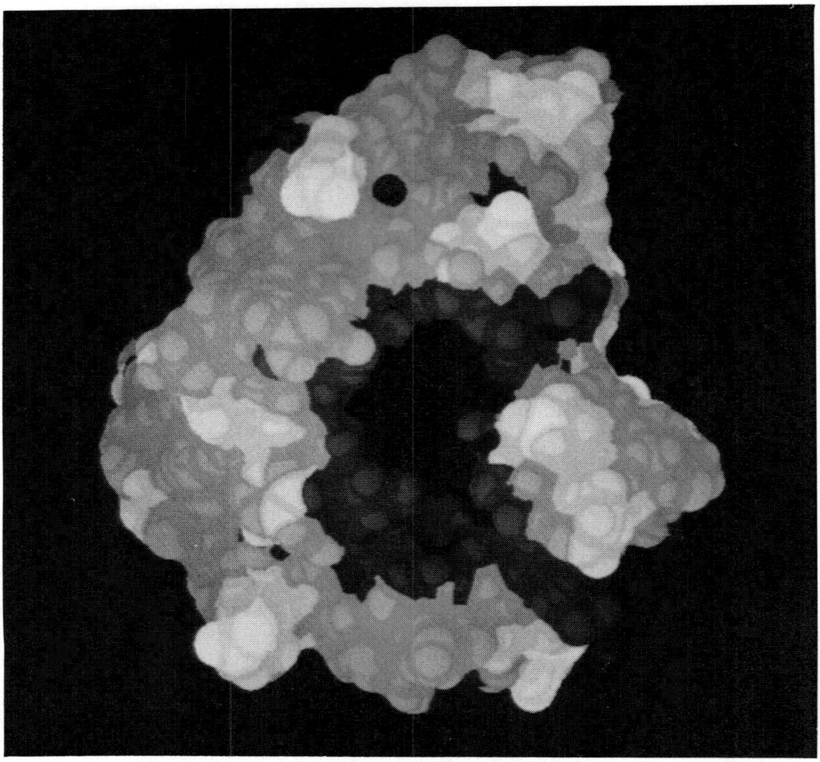

Fig. 5. The electrostatically color-coded, active site face of the SOD subunit, showing the highly positive potential around the active site channel (concave area, center). The electrostatic potential is divided into five categories coded by a radiating body brightness scale, with the most negative potential being white and the most positive potential being black (white < -21 kcal/mole; light gray -21 to -7 kcal/mole; medium gray -7 to $+7$ kcal/mole; dark gray $+7$ to 21 kcal/mole; black $>$ 21 kcal/mole). The solid external molecular surface was calculated with the programs AMS (Connolly, 1983a,b) and RAMS (Connolly, 1985) and coded by the electrostatic potentials seen by a water molecule. (Adapted from Getzoff *et al.*, 1983.)

surrounding the catalytic site. On the opposite side of the subunit, away from the active site, the distribution of electrostatic charge shows no definite pattern; i.e., there are no large concentrated patches of either positive or negative electrostatic potential. The characteristic positive area covering the active site channel represents approximately one-tenth of the surface area of the subunit and is composed almost exclusively of sequence-conserved amino acid residues.

Electrostatic field vectors calculated and displayed above the catalytic site indicate the direction in which a negative charge would flow in the field (Fig. 6). Vectors were calculated from the derivative of the electrostatic potential at points lying on concentric spheres radiating out from the open coordination site on the Cu. The direction of flow of the field is away from the sides and down into the center of the channel, where the reaction takes place. In cross section the channel funnels downward stepwise from a large shallow depression about 24 Å across to a deeper well about 10 Å wide and 5 Å deep, then narrows to less than 4 Å, forming a complementary pocket for the superoxide anion directly above the Cu ion (Fig. 7). At its greatest depth the channel is roughly 12 Å from top to bottom. The active site shape in SOD and other enzymes modulates intermolecular interactions directly by steric effects and indirectly by increased electrostatic effects due to the sequestering of water molecules from bulk solvent (Meyer *et al.*, 1985; Getzoff *et al.*, 1983). Guidance of the substrate O_2^- anion into the active site pocket by the electrostatic field of SOD appears to be significant at distances of 14 Å, thus coding for active site recognition prior to actual collision at the molecular surface (Fig. 6). The highly attractive portions of the electrostatic fields around the two active sites in the dimer cover a large solid angle, making the enzyme extremely efficient at attracting superoxide radicals. We dissected the electrostatic field to identify the contributing amino acid residues (Table II; Getzoff, 1982; Getzoff *et al.*, 1983); two of these have since been confirmed by biochemical studies using chemical modifications and exploring ionic-strength effects (Cudd and Fridovich, 1982). The major contributions to the electrostatic field are made by invariant residues.

In the SOD dimer interface, the electrostatic potential mapped on the buried surfaces is neutral in the center; only the edges show significant positive or negative electrostatic potentials. The surfaces are electrostatically complementary to one another and essentially seal the edges of the dimer contact at the protein–water interface (Getzoff *et al.*, 1986a). Electrostatic complementarity at molecular interfaces should increase both the efficiency with which the interacting molecules recognize and orient each other and the stability of the resultant complex. Site-directed mutation of amino acids contributing to electrostatic complementarity should therefore be able to change the specificity, kinetics, and stability of the complex. The tomato bushy stunt virus (TBSV) coat protein capsid is a good example of the important role of electrostatic complementarity in the stability of the protein complexes. Changing the electrostatic potential at interfaces between TBSV subunits from complementary to unfavorable by the removal of ions causes an almost explosive expansion of the viral capsid (Robinson and Harrison, 1982)

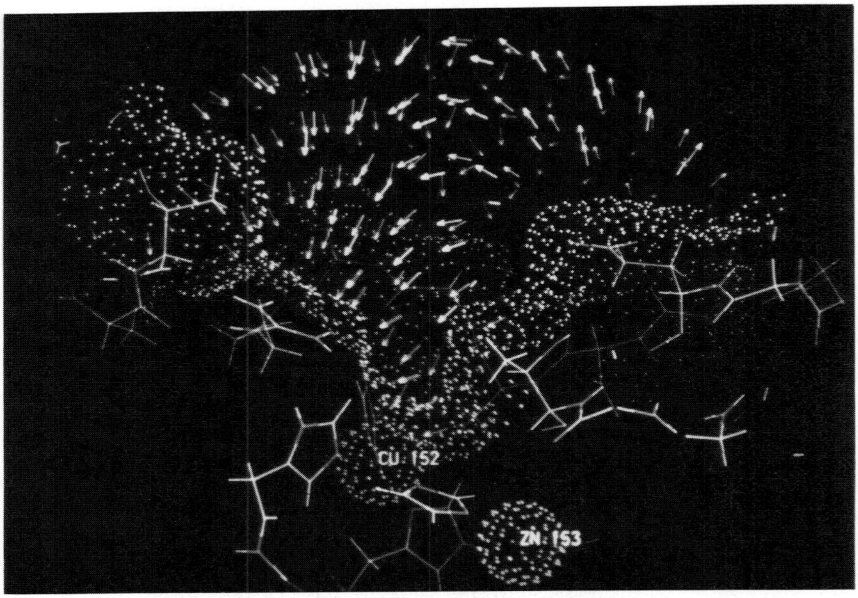

Fig. 6. Electrostatic field around the SOD active site channel. The arrows depict the electrostatic field that guides the negatively charged superoxide substrate to the catalytic Cu site (sphere, center), which lies in the most highly positive region of the molecular surface (shown by dots). The magnitude of the electrostatic potential is sufficient to allow recognition and alignment between enzyme and substrate prior to collision, thus increasing the reaction rate. In the original computer graphics representation, color coding was used to show the magnitudes of electrostatic potential for the molecular surface and field arrows (Adapted from Getzoff *et al.*, 1983).

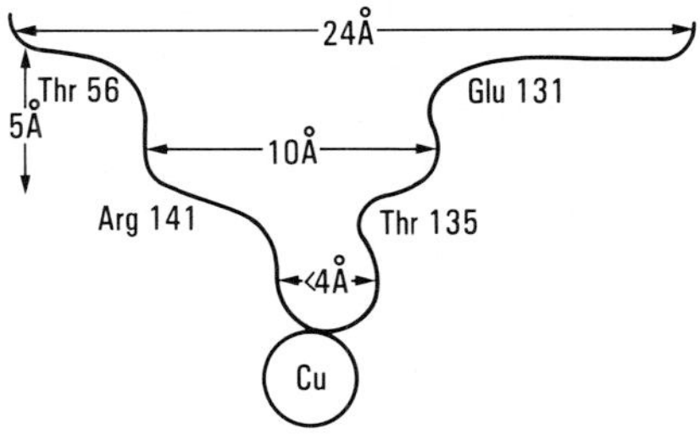

Fig. 7. Schematic diagram of the SOD active channel cross section. The diameter of the channel decreases stepwise to limit anion access to the catalytic Cu ion. Changes in residues forming the sides of the channel, such as Arg 141 and Thr 135, could be made to increase or decrease the availability of the Cu ion for interaction with other possible electron exchange partners (Adapted from Getzoff and Tainer, 1986.)

TABLE II

Effects of Individual Residues on the Electrostatic Field Direction[a]

Residue(s) neutralized[c]	Average angular change in electrostatic field direction produced in each of the eight shells[b]								
	0 Å	2 Å	4 Å	6 Å	8 Å	10 Å	12 Å	14 Å	All
Zn, His 61	1.3	4.1	1.9	2.1	3.1	3.9	4.3	5.8	4.4
Glu 119	0.6	1.7	3.1	5.5	7.1	8.8	8.6	8.7	8.1
Lys 120	0.7	2.0	5.3	7.5	18.0	22.6	26.3	24.5	22.4
Glu 130	0.2	0.5	1.0	2.0	3.9	6.1	9.7	19.0	10.7
Glu 131	0.6	3.0	10.1	23.1	30.7	37.4	37.9	36.9	34.7
Lys 134	0.4	1.2	2.7	5.6	11.9	20.3	31.0	37.3	26.7
Arg 141	7.4	16.0	31.9	29.3	21.7	23.9	24.5	20.7	23.2
Glu 119 and Lys 120	0.4	1.4	4.2	4.8	13.4	16.8	19.6	17.1	16.3
Glu 131 and Lys 134	0.3	2.1	7.9	17.5	25.0	34.2	43.4	44.2	37.3
$D = 80$[d]	4.8	6.4	13.6	9.7	12.8	14.3	16.7	18.2	15.8

[a] Adapted from Getzoff *et al.* (1983).

[b] Shell, named by the radial distance from the position of the axial water bound to the Cu(II) with the number of vectors calculated in each successive shell being: 1, 5, 10, 14, 32, 65, 105, and 107.

[c] Mathematically neutralized for calculating the electrostatic field.

[d] $D = 80$: changes resulting from the use of the bulk water dielectric constant of 80.

D. Probing Reaction Mechanisms

A family of superoxide dismutase enzymes having Cu and Zn, or Mn, or Fe at the active site protects against the toxic effects of the superoxide radical (O_2^-) by catalyzing its dismutation to molecular oxygen and hydrogen peroxide; this reaction occurs through the alternate reduction and oxidation of the active site metal ion (Cu, Mn, or Fe): $O_2^- + O_2^- + 2H^+ \rightarrow H_2O_2 + O_2$ (Fridovich, 1979). The catalytic rate of SOD is very rapid ($2 \times 10^{-9}\ M^{-1}\ sec^{-1}$; Klug *et al.*, 1972), suggesting the evolution of an optimal active site for the recognition and chemical catalysis of the substrate by the enzyme. Furthermore, in combination with biochemical studies, the atomic structure has implications for the mechanism of activity (Tainer *et al.*, 1983).

In Fig. 8 all of the active site residues including the liganding residues and the sequence-invariant Arg 141 (Malinowski and Fridovich, 1979) are shown skeletally along with the Cu and Zn ions. One unique feature of this enzyme active site is the bridge formed between the Zn and Cu ions

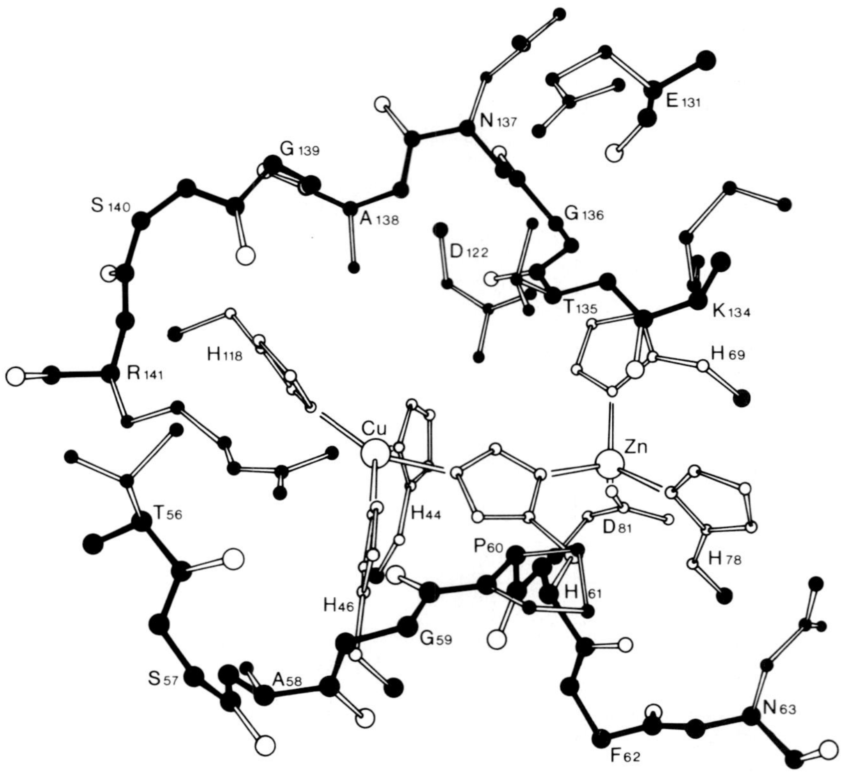

Fig. 8. Critical active site channel residues of SOD as viewed from the solvent. The main chain is shown by solid black bonds, the ligand side chains by open bonds, and the other side chains by solid atoms and open bonds. Residues are labeled with the one-letter amino acid code and residue number. Residues 131, 134–139, and 141 form one rim of the channel, residues 56, 58–60, and 63 form the other rim, and the metal ions and their ligands form the floor. Main-chain atoms only of residues 57, 62, and 140 have also been included in the drawing for continuity. The Cu(II) (left center) is liganded by His 118, 44, 61, and 46 (clockwise from upper left) with an uneven tetrahedral distortion from square planar geometry. The geometry of Zn ligands (right center) His 61, 69, and 78 and Asp 81 (clockwise from left) is tetrahedral with the largest distortion involving His 69. Asp 81 is behind the Zn and is buried. His 61 bridges between the Cu and Zn, with its ring almost planar to the metals but tilted toward the solvent by ~0.3 Å on the Cu side. In the metal-liganding residues, all five of the nonliganding imidazole nitrogens and the nonliganding Asp 81 carboxyl oxygens are hydrogen-bonded (Tainer *et al.*, 1982). Reduction of the Cu during catalysis should weaken the hydrogen bonds of the His rings bound to the Cu, allowing small shifts in ligand geometry that could give a more tetrahedral Cu site. The carboxyl group of solvent-inaccessible Asp 122, which forms a second, less direct bridge between the metals by hydrogen bonding to both Zn ligand His 69 and Cu ligand His 44, may couple the Cu oxidation state to the Zn geometry; coupled movement of Asp 122 and His 69 away from His 44 and toward His 78 in the Cu(I) enzyme would decrease the distortion in the Zn tetrahedral geometry. Based on

by the His 61 side chain. This bridge is made and broken as the reaction proceeds (McAdam *et al.*, 1977). Proton transfer from His 61 to the superoxide explains the pH independence of the reaction; i.e., this crucial proton does not come from bulk solvent.

The human SOD gene has been sequenced, cloned, and expressed (Hallewell *et al.*, 1985), allowing site-directed mutagenesis experiments. Specific changes already made include modification of apparently critical active site residues such as Arg 141. This type of experiment promises to directly probe the role of specific residues in the mechanism of enzyme reactions. Since enzymes have evolved to optimize a number of conflicting requirements, it may be possible to design improved catalytic sites. Probably it will be more interesting to apply an understanding of the detailed reaction mechanism to the redesign of enzyme active sites to perform different reactions.

IV. ENGINEERING PRINCIPLES

A. The Advantage of Failures

The technology now exists to make completely new proteins or to redesign known proteins to alter and improve their properties and functions. The missing step in achieving these goals is appropriate methodology for choosing the proper design. Our recursive approach is to use existing protein structural information for the prediction of the structural and functional changes arising from specific amino acid mutations, to mutate such sites, and to examine the structure of the modified protein to see if the changes are, in fact, those expected. Thus, the modeling of molecular structures and their chemical, physical, and functional parameters plays a critical role in the design of genetic experiments; and these experiments, in turn, are important for testing and improving appropriate models.

Given that the basic weakness in rational protein design is in the accuracy of the prediction from modeling, it is obvious that a recursive approach is not simply pragmatic, but required. More bluntly, current approaches to modeling protein interactions are unlikely to improve significantly without a data base of specific defined examples in which the model clearly fails. Predictive failures will provide the single most important source of data for improving our modeling technology. Patterns of

the interpretation of the active site atomic structure, the techniques of site-directed mutagenesis allow the specific role of each side chain to be tested. (Adapted from Tainer *et al.*, 1983.)

predictive error such as consistent problems with the design of turns will pinpoint problems in our approach, and isolated cases will point to error in our energy and force field parameters.

A corollary to our arguments is that rather than beginning from scratch, a more useful approach for protein design is to consider proteins as mosaics that can be changed by restructuring the component subdomains or by changing the nature of their interactions. This follows a kind of construction-by-parts approach, which may in some ways resemble the evolutionary recombination of introns to produce new proteins. It also provides a rapid scheme for accumulating sets of failures and successes in a number of defined and easily modeled subsystems. In addition, the results will be generally applicable to different proteins.

SOD is an especially attractive protein for our approach, because the β-barrel motif appears in proteins with many different functions, including electron transfer (plastocyanin and azurin), recognition and binding (immunoglobulin and concanavalin A), catalysis (serine proteases, pyruvate kinase, and staphylococcal nuclease), and stable structural assembly (icosahedral virus capsid proteins).

B. Prospects for Increased Stability

Mutations that decrease the stability or catalytic activity of an enzyme are usually easier to predict than those that improve these or other characteristics. However, based on a computer graphics analysis of the interdomain and intersubunit interactions in several proteins (Getzoff *et al.*, 1986a), prediction of specific mutations that may increase the stability of a protein appears to be a fairly realistic goal.

Redesigning SOD will involve changes in the structural subdomains already identified as affecting their behavior and stability. From Section II it is apparent that the primary interactions between these subdomains are hydrophobic complementarity, side chain-to-side chain hydrogen bonds, salt links between charged side chains, and the covalent disulfide bond. Improving and increasing these interactions should have a marked effect on stabilizing the protein fold. Additional information about stabilizing interactions may be obtained from work in progress on the characterization of SOD enzymes from thermophilic organisms.

Examples of changes to increase the enzyme's tertiary and quaternary stability can be inferred from the results of the analysis in Sections II and III. Within the β barrel most contacts are between adjacent strands with a few interactions across the β barrel. The integrity of the β-barrel subdomain can be altered by changing the number of cross-barrel contacts. This requires substituting long hydrophobic side chains in inward-facing positions on opposing β-barrel strands. The integrity of the loop regions depends primarily on main chain-to-side chain hydrogen bonds and hydrophobic packing interactions. The strength of these hydrophobic interactions can be increased by sequence changes to eliminate and fill gaps to as to allow increased shape complementarity within each loop. The stability of the dimer interaction depends on the continuous stripe formed by sequence-invariant and main-chain atoms, which accounts for two-thirds of the entire interacting surface area. Significant changes within this stripe will potentially disrupt the contact; modifications designed to improve the interactions in this area should improve dimer stability.

C. Current Limitations and Prospects

Knowledge of the interactions responsible for the shape and stability of the domains and subdomains forming SOD is a prerequisite for redesigning the enzyme for improved stability, changed substrate specificity, or coordination of different metal ions. Currently, there are simply too many unknowns to usefully apply structural information to more complex problems such as the design of multifunctional proteins having coupled domains with different enzymatic and binding properties. Yet, one can envision a second generation of redesigned SOD enzymes having coupled domains with additonal functions. The combination of SOD with peroxidase, for example, would join related functions in a single macromolecule. For now, the proposal that the mosaic nature of the protein domain be used to rebuild a protein by parts is not without problems. For short amino acid sequences, sequence identity does not guarantee secondary structural identity. For example, identical hexameric sequences that have different secondary structures have been found in pairs of known protein structures (Wilson *et al.*, 1985).

When the three-dimensional structure of the protein of interest is unknown or only partially known, it is often worthwhile to make some type of structural model to aid in the design of experiments. One approach is to find sequence and/or secondary structural homology to known protein structures and build a model from this information. Frequently this situation arises when genetic sequencing provides information about a protein that is not available in large enough amounts for structural studies. To address this problem, we use an empirically based approach involving four steps: (1) identification of the probable secondary structural elements, (2) selection of possible tertiary structural motifs based on the identity and placement of the secondary structural elements, (3) detailed search of sequences of the proteins known to fall in the same structural class(es) to find limited levels of sequence–structure homology, and (4) construction of a model consistent with the identified probable secondary structure, tertiary motif, and similarity to a solved structure within that class.

Since the level of sequence homology identified may not be statistically significant for a single pairwise matching, it is important to perform the multiple alignment in a way that gives a higher weight to the conserved and invariant sequence positions in both the known protein structure and the protein being modeled. In our approach, the development of a specific model is dependent on step 3 and is not attempted without the identification of a similar known structure. Once a tertiary structural model is obtained from this procedure, specific experimental questions regarding structure and function can be addressed. This approach has been applied to the pilin protein from *Neisseria gonorrhoeae* (Deal *et al.,* 1985) and the transposase protein A from bacteriophage Mu (Harshey *et al.,* 1985). Such predictive methods are limited and inaccurate but can aid rational design of experiments for proteins of unknown tertiary structure and efforts to redesign the conformation of a subdomain.

Even in solved protein structures, there are uncertainties due to the effects of mobility and solvation. Local flexibility of the polypeptide chain permits final adjustments to maximize the shape complementarity of interacting macromolecules through induced fit. Such flexibility is correlated with macromolecular shape; protrusions appear consistently to have high mobility (Tainer *et al.,* 1985a). However, the complexity of protein structure precludes accurate dynamic simulations over the time scales appropriate for intermolecular interaction. Consequently, predictions of macromolecular interactions will be difficult even when atomic structures are known for the individual molecules. A related problem is the proper treatment of bound water molecules, which may be exchanging extremely rapidly with bulk solvent, but consistently occupy certain sites. Some types of molecular interactions may be entropically driven by the release of bound water molecules. In others, bound water molecules act to alter the damping of electrostatic forces. Unfortunately, the dynamics of interaction of these water molecules with themselves and the protein have yet to be defined satisfactorily.

The great deal of structural information regarding macromolecular interactions and the set of extremely powerful genetic techniques available for implementing design changes are limited by large uncertainties due to the processes of induced fit and solvation. This combination of knowledge and ignorance recommends empirical experimentation in which probable desired changes identified by modeling are implemented by site-directed mutagenesis and then examined by X-ray crystallographic structure determination, as discussed above.

The long-term prospects for protein design seem extremely promising. Site-directed mutagenesis coupled with crystallography should provide the crucial data needed to improve our theoretical models for protein structure and interactions. Designed proteins are expected to have significant advantages over naturally occurring ones for may uses. Engineered proteins can be redesigned to optimize a single desired feature while ignoring other irrelevant functions; thus, in selected respects they can undoubtedly be made superior to naturally occurring proteins, whose structure reflects a compromise among the many constraints operating *in vivo*. Although the enzymatic capabilities of proteins may seem limited by the properties of amino acids, the binding of specially designed cofactors can be used to extend these capacities considerably.

ACKNOWLEDGMENTS

This chapter builds on a number of ideas from others, particularly David Richardson and Jane Richardson at Duke University, Michael Pique at the University of North Carolina at Chapel Hill, and Michael Connolly and Arthur Olson at the Research Institute of Scripps Clinic.

REFERENCES

Allison, S. A., Ganti, G., and McCammon, J. A. (1985). *Biopolymers* **24,** 1323–1336.

Bernstein, F. C., Koetzle, T. F., Williams, G. J. B., Meyer, E. F., Brice, M. D., Rodgers, J. R., Kennard O., Shimanouchi, T., and Tasumi, M. (1977). *J. Mol. Biol.* **112,** 535–542.

Chothia, C., and Janin, J. (1981). *Proc. Natl. Acad. Sci. U.S.A.* **78,** 4146–4150.

Cohen, F. E., Sternberg, M. J. E., and Taylor, W. R. (1981). *J. Mol. Biol.* **148,** 233–272.

Connolly, M. L. (1983a). *Science* **221,** 709–713.

Connolly, M. L. (1983b). *J. Appl. Crystallogr.* **16,** 548–558.

Connolly, M. L. (1985). *J. Mol. Graphics* **3,** 19–24.

Connolly, M. L., and Olson, A. J. (1985). *Comput. Chem.* **9,** 1–6.

Craik, C. S., Largman, S., Fletcher, T., Roczniak, S., Barr, P. J., Fletterick, R., and Rutter, W. J. (1985). *Science* **228,** 291–297.

Cudd, A., and Fridovich, I. (1982). *J. Biol. Chem.* **257,** 1143–1147.

Deal, C. D., Tainer, J. A., So, M., and Getzoff, E. D. (1985). *In* "The Pathogenic Neisseriae" (G. K. Schoolnik, ed.), pp. 302–308. Am. Soc. Microbiol., Washington, D.C.

Forman, H. J., and Fridovich, I. (1973). *J. Biol. Chem.* **248,** 2645–2649.

Fridovich, I. (1979). *Adv. Inorg. Biochem.* **1,** 67–90.

Getzoff, E. D. (1982). Ph.D. Dissertation, Duke University, Durham, North Carolina.

Getzoff, E. D., and Tainer, J. A. (1986). *In* "Ion Channel Reconstitution" (C. Miller, ed.), pp. 57–73. Plenum, New York.

Getzoff, E. D., Tainer, J. A., Weiner, P. K., Kollman, P. A., Richardson, J. S., and Richardson, D. C. (1983). *Nature (London)* **306,** 287–291.

Getzoff, E. D., Tainer, J. A., and Lerner, R. A. (1985). *In* "Immune Regulation" (M. Feldmann, ed.), p. 243–258. Humana Press, Clifton, New Jersey.

Getzoff, E. D., Olson, A. J., and Tainer, J. A. (1986a). *Biophys. J.* **49,** 191–206.

Getzoff, E. D., Olson, A. J., and Tainer, J. A. (1986b). *In* "Superoxide and Superoxide Dismutase in Chemistry, Biology and Medicine" (G. Rotilio, ed.), pp. 135–140. Elsevier, Amsterdam.

Goldenberg, D., Smith ,D., and King, J. (1983). *Proc. Natl. Acad. Sci. U.S.A.* **80,** 7060–7064.

Hallewell, R. A., Masiarz, F. R., Najarian, R. C., Puma, J. P., Quiroga, M. R., Randolph, A., Sanchez-Pescador, R., Scandella, C. J., Smith, B., Steimer, K. S., and Mullenbach, G. T. (1985). *Nucleic Acids Res.* **13,** 2017–2034.

Harshey, R. M., Getzoff, E. D., Baldwin, D. L., Miller, J. L., and Chaconas, G. (1985). *Proc. Natl. Acad. Sci. U.S.A.* **82,** 7676–7680.

Holbrook, S. R., and Kim, S.-H. (1984). *J. Mol. Biol.* **173,** 361–388.

Karplus, P. A., and Schultz, G. E. (1985). *Naturwissenschaften* **72,** 212–213.

Klug, D., Rabani, J., and Fridovich, I (1972). *J. Biol. Chem.* **247,** 4839–4842.

Koppenol, W. H. (1982). *In* "Oxidases and Related Redox Systems" (T. E. King, H. S. Mason, and M. Morrison, eds.), pp. 127–136. Pergamon, Oxford.

McAdam, M. E., Fielden, E. M., Lavelle, F., Calabrese, L., Cocco, D., and Rotilio, G. (1977). *Biochem. J.* **167,** 271.

Malinowski, D. P., and Fridovich, I (1979). *Biochemistry* **18,** 5055–5060.

Meyer, T. E., Cheddar, G., Bartsch, R. G., Getzoff, E. D., Cusanovich, M. A., and Tollin, G. (1985). *Biochemistry* **25,** 1383–1390.

Moore, G. R., and Williams, A. J. P. (1980). *Eur. J. Biochem.* **103,** 543–550.

O'Donnell, T. J., and Olson, A. J. (1981). *Comput. Graphics* **15,** 133.

Olson, A. J., Getzoff, E. D., and Tainer, J. A. (1984). "Terms of Entrapment: Structure and Function of Superoxide Dismutase" (computer animated film). Copyright Scripps Clinic and Research Foundation, La Jolla, California.

Richards, F. M. (1977). *Annu. Rev. Biophys. Bioeng.* **6,** 151–176.

Richardson, J. S. (1981). *Adv. Protein Chem.* **34,** 168–339.

Richardson, J. S., Getzoff, E. D., and Richardson, D. C. (1978). *Proc. Natl. Acad. Sci. U.S.A.* **75,** 2574–2578.

Robinson, I. K., and Harrison, S. C. (1982). *Nature (London)* **297,** 563–568.

Sheriff, S., Hendrickson, W. A., Stenkamp, R. E., Sieker, L. C., and Jensen, L. H. (1985). *Proc. Natl. Acad. Sci. U.S.A.* **82,** 1104–1107.

Stellwagen, E., and Wilgus, H. (1978). *In* "Biochemistry of Thermophily" (S. Friedman, ed.), pp. 223–232. Academic Press, New York.

Swaminathan, S., Ichiye, T., van Gunsteren, W., and Karplus, M. (1982). *Biochemistry* **21,** 5230–5241.

Tainer, J. A., Getzoff, E. D., Beem, K. M., Richardson, J. S., and Richardson, D. C. (1982). *J. Mol. Biol.* **160,** 181–217.

Tainer, J. A., Getzoff, E. D., Richardson, J. S., and Richardson, D. C. (1983). *Nature (London)* **306,** 284–287.

Tainer, J. A., Getzoff, E. D., Alexander, H., Houghten, R. A., Olson, A. J., Lerner, R. A., and Hendrickson, W. A. (1984). *Nature (London)* **312,** 127–133.

Tainer, J. A., Getzoff, E. D., and Olson, A. J. (1985a). *In* "*Molecular Dynamics and Protein Structure*" (J. Hermans, ed.), pp. 110–115. University of North Carolina, Chapel Hill.

Tainer, J. A., Getzoff, E. D., Paterson, Y., Olson, A. J., and Lerner, R. A. (1985b). *Annu. Rev. Immunol.* **3,** 501–535.

Tainer, J. A., Getzoff, E. D., Sayre, J., and Olson, A. J. (1985c). *J. Mol. Graphics* **3,** 103–105.

Thornton, J. M., Edwards, M. S., Taylor, W. R., and Barlow, D. J. (1985). *J. Mol. Graphics* **3,** 105–106.

Valentine, J. S., and Pantoliano, M. W. (1981). *In* "Copper Proteins" (J. G. Spiro, ed.), Met. Ions Biol. Vol. 3, pp. 291–358. Dekker, New York.

Weiner, P. K., and Kollman, P. A. (1981). *J. Comput. Chem.* **2,** 287–303.

Weiner, S. J., Kollman, P. A., Case, D. A., Singh, U. C., Ghio, C., Alagona, G., Profeta, S., Jr., and Weiner, P. (1984). *J. Am. Chem. Soc.* **106,** 765–784.

Westhof, E., Altschuh, D., Moras, D., Bloomer, A. C., Mondragon, A., Klug, A., and Van Regenmortal, M. H. V. (1984). *Nature (London)* **311,** 123.

Wilson, I. A., Haft, D. H., Getzoff, E. D., Tainer, J. A., Lerner, R. A., and Brenner, S. (1985). *Proc. Natl. Acad. Sci. U.S.A.* **82,** 5255–5259.

5

The Design and Construction of Biologically Active Peptides, Including Hormones

EMIL THOMAS KAISER

Laboratory of Bioorganic Chemistry and Biochemistry
The Rockefeller University
New York, New York 10021-6399

I. INTRODUCTION

In our laboratory several approaches to the design of biologically active proteins have been pursued. In one approach, starting with an existing tertiary structure such as that present in a naturally occurring enzyme and using either chemical modification (1) or recombinant DNA methodology (2,3), we have introduced a new active site group in the existing enzyme system. Examples of the chemical modification approach include the preparation of the semisynthetic enzymes flavopapain and flavoglyceraldehyde-3-phosphate dehydrogenase. In the case of papain, by covalent modification of a hydrolytic enzyme with appropriate flavin derivatives we have succeeded in converting the enzyme into a new semisynthetic enzyme with oxidoreductase activity (1,4–6). With glyceraldehyde-3-phosphate dehydrogenase (7), taking an enzyme which normally uses NAD/NADH as the cofactor, by chemical modification of a sulfhydryl group in each subunit with an appropriate flavin analog we have succeeded in converting the enzyme into one which uses a flavin as the prosthetic group. Employing recombinant DNA methodology, we have succeeded in converting the active site hydroxyl of the Ser 102 residue in *Escherichia coli* alkaline phosphatase into a sulfhydryl group (2,3). The resultant mutant Cys 102 alkaline phosphatase shows catalytic

PROTEIN ENGINEERING
Applications in Science,
Medicine, and Industry

activity toward a series of aryl phosphate monoesters. Although our work with these approaches has demonstrated that it is possible to take an existing tertiary structure which contains an active site and convert that active site into one with its own catalytic activity, from the standpoint of design and construction of entire enzyme systems, a major goal which remains to be achieved is to develop our ability to predict the way in which primary amino acid sequences will fold into tertiary structure. In other words, if we are to develop our ability to design and to carry out the total synthesis of enzymes from the constituent amino acids, it will be necessary to establish methodology for the prediction of how protein chains fold. Unfortunately, at present our knowledge of the basis of tertiary structure is still rather fragmentary. For this reason, a number of years ago we decided to initiate another approach to protein design involving the construction of those regions of peptide and protein systems which are outside the active site areas (8,9). In setting ourselves this target we decided that it would be best to start with systems where to a first approximation we could neglect tertiary structure. Our argument was that if we were to focus primarily on secondary structural features we would have an excellent chance to be able to build reasonable models, since our ability to predict amino acid sequences which form given secondary structures is really quite good.

When searching for peptide and protein systems which we might model in terms of their secondary structures, we were drawn to those molecules which bind to biological interfaces such as membranes. Through the work to be described in part in this chapter we have shown that amphiphilic secondary structures play a major role in the biological activity and physical properties of many peptides and proteins which bind on biological surfaces. In structures such as the amphiphilic α helix one face of the structural region is hydrophilic while the other is hydrophobic. We have shown that for surface-active peptide and protein molecules ranging in size from the apolipoproteins, which are large, to peptides of more modest length such as the bee venom toxin melittin and various peptide hormones, amphiphilic secondary structural regions appear to be important to the biological and physical properties (10–17). With these examples, in a first approximation the possible formation of tertiary structures can be neglected in designing models. Therefore, we have found it feasible to proceed with a rational approach through the design of new amino acid sequences which possess many of the important properties of major amphiphilic secondary structural regions of the natural systems. The modeling approach that we have developed has allowed us to make model peptides with activities similar to if not greater than those of the corresponding naturally occurring peptides and proteins.

II. DEVELOPMENT OF PRINCIPLES FOR THE DESIGN OF MODELS FOR SURFACE-ACTIVE PEPTIDES AND PROTEINS

In addition to the use of several computer programs which have been developed for the recognition of amphiphilic secondary structures, we have found it useful to search for such secondary structures in peptides and proteins by means of relatively simple visual or graphical methods. It is easy to discern potential amphiphilic β sheets since they consist of alternating hydrophilic and hydrophobic residues. Furthermore, if it is suspected that a given region of a peptide or protein might have the potential to form a helical structure, the possibility that this structure might be an amphiphilic one induced in an appropriate amphiphilic environment can be assessed through the use of a drawing of a helical wheel having the appropriate pitch (8,9). Once we find that such amphiphilic secondary structures may be formed in peptides and proteins, we generally use several experimental approaches to test the importance of these conformations. Several indirect methods can be employed, including studies of the binding of the peptides and proteins to amphiphilic surfaces such as those encountered with polystyrene beads coated with phospholipid (18). Also, the formation of stable monomolecular layers at the air–water interface can be an important diagnostic for testing the proposal that given peptides and proteins are induced to form regular structures under these conditions. The most powerful approach for determining the importance of amphiphilic structures in biologically active peptides and proteins involves the use of peptide synthesis to construct appropriate models.

The basis of our modeling approach is the hypothesis that when the biological and physical properties of a peptide or protein depend on the secondary structural features of a given region, then through the construction of models for the region under examination we should be able to design peptides with minimal homology to the natural system but with the potential to form a very closely related secondary structure. In essence, if a particular amino acid sequence is not crucial for a given region of a peptide or protein but instead a given type of secondary structural feature is crucial, the secondary structure should be reproduced by a model comprised of an amino acid sequence which has relatively little homology to the natural sequence. In the work that is to be described here briefly we have made use of solid-phase synthesis, but the design principles we have applied could just as well be employed in recombinant DNA methodological approaches to this problem.

III. CONSTRUCTION OF AMPHIPHILIC HELICAL MODELS OF APOLIPOPROTEINS, PEPTIDE TOXINS, AND HORMONES

The first surface-active protein for which we undertook modeling was apolipoprotein A-I (apo A-I), the principal protein in high-density lipoprotein (HDL) (8–11). It had been pointed out by Segrest *et al.* on the basis of the construction of molecular models that there appear to be segments of the apo A-I molecule that have the potential to form amphiphilic α helices (19). Subsequently, working independently, Fitch (20) and McLachlan (21) suggested that in apo A-I there might be a repeating pattern of about six such amphiphilic α-helical regions in which the helices are punctuated at regular intervals by residues like Pro. This analysis of the apo A-I molecule suggested to us a picture of the binding of apo A-I to HDL in which the amphiphilic helices lie with their axes roughly tangential to the surface and their hydrophobic faces penetrating between the phospholipid head groups, coming in contact with the long aliphatic chains of the phospholipid, while their hydrophilic faces are oriented toward the aqueous phase.

To test this picture of the binding of apo A-I to HDL we synthesized a 22-residue model peptide designed to reproduce in an idealized fashion the principal structural feature, the amphiphilic α helix postulated to be an essential structural component of apo A-I. The design of the amino acid sequence of the model was based on the following four criteria. First, the amino acids used in the preparation of the model were ones having a high potential to form α helices as predicted by empirical methods such as that described by Chou and Fasman (22). Second, the ratios of charged to uncharged residues and acidic to basic residues should approximate the balance found in the natural domains of apo A-I. Third, when in the predicted helical conformation the model peptide should show amphiphilic character corresponding to that expected for an apo A-I domain in an idealized fashion. And finally, the linear arrangement of the amino acids in the model should have a minimal resemblance to any segment of the natural sequence of apo A-I. The amino acid sequence of the resulting model is shown as follows. When an α-helical projection of this model is made, it can readily be seen that the distribution of hydrophobic and hydrophilic amino acids corresponds to an idealized structure for one of the amphiphilic helical segments of naturally occurring apo A-I (8,9). With the exception of the residues at the N terminus and the C terminus, only three amino acids were utilized in the synthesis: leucine was employed as the hydrophobic residue, glutamic acid as the acidic residue, and lysine as the basic residue. These amino acids are believed to have a

```
 1                                           10
Pro-Lys-Leu-Glu-Glu-Leu-Lys-Gly-Lys-Leu-
 11                                          20
Lys-Glu-Leu-Leu-Glu-Lys-Leu-Lys-Glu-Lys-
 21
Leu-Ala
```

Peptide model for apo A-I

high helix-forming potential (22). Further, when we compared the model sequence with representative amphiphilic helical domains of apo A-I it appeared that the charge balance of natural segments was retained in the model, while from inspection of a helical wheel projection of the model one could see clearly that the amphiphilicity of the synthetic peptide corresponded to an idealized version of the regions in the natural protein. Thus, whereas in the putative amphiphilic α-helical segments of the naturally occurring peptides there are several "mistakes," which correspond, for instance, to a polar amino acid residing on the nonpolar helix face, such mistakes do not exist in the model, which has a clear separation of the hydrophobic from the hydrophilic regions.

Although the synthetic model peptide was only 1/11 the length of apo A-I itself, the model not only exhibited the expected helical structure, as determined by circular dichroism measurements on solutions at high concentrations of peptide or in structure-forming solvents such as trifluoroethanol, but also reproduced many of the fundamental properties of apo A-I (8–11). In particular, the model peptide behaved in a manner similar to that of the natural protein at the air–water and phospholipid–water interfaces. Furthermore, the model peptide was able to participate in the activation of the important enzyme lecithin–cholesterol acyltransferase much as does apo A-I itself. From our studies we concluded that the physical and biological characteristics of apo A-I are, indeed, highly dependent on the presence of amphiphilic α-helical structures which can be induced on binding to surfaces. The successful development of our model peptide shows that, at least for the properties we have investigated, specific side-chain interactions and complex tertiary folding of the protein do not play major roles in apo A-I (23). Most recently, we found that a 44-amino-acid peptide which corresponds to a covalent dimer of the first model peptide for apo A-I is even more effective in mimicking the physical properties of apo A-I. From our work on the 44-peptide we concluded that the most likely fundamental structural unit in apo A-I may consist of two amphiphilic helical regions flanking a bend which is formed by a short sequence including the Pro residue.

In the next stage of our modeling approach we developed a model for melittin, a hemolytic peptide consisting of 26 residues in its principal form

```
1                                      10
Gly-Ile-Gly-Ala-Val-Leu-Lys-Val-Leu-Thr-
11                                     20
Thr-Gly-Leu-Pro-Ala-Leu-Ile-Ser-Trp-Ile-
21
Lys-Arg-Lys-Arg-Gln-Gln
```

Amino acid sequence of melittin

```
1                                      10
Leu-Leu-Gln-Ser-Leu-Leu-Ser-Leu-Leu-Gln-
11                                     20
Ser-Leu-Leu-Ser-Leu-Leu-Leu-Gln-Trp-Leu-
21
Lys-Arg-Lys-Arg-Gln-Gln
```

Amino acid sequence of model for melittin

and acting as the main toxic component of honeybee venom (12). According to our structural hypothesis, melittin consists of two sections, with the first 20-amino-acid portion forming an amphiphilic α helix which is rather hydrophobic in nature and which is joined to a C-terminal hexapeptide region containing basic residues. Neither the 1–20 proposed amphiphilic α helix nor the C-terminal cluster hexapeptide is hemolytically active in itself (24,25). We have proposed that melittin's ability to lyse cells involves interaction of the amphiphilic α helix with the cell surface which positions the very basic C-terminal region in such a way that lysis can be initiated. To test our structural model for melittin we designed a synthetic peptide toxin, using principles similar to those employed in the design of the apo A-I model (12). For comparison, the sequences of melittin and the model peptide are shown above. In the region from residues 1 through 20 the model has approximately two-thirds hydrophobic residues and one-third hydrophilic residues, roughly a balance comparable to that in melittin itself, but the model shares only five residues in this region with melittin. In the basic hexapeptide at the C terminus the sequence of the model was not changed from that of melittin. The model for melittin duplicated all of the physical properties of melittin, including its ability to self-associate as a tetramer with high α-helical content. It was also found to activate phospholipase A_2 similarly to melittin and to lyse human erythrocytes somewhat more effectively than melittin (12,26).

We have also extended the modeling approach developed with apo A-I and melittin to a number of peptide hormones which are in the range of 10 to 50 amino acids in length and which do not have extensive disulfide cross-linking (9,13-17). In many such hormones there appear to be significant amphiphilic helical regions, and in focusing on such regions we have

$(\text{Val-Glu-Val-(TFA)Orn})_3\text{-Val}$	Protected 13-peptide
$(\text{Val-Glu-Val-Orn})_3\text{-Val}$	13-Peptide
$(\text{Val-Glu-Val-(TFA)Orn})_2\text{-Val}$	Protected 9-peptide
$(\text{Val-Glu-Val-Orn})_2\text{-Val}$	9-peptide
$(\text{Val-Glu-Val-(TFA)Lys})_1\text{-Val}$	Protected 5-peptide
$(\text{Val-Glu-Val-Lys})_1\text{-Val}$	5-peptide

designed many new models to test the importance of the amphiphilic helical structures. We have been able to design models which differ markedly in their sequences in the amphiphilic helical regions from the natural peptide hormones and yet have comparable abilities to carry out receptor binding as well as to cause the physiological effects of the corresponding hormones. The hormones for which we have prepared models include calcitonin, β-endorphin, and glucagon (9). In addition, strong evidence has been obtained that corticotropin-releasing hormone (27) and growth hormone-releasing hormone contain important amphiphilic secondary structural regions.

IV. THE DESIGN OF AMPHIPHILIC β STRANDS

We have undertaken the characterization of peptides capable of forming amphiphilic β-strand structures as functions of chain length, charge, and environment (28,29). The sequences of some of the peptides that we have synthesized are shown above. They have sequences of repeating units of alternating hydrophobic and hydrophilic residues. In addition, we alternated the acidic and basic residues on the hydrophilic face to minimize charge repulsion. The number of the repeat units in the peptide was varied from 1 to 3 to allow us to determine the minimum chain length necessary to form an ordered structure. We utilized the Merrifield solid-phase peptide synthesis in order to obtain a more homogeneous product than would normally be obtained with polymerization methods. The peptides were synthesized with their basic groups protected with the trifluoroacetyl (TFA) group, a function which is not labile to liquid HF treatment, and the acidic groups were protected in the synthesis with benzyl residues, which are removed by the HF treatment. The peptide as recovered from HF cleavage possesses only negatively charged side-chain functional groups and a free N terminus. By treatment with 1 *M* piperidine we were able to remove the TFA protective groups. This allowed us to determine the effect of different charge balances on these peptides. While our peptides as designed did not have a high potential for β-sheet formation, based on the Chou and Fasman parameters (22), measurements on their solutions by circular dichroism spectroscopy showed significant β-sheet content, with up to 90% β sheet calculated for the 13-residue protected peptide. The potential amphiphilicity of the β-sheet conformation is the driving force for its formation. We found that the β-strand conformation of the 13-residue deprotected peptide was stable in trifluoroethanol, 6 *M* guanidine hydrochloride, and octanol. Because the peptides are strongly self-associating in water, the unfavorable contacts of the hydrophobic residues with water are reduced. Our work clearly shows that even small peptides such as nonapeptides can be designed to form stable β-strand conformations.

Amphiphilic β-strand conformations occur in a variety of surface-active proteins including the principal protein component of low-density lipoprotein (LDL), apolipoprotein B (apo B) and fibronectin. In sonication experiments with mixtures of cholesterol oleate, triolein, and egg lecithin approximating the ratios found in low-density lipoprotein, the nonapeptide and the tridecapeptide were found to be capable of forming large particles with molecular weights of approximately 2.5×10^6, a size comparable to that of LDL (30). In other words, we are able to produce synthetic LDL particles in these sonication experiments. In further work we hope to determine not only the physical but also the biological properties of such particles.

In conclusion, through work on amphiphilic helical and β-strand arrangements we have been able to show that it is possible to design in a rational way models for surface-active peptides and proteins where to a first approximation tertiary structural considerations can be neglected. In our current work we are attempting to incorporate many of the concepts and experimental approaches developed in the secondary structural systems to introduce important segments of proteins which we have designed where tertiary structures must be considered seriously.

ACKNOWLEDGMENT

The work presented here was supported in part by Public Health Service Program Project HL-18577 and by a grant from the Dow Chemical Company Foundation.

REFERENCES

1. Lawrence, D. S., and Kaiser, E. T. (1984). *Science* **226,** 505.
2. Kaiser, E. T. (1985). *Nature* (*London*), **313,** 630.

3. Ghosh, S. S., Bock, S. C., Rokita, S. E., and Kaiser, E. T. (1986). *Science* **231,** 145–158.
4. Levine, H. L., and Kaiser, E. T. (1978). *J. Am. Chem. Soc.* **100,** 7670.
5. Slama, J. T., Radziejewski, C., Oruganti, S. R., and Kaiser, E. T. (1981). *J. Am. Chem. Soc.* **106,** 6778.
6. Radziejewski, C., Ballou, D. P., and Kaiser, E. T. (1985). *J. Am. Chem. Soc.* **107,** 3352.
7. Hilvert, D., and Kaiser, E. T. (1986). *J. Am. Chem. Soc.* (in press).
8. Kaiser, E. T., and Kezdy, F. J. (1984). *Science* **223,** 249.
9. Kaiser, E. T., and Kezdy, F. J. (1983). *Proc. Natl. Acad. Sci. U.S.A.* **80,** 1137.
10. Kroon, D. J., Kupferberg, J. P., Kaiser, E. T., and Kezdy, F. J. (1978). *J. Am. Chem. Soc.* **100,** 5975.
11. Yokoyama, S., Fukushima, D., Kezdy, F. J., and Kaiser, E. T. (1980). *J. Biol. Chem.* **255,** 7333.
12. DeGrado, W. F., Kezdy, F. J., and Kaiser, E. T. (1981). *J. Am. Chem. Soc.* **103,** 6965.
13. Taylor, J. W., Miller, R. J., and Kaiser, E. T. (1981). *J. Am. Chem. Soc.* **103,** 679.
14. Moe, G. R., Miller, R. J., and Kaiser, E. T. (1983). *J. Am. Chem. Soc.* **105,** 4100.
15. Moe, G. R., and Kaiser, E. T. (1985). *Biochemistry* **24,** 1971.
16. Blanc, J., Miller, R. J., and Kaiser, E. T. (1983). *J. Biol. Chem.* **258,** 8277.
17. Blanc, J., and Kaiser, E. T. (1984). *J. Biol. Chem.* **259,** 9549.
18. Retsinger, G. Private communication.
19. Segrest, J. P., Jackson, R. L., Morrisett, J. D., and Gotto, A. M., Jr., (1974). *FEBS Lett.* **38,** 247.
20. Fitch, W. M. (1977). *Genetics* **86,** 623.
21. McLachlan, A. D. (1977). *Nature* (*London*) **267,** 465.
22. Chou, P. Y., and Fasman, G. D. (1978). *Annu. Rev. Biochem.* **47,** 251.
23. Nakagawa, S. H., Lau, H. S. H., Kezdy, F. J., and Kaiser, E. T. (1985). *J. Am. Chem. Soc.* **107,** 7087–7092.
24. Habermann, E. (1972). *Science* **177,** 314.
25. Schroeder, E., Lubke, K., Lehmann, M., and Bertz, I. (1971). *Experientia* **27,** 764.
26. DeGrado, W. F., Musso, G. F., Lieber, M., Kaiser, E. T., and Kezdy, F. J. (1982). *Biophys J.* **37,** 329.
27. Lau, S. H., Rivier, J., Vale, W., Kaiser, E. T., and Kezdy, F. J. (1983). *Proc. Natl. Acad. Sci. U.S.A.* **80,** 7070.
28. Osterman, D., Mora, R., Kezdy, F. J., Kaiser, E. T., and Meredith, S. C. (1984). *J. Am. Chem. Soc.* **106,** 6845.
29. Osterman, D., and Kaiser, E. T. (1985). *J. Cell Biochem.* **29,** 57–72.
30. Meredith, S., and Mora, R. Private communication.

6

Structural and Functional Analysis of Thermophile Proteins

TAIRO OSHIMA

Department of Life Science
Tokyo Institute of Technology
Nagatsuta, Yokohama 227, Japan

I. TACTICS OF THERMOPHILES

Thermophiles are fascinating microorganisms for those who are interested in protein design. Enzymes and proteins of thermophiles are more resistant to heat than their counterparts from mesophilic organisms. Studies of their unusual stability will provide clues for designing proteins possessing more than usual heat resistance. In the near future, it should be possible to confer extra stability on mesophilic proteins based on knowledge obtained from biochemical studies of cell constituents of thermophiles.

The author and his colleagues (Oshima *et al.*, 1976; Ohno-Iwashita *et al.*, 1976; Oshima, 1978a, 1979, 1982) have studied the molecular basis of thermophily by using extremely thermophilic bacteria, *Thermus thermophilus,* isolated from a hot spring (Oshima and Imahori, 1974). Although many questions concerning protein stability remain unsolved, the results of many research groups suggested two general principles related to the unusual stability of cell constituents of thermophiles: (1) the structural changes important for improved stability are rather subtle, and (2) a variety of mechanisms are used to make biopolymers heat stable, and often a molecule is stabilized by a combination of two or more different mechanisms.

Although the mechanisms used are diverse, they can be classified in three categories:

PROTEIN ENGINEERING
Applications in Science,
Medicine, and Industry

(1) Intrinsic mechanisms, in which intramolecular interactions are increased by changes in internal structure. Many enzymes and proteins of thermophiles are stabilized by increased intramolecular interactions caused by replacements of amino acid residues. For instance, phosphoglycerate kinase from *T. thermophilus* is a simple protein without any prothetic group and is unusually stable at high temperatures even in the absence of substrates, metal ions, or other protectors (Nojima *et al.*, 1979). Likewise, polypeptide elongation factors Tu and Ts of *T. thermophilus* are simple proteins and are much more stable than the corresponding factors of *Escherichia coli* in the absence of GDP or Mg^{2+} (Arai *et al.*, 1978).

(2) Protector mechanisms, in which a biopolymer (or cellular component) is stabilized by binding with a stabilizing cofactor. For instance, a thermophile α-amylase is stable at high temperature only in the presence of Ca^{2+} and Na^{+} ions (Hasegawa *et al.*, 1976). Similarly, a thermophile protease (thermolysin) is stable in the presence of Ca^{2+} (Weaver *et al.*, 1976). For catalytic activity, Zn^{2+} is required but Ca^{2+} is not directly involved in the active site of thermolysin.

(3) Biochemical modifications. As described later, tRNA from *T. thermophilus* is stabilized in part by the thiolation of a ribothymidine residue (Watanabe *et al.*, 1976b). From this observation, one can imagine a thermophilic enzyme which is stabilized by posttranslational modification(s) on some amino acid residue(s). However, so far there has been no report of a thermostable enzyme which is stablized by the biochemical modification mechanism.

In some cases, a biopolymer is stabilized by a combination of two or three mechanisms. For instance, as described later, tRNA from *T. thermophilus* is stabilized by a combination of the intrinsic and biochemical modification mechanisms.

II. CATALYTIC PROPERTIES OF THERMOPHILIC ENZYMES

In many cases the catalytic properties of a thermophile enzyme are similar to those of its mesophilic counterpart except for stability to heat (Zuber, 1976; Heinrich, 1976; Friedman, 1978; Oshima, 1978a; Amelunxen and Murdock, 1978; Ljungdahl, 1979; Zuber, 1979). For instance, molecular weight, subunit structure, kinetic parameters, amino acid composition, secondary structure contents, and the presence of reactive cysteine in the active site of glyceraldehyde-3-phosphate dehydrogenase from *T. thermophilus* are identical or similar to those from mesophilic organisms (Fujita *et al.,* 1976). These findings suggest that the changes important for the unusual stability do not affect the molecular architecture of the active site of thermophile enzymes.

Many thermophile enzymes have been sequenced, and some have been studied by X-ray crystallography. Up to now, however, it has not been possible to explain the unusual stability of any enzyme completely on the basis of molecular structure. In some cases, increased intramolecular interactions were pointed out and thought to be important for the increased stability (Perutz and Raidt 1975; Perutz, 1978). For instance, based on the three-dimensional structure of glyceraldehyde-3-phosphate dehydrogenase from *Bacillus stearothermophilus,* additional interactions in the subunit interface were found and thought to be important for the stability of the enzyme (Biesecker *et al.,* 1977). However, the increased interactions reported may be only some of the intramolecular interactions important for the increased stability, and there was no quantitative study of the implications of these interactions for the unusual stability of the thermophile glyceraldehyde-3-phosphate dehydrogenase. Such extra interactions between subunits were not found in the same enzyme from *Thermus aquaticus* (Walker *et al.,* 1980).

Amino acid replacements found in thermophile proteins have been statistically investigated and some general rules have been postulated (Argos *et al.,* 1979). The most abundant replacements are Gly→Ala, Ser→Ala or Thr, Lys→Arg, and Asp→Glu. Often thermophile proteins have high Glx/Asx and Arg/Lys ratios compared to the corresponding proteins from mesophilic organisms (Singleton et al., 1977). A low Cys content in thermophile enzymes has often been reported. However, no convincing theory has been published to support these empirical rules.

The hypothesis that thermophile proteins are stabilized by subtle changes in chemical structure is supported by studies with mutant proteins (Elwell and Schellman, 1977; Yutani *et al.,* 1977). For instance, Yutani *et al.* (1977) investigated the thermal inactivation of the α subunit of tryptophan synthase produced by *E. coli* mutants. They found that replacement of single amino acids at specific sites considerably changed the thermal stability of the protein. A mutant protein in which Glu 49 was replaced with Met showed a higher thermal stability than the wild protein. Another mutant protein in which Glu 49 was replaced with Gln was less stable to heat than the wild one. The results clearly support the idea that the conformational stability of a protein can be changed by changing a small number of amino acid residues without loss of activity. Position 49 of tryptophan synthetase α subunit seems to be one of the key corners for its conformational stability.

In this case, the free energy change of heat denaturation increased with

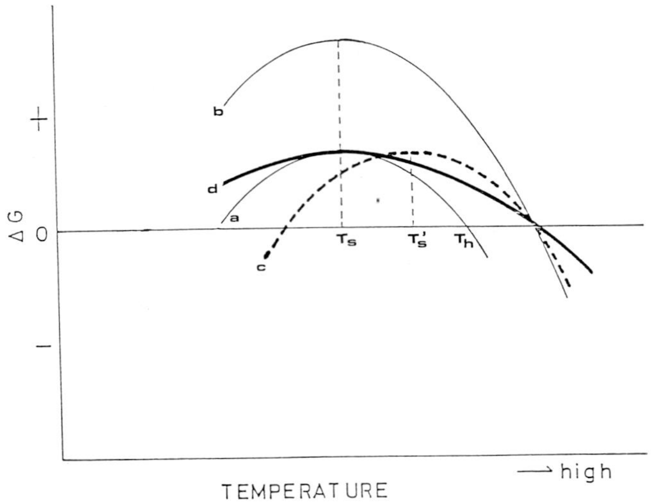

Fig. 1. Hypothetical free energy changes (ΔG) of (a) normal and (b to d) thermostable globular proteins. T_s, Temperature of maximum stability; T_h, thermal denaturation temperature of the normal protein.

the increase in thermal stability on replacing amino acid residue 49 (Yutani *et al.*, 1982). As described in the next section, the stable protein with Met 49 seemed to be stabilized by a mechanism expressed by curve b in Fig. 1

III. PHYSICOCHEMICAL STUDIES

The temperature dependence of the free energy change ΔG of heat denaturation of a protein is represented by curve a in Fig. 1 (Brandt, 1969). The protein has maximum stability at T_s and a thermal denaturation temperature of T_h. There are three basic ways to make the protein stable (Nojima *et al.*, 1978b): (1) lifting curve a to curve b, (2) shifting curve a to curve c, and (3) flattening curve a to curve d.

In the first mechanism (curve b), the ΔG value at maximum stability is much larger than that of curve a. The heat-stable protein (curve b) may have more hydrophobic interactions, more hydrogen bonds, and/or more disulfide bonds than the protein of curve a. The heat-stable protein represented by curve b is also stable at low temperatures.

The protein represented by curve c is stable at a higher temperature than T_h but unstable at low temperatures. This protein has a much higher temperature of maximum stability (T_s'), but the ΔG at T_s' is not larger than the ΔG at T_s for the protein of curve a. Both the temperature of maximum stability and the ΔG value at T_s do not increase in a heat-stable protein of the third type (curve d). This protein is stable at low temperatures too.

Nojima *et al.* (1977) studied the thermodynamic parameters of reversible thermal denaturation of phosphoglycerate kinases of a thermophilic (*T. thermophilus*) and a mesophilic (yeast) microorganism in the presence of guanidine hydrochloride. They concluded that *T. thermophilus* phosphoglycerate kinase is stabilized by a combination of the mechanisms corresponding to curves b and d in Fig. 1. Although the unusual stability of the thermophile phosphoglycerate kinase is mainly conferred by mechanism d, the first mechanism (curve b) is also involved to some extent since the free energy change at the temperature of maximum stability of the thermophile phosphoglycerate kinase is larger (50 kJ/mol at 25°C) than that (22.5 kJ/mol) of the yeast enzyme.

In contrast to *T. thermophilus* phosphoglycerate kinase, cytochrome *c* of the same organism is stabilized by a combination of the first and second mechanisms since T_s and ΔG at T_s are higher than their counterparts in mesophilic organisms (Nojima *et al.*, 1978a). Similarly *E. coli* tryptophan synthase α-subunit stabilized by replacing Glu 49 with hydrophobic amino acid residues showed larger ΔG values than that of the native protein (Yutani *et al.*, 1982).

Polypeptide elongation factor Tu (EF-Tu) is stabilized by binding with GDP. In the case of *T. thermophilus* EF-Tu, it was found that the temperature dependence of ΔG was flattened by GDP binding (H. Nojima, unpublished data). This observation suggests that a change in the thermodynamic properties of a protein can be induced by binding of a relatively small ligand.

Conformation of a globular protein is maintained by a relatively small ΔG value, roughly 50 kJ/mol. Thus a subtle change in amino acid sequence or binding with a ligand can considerably change the thermodynamic properties and the protein stability.

Kinetic studies on thermal denaturation were also carried out with a few proteins of thermophiles. Fujita and Imahori (1974) showed that a thermophile glyceraldehyde-3-phosphate dehydrogenase can be stabilized by small changes in $\Delta S\ddagger$ and/or $\Delta H\ddagger$. The enzyme will be stable up to 90°C if $\Delta S\ddagger$ and/or $\Delta H\ddagger$ values of a mesophile glyceraldehyde-3-phosphate dehydrogenase are increased to 110%.

Nojima and Noda (1979) measured the kinetic parameters of heat denaturation of phosphoglycerate kinases of a thermophile and a mesophile by using changes in circular dichroism. At 25°C, $\Delta G\ddagger$ of the yeast enzyme was 117 kJ/mol and that of *T. thermophilus* was 137 kJ/mol.

We analyzed the thermal denaturation of phosphoglycerate kinase from

T. thermophilus in comparison with that of yeast enzyme by using a differential scanning calorimeter (J. Kimura and T. Oshima, unpublished). The $\Delta H\ddagger$ values for the thermophile and yeast enzymes were 1550 kJ/mol at 94°C and 840 kJ/mol at 52.5°C, respectively, and the $\Delta H\ddagger$ values were 575 kJ/mol at 94°C and 588 kJ/mol at 52.5°C, respectively. The difference between the $\Delta \mathrm{H}^{\ddagger}$ values of the thermophile and mesophile enzymes was small and was consistent with the results reported by Fujita and Imahori (1974) and Nojima and Noda (1979).

These thermodynamic studies also support the idea that the thermophile proteins are stabilized by subtle changes in their chemical structure, and the mechanisms used are diverse. However the thermodynamic and kinetic properties for heat denaturation of a thermophile protein have not yet been satisfactorily interpreted on the basis of structure.

Attempts have been made to correlate thermal stability with structural or physicochemical parameters of a protein. Bull and Breese (1973) reported that thermal stability correlates with the total volume of amino acid side chains in a protein. Stellwagen and Wilgus (1978) pointed out a linear relation between the denaturation temperature and the ratio of the surface area to the total volume of a protein. Ikai (1980) proposed a positive correlation between thermal stability and the volume occupied by hydrophobic, aliphatic amino acid side chains. Often these hypotheses have exceptions, and it seems that a large number of thermostable as well as thermolabile proteins should be analyzed to judge their feasibility.

IV. tRNA AS A MODEL

The tRNA from *T. thermophilus* has a higher melting temperature than tRNAs from mesophilic organisms. The thermophile tRNA can be used instead of mesophile tRNA in *E. coli* cell-free protein synthesis without an increase in miscoding (Ohno-Iwashita *et al.*, 1975). Like the thermophile enzymes, the sites important for the biological functions of the thermophile tRNA seem to be preserved during adaptation to high temperature. So far, thermophile tRNAs are the only biopolymers of thermophilic organisms whose unusual stability is clearly understood on the basis of molecular structure.

The chemical structure of *T. thermophilus* $\mathrm{tRNA}_{\mathrm{f}}^{\mathrm{Met}}$ is identical to that of *E. coli* $\mathrm{tRNA}_{\mathrm{f}}^{\mathrm{Met}}$ except for one base replacement and three modifications (Watanabe *et al.*, 1979). A G-U pair (connected by two hydrogen bonds like a A-U pair) in *E. coli* tRNA was replaced by a G-C pair in the thermophile tRNA; that is, a uracil residue in the T stem was replaced by a cytosine. The empirical data suggest that an increase of a hydrogen bond in the base-paired region of tRNA brings about a 1.5°C increase in the melting temperature. However, $tRNA_f^{Met}$ from *T. thermophilus* showed a higher melting temperature than that expected from its G+C content in the base-paired region (Watanabe *et al.*, 1976b). This observation suggested that the modified base(s) confers the additional stability to the thermophile tRNA.

Except for the base replacement mentioned above, three modified nucleosides were found in the thermophile tRNAs which were not found in the *E. coli* counterpart: 2-thioribothymidine at position 54 in the T loop, 1-methyladenosine at position 58 in the T loop, and 2′-0-methylguanosine at position 18 in the D loop.

Extensive studies showed that the thiolation at ribothymidine 54 and the additional hydrogen bond resulting from replacement of a G-U pair with a G-C pair are important for the increased melting temperature of $tRNA_f^{Met}$ of *T. thermophilus* (Watanabe *et al.*, 1974, 1976a,b, 1979, 1980, 1983; Davanloo *et al.*, 1979; Watanabe, 1980; Yamamoto *et al.*, 1983). It was speculated that the thiolation at the position 2 carbon of uracil (or thymine) strengthens the stacking interaction between neighboring bases in a helix (Scheit and Faerber, 1975). However, a more recent study has shown the highly restricted conformation of the thiolated nucleotide to give extra stability to the thermophile tRNA (Yamamoto *et al.*, 1983). Perhaps both mechanisms contribute to the increased stability.

Methylations at G-18 and A-58 did not change the melting temperature of tRNA. The thermophile tRNA was resistant to ribonuclease digestions, and these methylations were thought to be important to the ribonuclease resistance. These modifications may restrict local conformational fluctuations and may contribute to the unusual stability to some extent. To confirm this speculation a modification enzyme, guanosine-2′-*O*-methylase EC 2.1.1.34, was purified from a cell-free extract of *T. thermophilus* (Kumagai *et al.*, 1982). The heat-stable enzyme catalyzes the ribose methylation of a specific guanosine residue in the D loop of tRNA, using *S*-adenosylmethionine as the methyl donor. When tRNA from yeast was modified with the thermophile methylase and *S*-adenosylmethionine, only one methyl group per molecule was introduced in the specific guanosine residue of the yeast tRNA. The resulting methylated tRNA was as resistant to ribonuclease A and ribonuclease T_1 as thermophile tRNA. Some of the results are shown in Fig. 2. To our knowledge, this is the first example of artificial stabilization of a biopolymer by the same means as those employed by thermophiles during their adaptation to high temperatures.

The studies on thermophile tRNA showed that the molecule is stabilized by subtle changes in chemical structure and by a combination of two different means. The molecule is stabilized by an extra intramolecular

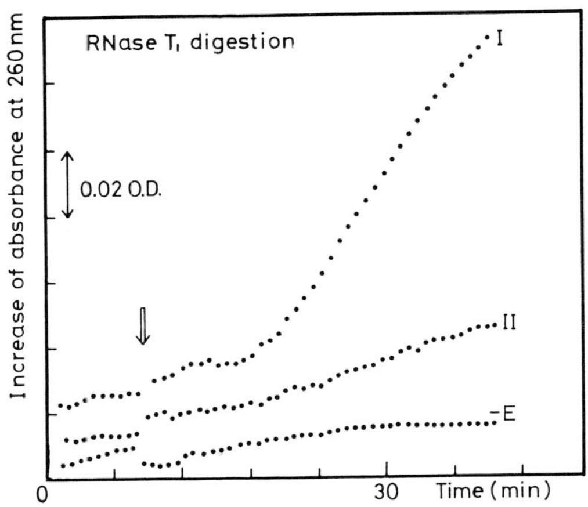

Fig. 2. Resistance to RNase T_1 digestion of yeast (I) $tRNA^{Phe}$ and (II) 2′-O-methylated (G-19) $tRNA^{Phe}$. The rate of hydrolysis was followed by absorbance change at 260 nm at 43°C. The arrow indicates the addition of RNase T_1. (Reproduced from Kumagai *et al.*, 1982.)

hydrogen bond formed by replacing a uridyl residue with a cytidyl residue, and by the increase in stacking interaction and/or conformational rigidity brought about by thiolation of a ribothymidine residue in the T loop. These changes are quite subtle compared to the sum of the intramolecular hydrogen bonds (more than 60) in the molecule or the number of atoms composing the tRNA molecule, one (oxygen atom) of which is replaced by a sulfur atom. It is interesting that these changes are concentrated at a location quite far from the anticodon loop and the amino acid accepting terminal, both sites of direct importance to the biological functions of tRNA. These observations from our studies on thermophile tRNA should provide useful guidelines for the basic design of protein stabilization.

A psychrophilic shrimp, *Euphausia,* produces a tRNA less stable than the mesophile tRNA. Based on the observations mentioned above, we first expected that the psychrophile tRNA would contain a low G-C content in the base-paired region and/or a modified base that destabilizes the molecule. This speculation was wrong. The molecule contains unusual base pairs or mismatched pairs such as G-U or A-C (Watanabe *et al.*, 1981). Except for these unusual pairs, no significant difference was found between the psychrophilic and mesophilic eukaryote tRNAs; therefore these pairs seemed to be the only structural basis for the destabilizing mechanism. This finding suggests that the means for changing the stability of a biopolymer are diverse and differ from organism to organism.

V. MOLECULAR CLONING

DNA cloning and site-directed mutagenesis have opened a new era in the study of mechanisms for the thermostability of proteins from thermophiles. The role of an amino acid residue at a specific site in conformational stability can be examined by replacing the residue with other amino acids through changes in the DNA sequence. However, up to now only a small number of genes of extreme thermophiles have been cloned.

The gene for 3-isopropylmalate dehydrogenase of *T. thermophilus* strain HB8, which corresponds to the *E. coli leuB* gene, was cloned into *E. coli* (Nagahari *et al.,* 1980; Tanaka *et al.,* 1981). The *E. coli* cells harboring a recombinant plasmid containing the thermophile *leuB* gene produced a heat-stable 3-isopropylmalate dehydrogenase. The optimum temperature of the dehydrogenase produced by *E. coli* harboring the recombinant plasmid is close to that of the dehydrogenase produced by *T. thermophilus* within experimental error. This observation clearly indicates that the thermophile 3-isopropylmalate dehydrogenase is stabilized by the intrinsic mechanism. Similarly, a gene for a thermostable malate dehydrogenase of *T. flavus* was cloned and expressed in *E. coli*. The gene product expressed in *E. coli* was also stable to heat (S. Iijima, personal communication).

Gene cloning is a new way to purify a thermophile enzyme. When a cell-free extract of *E. coli* harboring the thermophile gene was heated at 70°C, most of the proteins coded by the *E. coli* genes were denatured and precipitated. Roughly 80% of the proteins in the extract were removed by this treatment. Thus the thermophile enzyme was easily purified. Recently, the thermophile 3-isopropylmalate dehydrogenase was crystallized. (T. Yamada, personal communication).

The thermophile DNA coding for 3-isopropylmalate dehydrogenase was sequenced along with a 0.6-kilobase (kb) flanking region (Kagawa *et al.,* 1984). The base sequence reported was recently corrected (T. Oshima *et al.,* manuscript in preparation). The amino acid composition deduced from the corrected base sequence is shown in Table I, along with codon usage data. The amino acid composition showed features characteristic of many thermophile proteins. No cysteine residue was found in the 339 amino acid residues. The Glu/Asp and Arg/Lys ratios were high, as reported for many other thermophile enzymes (Singleton *et al.,* 1977). The contents of hydrophobic amino acid residues such as Ala, Pro, Val, and Leu were conspicuously high.

The DNA sequence also has interesting features. The use of synonymous codons is highly restricted to those whose third letter is G or C, as shown in Table I. The G + C content of *T. thermophilus* DNA is about

TABLE I

Amino Acid Composition of *Thermus thermophilus* 3-Isopropylmalate dehydrogenase and Codon Usage in Its Structural Gene[a]

First position	Second position: U			C			A			G			Third position
	Phe	x	3	Ser	o	1	Tyr	x	1	Cys		0	U
U	Phe	o	9	Ser	x	8	Tyr	o	5	Cys		0	C
	Leu	x	1	Ser	x	0	Term[b]		1	Term[b]		0	A
	Leu	x	3	Ser	x	1	Term[b]		0	Trp		2	G
C	Leu	x	5	Pro	x	3	His		0	Arg	o	1	U
	Leu	x	15	Pro	x	18	His		5	Arg	o	9	C
	Leu	x	1	Pro	o	0	Gln	x	0	Arg	x	2	A
	Leu	o	10	Pro	o	6	Gln	o	3	Arg	x	10	G
A	Ile	x	0	Thr	o	0	Asn	x	0	Ser		0	U
	Ile	o	8	Thr	o	6	Asn	o	6	Ser		5	C
	Ile	x	1	Thr	x	0	Lys	o	1	Arg	x	0	A
	Met		6	Thr	x	7	Lys	x	15	Arg	x	6	G
G	Val	o	0	Ala	o	1	Asp		0	Gly	x	0	U
	Val	x	12	Ala	x	28	Asp		15	Gly	o	11	C
	Val	o	0	Ala	o	1	Glu	o	2	Gly	x	6	A
	Val	x	20	Ala	o	12	Glu	x	28	Gly	x	19	G

[a] Optimal codons in *E. coli* are marked with o and nonoptimal codons with x.
[b] Termination codon.

69% and that of the coding region (for 3-isopropylmalate dehydrogenase) is 70.2%. High Arg/Lys ratios in amino acid composition of a thermophile protein can be the result of a high G + C content of DNA coded for the protein and may not correlate directly to unusual stability of the protein. The first letters have a high guanine content. This leads to high Val, Ala, Glu, and Gly contents in the amino acid composition.

The second letters showed no significant preference for any of four bases. The third letters have a high G + C content, about 91%. Since the third letters are highly restricted to G or C, the codon usage differs from that in *E. coli* genes. For instance, GUU and GUA are the optimal codons for Val in *E. coli* genes (Ikemura, 1981) but are not used in the thermophile isopropylmalate dehydrogenase gene, whereas GUC and GUG, which are nonoptimal codons in *E. coli,* are used 12 and 20 times in the thermophile gene, respectively.

Studies on stability–structure relationship as well as structure–function studies of 3-isopropylmalate dehydrogenase of *T. thermophilus* have been started in the author's laboratory. We have found that the *T. thermophilus* strain 110 produces less stable 3-isopropylmalate dehydrogenase than that produced by *E. coli* harboring a recombinant plasmid containing *leuB* gene from *T. thermophilus* strain HB8. The *leuB* gene of strain 110 was cloned, and, at present, 85% of the coding region is sequenced. So far four substitutions and one insertion have been found in the amino acid sequence deduced from the base sequence of *leuB* gene of strain 110 when compared with that of strain HB8. Further studies using site-directed mutagenesis will soon clarify the substitutions which are directly responsible for the difference in thermal stability.

In our department, Professor Noso and co-workers (Y. Noso *et al.*, unpublished data) cloned the 3-isopropylmalate dehydrogenase genes from extremely thermophilic, moderately thermophilic, and mesophilic bacilli. The sequencing of these cloned genes is in progress. Comparisons of amino acid sequence data of these dehydrogenases, which showed a wide variety of thermal resistance, should suggest the key corners of the enzyme protein that are critical to conformational stability and will give clues to understanding the molecular basis of unusual stability of thermophilie enzymes. Eventually, we will be able to design 3-isopropylmalate dehydrogenase possessing the desired stability.

REFERENCES

Amelunxen, R. E., and Murdock, A. L. (1978). *CRC Crit. Rev. Microbiol.* **6,** 343–393.

Arai, K., Ota, Y., Arai, N., Nakamura, S., Hennecke, C., Oshima, T., and Kaziro, Y. (1978). *Eur. J. Biochem.* **92,** 509–519.

Argos, P., Rossman, M. G., Grau, U. M., Zuber, H., Frank, G., and Tratschin, J. D. (1979). *Biochemistry* **18,** 5698–5703.

Biesecker, G., Harris, J. I., Thierry, J. C., Walker, J. E., and Wonacott, A. J. (1977). *Nature (London)* **226,** 328–333.

Brandt, J. F. (1969). "Structure and Stability of Biological Macromolecules." Dekker, New York.

Bull, H. B., and Breese, K. (1973). *Arch. Biochem. Biophys.* **158,** 681–686.

Davanloo, P., Sprinzl, M., Watanabe, K., Albani, M., and Kersten, H. (1979). *Nucleic Acids Res.* **6,** 1571–1581.

Elwell, M. L., and Schellman, J. A. (1977). *Biochim. Biophys. Acta* **494,** 367–383.

Friedman, S. M., ed. (1978) "Biochemistry of Thermophily." Academic Press, New York.

Fujita, S. C., and Imahori, K. (1974). *In* "Peptides, Polypeptides, and Proteins" (E. R. Blout, F. A. Bovey, M. Goodman, and N. Lotan, eds.), pp. 217–229. Wiley, New York.

Fujita, S. C., Oshima, T., and Imahori, K. (1976). *Eur. J. Biochem.* **64,** 57–68.

Hasegawa, A., Miwa, M., Oshima, T., and Imahori, K. (1976). *J. Biochem. (Tokyo)* **79,** 35–42.

Heinrich, M. R., ed. (1976). "Extreme Environments: Mechanisms of Microbial Adaptation." Academic Press, New York.

Ikai, A. (1980). *J. Biochem. (Tokyo)* **88,** 1895–1898.

Ikemura, T. (1981). *J. Mol. Biol.* **151,** 389–409.

Kagawa, Y., Nojima, H., Nukiwa, N., Ishizuka, M., Nakajima, T., Yasuhara, T., Tanaka, T., and Oshima, T. (1984). *J. Biol. Chem.* **259,** 2956–2960.
Kumagai, I., Watanabe, K., and Oshima, T. (1982). *J. Biol. Chem.* **257,** 7388–7395.
Ljungdahl, L. G. (1979). *Adv. Microb. Physiol.* **19,** 149–243.
Nagahari, K., Koshikawa, T., and Sadaguchi, K. (1980). *Gene* **10,** 137–145.
Nojima, H., and Noda, H. (1979). *J. Biochem.* (*Tokyo*) **86,** 1055–1065.
Nojima, H., Ikai, A., Oshima, T., and Noda, H. (1977). *J. Mol. Biol.* **116,** 429–442.
Nojima, H., Hon-nami, K., Oshima, T., and Noda, H. (1978a). *J. Mol. Biol.* **122,** 33–42.
Nojima, H., Ikai, A., Noda, K., Hon-nami, K., and Oshima, T. (1978b). *In* "Biochemistry of Thermophily" (S. M. Friedman, ed.), pp. 305–323. Academic Press, New York.
Nojima, H., Oshima, T., and Noda, H. (1979). *J. Biochem.* (*Tokyo*) **85,** 1509–1517.
Ohno-Iwashita, Y., Oshima, T., and Imahori, K. (1975). *Z. Allg. Mikrobiol.* **15,** 129–132.
Ohno-Iwashita, Y., Oshima, T., and Imahori, K. (1976). *In* "Enzymes and Proteins from Thermophilic Microorganisms" (H. Zuber, ed.), pp. 333–345. Birkhaeuser, Basel.
Oshima, T. (1978a). *Enzyme Eng.* **4,** 41–46.
Oshima, T. (1978b). *In* "Biochemistry of Thermophily" (S. M. Friedman, ed.), pp. 211–220. Academic Press, New York.
Oshima, T. (1979). *In* "Strategies of Microbial Life in Extreme Environments" (M. Shilo, ed.), pp. 455–469. Dahlem Konferenzen, Berlin.
Oshima, T. (1982). *In* "Plant Cold Hardiness and Freezing Stress" (P. H. Li and A. Sakai, eds.), Vol. 2, pp. 661–670. Academic Press, New York.
Oshima, T., and Imahori, K. (1974). *Int. J. Syst. Bacteriol.* **24,** 102–112.
Oshima, T., Sakaki, Y., Wakayama, N., Watanabe, K., Ohashi, Z., and Nishimura, S. (1976). *In* "Enzymes and Proteins from Thermophilic Microorganisms" (H. Zuber, ed.), pp. 317–331. Birkhaeuser, Basel.
Perutz, M. F. (1978). *Science* **201,** 1187–1191.
Perutz, M. F., and Raidt, H. (1975). *Nature; (London)* **255,** 256–259.
Scheit, K. H., and Faerber, P. (1975). *Eur. J. Biochem.* **50,** 549–555.
Singleton, R., Jr., Middaugh, C. R., and MacElroy, R. D. (1977). *Int. J. Peptide Res.* **10,** 39–50.
Stellwagen, E., and Wilgus, H. (1978). *Nature* (*London*) **275,** 342–343.
Tanaka, T., Kawano, N., and Oshima, T. (1981). *J. Biochem.* (*Tokyo*) **89,** 677–682.
Walker, J. E., Wonacott, A. J , and Harris, J. I. (1980). *Eur. J. Biochem.* **108,** 581–186.
Watanabe, K. (1980). *Biochemistry* **19,** 5542–5549.
Watanabe, K., Oshima, T., Saneyoshi, M., and Nishimura, S. (1974). *FEBS Lett.* **43,** 59–63.
Watanabe, K., Oshima, T., and Nishimura, S. (1976a). *Nucleic Acids Res.* **3,** 1703–1713.
Watanabe, K., Shimma, M., and Oshima, T. (1976b). *Biochem. Biophys. Res. Commun.* **72,** 1137–1144.
Watanabe, K., Kuchino, Y., Yamaizumi, Z., Kato, M., Oshima, T., and Nishimura, S. (1979). *J. Biochem.* (*Tokyo*) **86,** 893–905.
Watanabe, K., Oshima, T., Iijima, K., Yamaizumi, Z., and Nishimura, S. (1980). *J. Biochem.* (*Tokyo*) **87,** 1–13.
Watanabe, K., Asai, K., Oshima, T., and Kuchino, Y. (1981). *J. Biochem.* (*Tokyo*) **90,** 1259–1266.
Watanabe, K., Oshima, T., Hansske, F., and Ohta, T. (1983). *Biochemistry* **22,** 98–102.
Weaver, L. H., Kester, W. R., Ten Eyck, L. F., and Matthews, B. W. (1976). *In* "Enzymes and Proteins from Thermophilic Microorganisms" (H. Zuber, ed.), pp. 31–39. Birkhaeuser, Basel.
Yamamoto, Y., Yokoyama, S., Miyazawa, T., Watanabe, K., and Higuchi, S. (1983). *FEBS Lett.* **157,** 95–99.

Yutani, K., Ogasahara, K., Sugino, Y., and Matsushiro, A. (1977). *Nature (London)* **267,** 274–275.

Yutani, K., Ogasahara, K., Kimura, A., and Sugino, Y. (1982). *J. Mol. Biol.* **160,** 387–390.

Zuber, H., ed. (1976). "Enzymes and Proteins from Thermophilic Microorganisms." Birkhaeuser, Basel.

Zuber, H. (1979). *In* "Strategies of Microbial Life in Extreme Environments" (M. Shilo, ed.), pp. 393–415. Dahlem Konferenzen, Berlin.

7

The Conformation of Diphtheria Toxin: A Protein That Penetrates Membranes at Low pH

ERWIN LONDON,* MICHAEL G. BLEWITT,*
AMITABHA CHATTOPADHYAY,†
LAURA A. CHUNG,* AND JIAN-MIN ZHAO*

** Department of Biochemistry*
† Department of Chemistry
State University of New York at Stony Brook
Stony Brook, New York 11794

There are two important reasons for considering the conformation of membrane-penetrating toxin proteins in the context of protein engineering. First, the "natural engineering" of such toxins, i.e., their conformation, involves some of the more fascinating designs found in nature, as they must undergo large conformational changes in order to function. Therefore, detailed study of toxin conformation will add appreciably to our understanding of the principles of protein folding. The second reason arises from the interest in artifical immunoglobulin–toxin covalent hybrids ("immunotoxins") as therapeutic agents targeted specifically against tumor cells. To design improved immunotoxin agents by protein engineering techniques it will be necessary to understand the mechanism of toxin action in detail.

I. DIPHTHERIA TOXIN STRUCTURE AND FUNCTION

The molecule we have been studying is diphtheria toxin. The toxin is a medium-sized protein (M_r 58,348) composed of two subunits: A (M_r

PROTEIN ENGINEERING
Applications in Science,
Medicine, and Industry

21,167) and B (M_r 37199) (Greenfield *et al.*, 1983; Kaczorek *et al.*, 1983; Ratti *et al.*, 1983). The two subunits are held together by an exposed polypeptide link that is very sensitive to proteolysis. The sequence of the protein has certain noteworthy features. There are only two disulfide bonds, one within the B subunit and one connecting the A and B subunits. The protein has no free Cys residues. There are a total of five Trp residues scattered throughout the protein (two in A and three in B). This provides convenient intrinsic fluorescence probes at various regions in the molecule. There is a curious lack of His in the A subunit (there is only one residue, at position 21), while the B subunit has 15 residues. This may reflect the enzymatic function of the A subunit, which involves derivatization of a specific modified His, as described below.* Also important are several relatively hydrophobic strings of residues concentrated in the N-terminal half of the B subunit, likely candidates for membrane-penetrating regions (Greenfield *et al.*, 1983; Lambotte *et al.*, 1980).

The functionally important enzymatic reaction catalyzed by the A subunit is the transfer of ADP-ribose from NAD^+ to elongation factor 2 (EF-2), which thereby is inactivated. In this way the toxin shuts off protein synthesis. The ADP-ribose is attached to the amino acid diphthamide, a product of posttranslational modification of a His residue unique to EF-2 (Van Ness *et al.*, 1980). Other activities catalyzed by toxin are two apparent side reactions: (1) a slow NAD^+ glycohydrolase (NAD^+ nucleosidase) reaction, i.e., the splitting of NAD^+ into ADP-ribose and nicotinamide, and (2) self ADP-ribosylation. Interestingly, both the whole toxin and the A subunit have the NAD^+ glycohydrolase activity, suggesting that at least part of the A subunit is folded into its active conformation in the while toxin. However, the whole toxin does not catalyze ADP-ribosylation of EF-2, perhaps, as has been suggested, due to the B subunit sterically blocking the EF-2 binding site (Collier, 1982).

Considerable effort has been directed toward identifying the active sites and ligand binding sites. The NAD^+ binds to a site on the A subunit. The binding site includes Glu 148, as judged from photoaffinity cross-linking studies (Carroll and Collier, 1984). Isolated toxin is largely associated with dinucleotide ligands (80% ApUp, 15% ApGp) that bind extremely tightly, probably at this site (Barbieri *et al.*, 1981; Collins and Collier, 1984; Collins *et al.*, 1982). EF-2 evidently binds to the A subunit subsequent to NAD^+, but little else is known (Collier, 1982). Another site that has been identified is the so-called P site, which binds many anionic

* In fact, we find that the sequence Tyr-His-Gly-Thr around His 21 of subunit A is also found at His 440 of the ADP-ribosylating fragment of exotoxin A (Gray *et al.*, 1984) and there is some further limited homology on either side of the tetrapeptide. Therefore we suggest that it is possible that the tetrapeptide forms part of the active site.

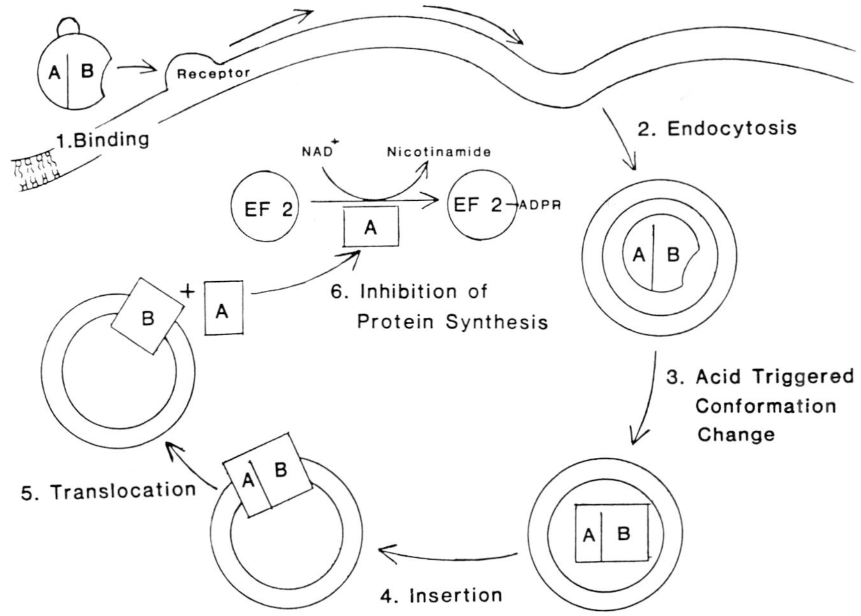

Fig. 1. Entry of diptheria toxin into cells; ADPR is ADP-ribose.

ligands (Collier, 1982). This site involves the C-terminal portion of the B subunit.

Spatially, this site is very likely to be adjacent to the NAD^+ site because of competitive inhibition between NAD^+ binding and P-site ligand binding (Collins and Collier, 1984). The C-terminal region of the B subunit also contains the binding site for the cell surface receptor, and the P and receptor sites may be closely related (Eidels *et al.*, 1982).

Figure 1 schematically illustrates the entry of diphtheria toxin into cells. Several studies now indicate that subsequent to receptor binding the toxin enters cells by receptor-mediated endocytosis. There has been much interest in the nature of the cell surface receptor. The bulk of the evidence supports the existence of a glycoprotein receptor for the toxin (Eidels *et al.*, 1983). However, based on the interaction with phosphorylated molecules, it has been suggested that certain lipids could act as receptors, although perhaps only of a secondary nature (Alving *et al.*, 1980). Endocytosis delivers the toxin to an acidic organelle (Sandvig and Olsnes, 1980; Draper and Simon, 1980), most likely an endosome (Marnell *et al.*, 1984), which is the name given to certain vesicles that are believed to be "intermediate" on the endocytosis pathway to lysosomes. The low pH in the lumen of these acidic organelles apparently triggers a hydrophilic-to-

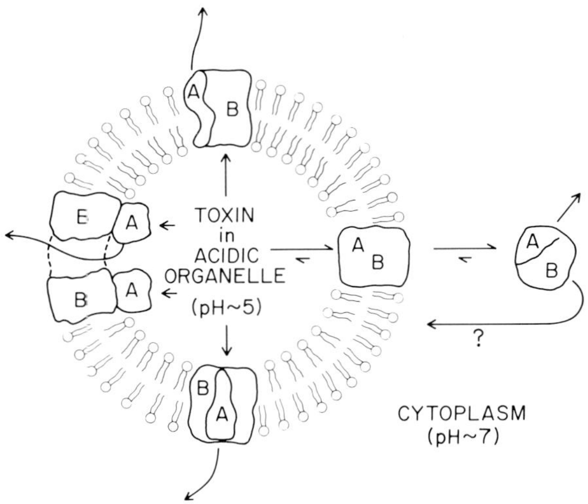

Fig. 2. Hypothetical mechanisms for membrane penetration.

hydrophobic change in toxin structure (Sandvig and Olsnes, 1981; Blewitt *et al.*, 1984, 1985). This is believed to result in insertion of the toxin into the lipid bilayer of the organelle membrane and, indeed, exposure to low pH results in insertion into model membranes *in vitro* (Donovan *et al.*, 1981; Kagan *et al.*, 1981; Zalman and Wisnieski, 1984; Hu and Holmes, 1984). In the next step the A subunit is released into the cytoplasm. It then turns off protein synthesis by the enzymatic ADP-ribosylation of EF-2 described above.

The least well-understood steps in diphtheria toxin entry involve the behavior of the toxin between the time of exposure to low pH and release of the A subunit. These steps require pronounced conformational changes. Already, several proposals have been made for the structure of membrane-inserted toxin, and these are shown in Fig. 2. Starting clockwise from the bottom, the first mechanism proposes that the toxin inserts such that the hydrophobic sites of the B subunit contact the bilayer and the hydrophilic A subunit is protected from contact with the bilayer. Upon release of the A subunit, a pore through the B subunit remains (Misler, 1984). In the second model this pore is formed by an oligomer of inserted B subunits. Indeed, toxin and isolated B subunit form pores in model membranes at low pH (Donovan *et al.*, 1981; Kagan *et al.*, 1981; Zalman and Wisnieski, 1984), but that does not prove they are important *in vivo*. One critical clue to their role is pore size. Any functional pore must be at least large enough to accomodate the A subunit in an unfolded state. The reason subunit A need not remain folded in the membrane is that it can efficiently refold after exposure to strongly denaturing conditions (Collier, 1982). Unfortunately, studies attempting to determine the size of the pores have given different answers (Donovan *et al.,* 1981; Zalman and Wisnieski, 1984). Another possible model is that hydrophobicity of the toxin B subunit may be sufficient to "drag" the hydrophilic A subunit into the membrane so that it is in contact with, rather than shielded from, the bilayer. In fact, photocrosslinking studies have suggested that subunit A does contact the bilayer on insertion (Hu and Holmes, 1984; Zalman and Wisnieski 1984). The membrane-inserted A subunit would perhaps take on an unnatural conformation, but again, on release it could refold. Unfortunately, the presence of reactive amino acid residues can distort cross-linking results even when "nonspecific" reagents are used (Ross *et al.,* 1982), so the exact degree of exposure of A subunit to the bilayer is uncertain. Another possible mechanism involves translocation of the whole toxin. If the pH-sensitive sites which trigger a change to a hydrophobic conformation at low pH became exposed to the neutral pH cytoplasm after insertion, then the toxin might switch back into the hydrophilic conformation and dissolve in the cytoplasm. In fact, starting at pH 3, which avoids excessive aggregation (Blewitt *et al.,* 1985), low-pH conformational changes are mostly reversed on neutralizing pH. However, this model assumes that the sites controlling the conformational transition become exposed to the cytoplasm upon insertion, and there is no supporting evidence for this assumption. A final possibility, not shown in Fig. 2, is that the whole toxin breaks endosomes open, resulting in release of the toxin. Indeed, recent studies showing that toxin can induce vesicle fusion at low pH suggest that the toxin has the potential to disrupt membranes (Cabiaux *et al.,* 1984), and it has been proposed that toxin enters the cytoplasm in a burst involving a number of toxin molecules (Hudson and Neville, 1985). At present we cannot say which model is correct.

Once the A subunit is exposed to the cytoplasm it can presumably be released from the B subunit if the disulfide bond between the subunits has been cleaved by reduction. Reduction of disulfide bonds by natural sulfhydryl agents such as glutathione is unlikely to take place in the lumen of an acidic organelle. The low pH of the lumen in an acidic organelle will prevent reduction because such reactions depend on the S^- form of the thiol (Torchinsky, 1981). Instead, previously proposed models assume that this cleavage occurs in the reducing environment of the cytoplasm, which places reduction subsequent to insertion, although it has also been suggested that membrane proteins could interact with toxin disulfides (Wright *et al.,* 1984).

II. THE HYDROPHILIC-TO-HYDROPHOBIC SWITCH: TRANSITION pH

The preceding discussion demonstrates that it will be difficult to characterize the true mechanism by which the toxin penetrates membranes and accomplishes translocation of the A subunit. Our approach is to use biochemical and biophysical techniques to dissect each step. Our first studies have concentrated on the nature of the switch to a hydrophobic conformation (Blewitt *et al.*, 1984, 1985). The experiments were done with 10^{-7} *M* toxin. This concentration is close to the physiological concentration expected in acidic organelles because the concentration of a single toxin molecule in a spherical acidic organelle of diameter 0.1–1.0 μm would be 3×10^{-6} to 3×10^{-9} *M*, respectively, as calculated from the number of moles of toxin divided by the internal volume. It may even be that this is an underestimate of concentrations *in vivo* because the concentration of toxin molecules may be higher than one per organelle (Hudson and Neville 1985).

Our observations revealed that the change induced by low pH, and detected by Trp fluorescence intensity, is a highly cooperative transition, which occurs over a pH range of only 0.2 unit (Fig. 3). The midpoint, at 23°C, is close to pH 5. This falls within the range of pH encountered in acidic organelles, about pH 4.8 in lysosomes (Geisow, 1984), and pH 5–5.5 in endosomes (Maxfield, 1982; Geisow and Evans, 1984), and therefore suggests that the transition observed *in vitro* could be the same as that which occurs *in vivo*. This conclusion is reinforced by the kinetics of the transition, which is very fast at pH 3 or 4.2 ($t_{1/2} < 30$ sec). This is important as the time for cytotoxicity (i.e., EF-2 inactivation) to appear is on the order of minutes at high toxin concentrations, and therefore any changes requiring a long time are unlikely to be physiologically significant.

Several environmental conditions and structural variations may affect the transition pH. Ionic strength is one such factor. In low salt concentrations the transition shifts to pH 4, while in the presence of 150 m*M* monovalent salts the transition pH is consistently 5. This sensitivity to ionic strength is an important clue to the mechanism of the transition, as discussed later. On the other hand, the transition pH is not influenced by the form of toxin used. Bound toxin (which contains one molecule of tightly bound dinucleotide), free toxin, toxin nicked between A and B or with an intact polypeptide, monomer, or dimer, all have very similar pH transitions. One factor we have not been able to test is the effect of toxin binding to receptor on the transition pH. However, the studies of Draper and Simon (1980) indicate that receptor binding does not have a major effect. They found that receptor-bound toxin directly penetrated the

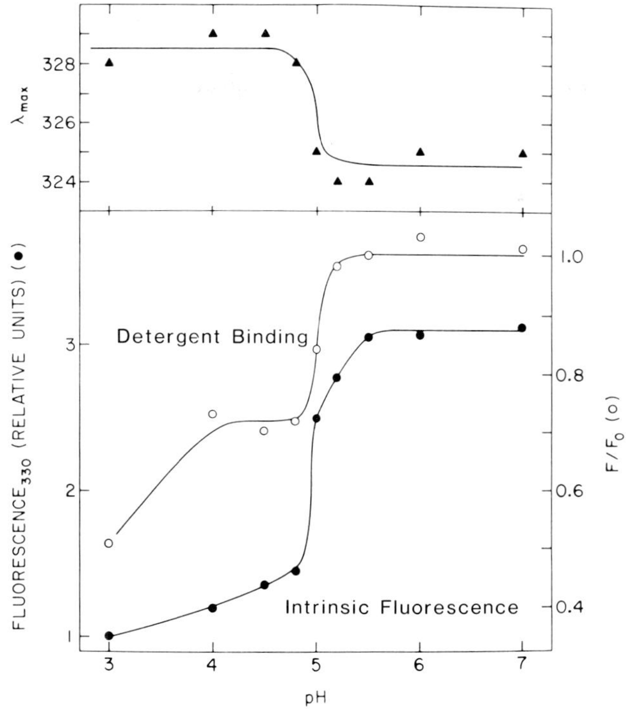

Fig. 3. Effect of pH on toxin fluorescence and detergent binding. F/F_0, Ratio of fluorescence in the presence of quenching detergent micelles to that in the presence of micelles without quencher. An F/F_0 value less than 1 indicated detergent binding. (Adapted from Blewitt *et al.*, 1985. Copyright 1985, American Chemical Society. Reprinted with permission.)

plasma membrane when cells were incubated below pH 5. The implication is that below pH 5 the receptor-associated toxin becomes hydrophobic. Since this is the same pH at which "receptor-free" toxin becomes hydrophobic, one must assume that receptor binding is not a critical regulatory factor. It is not even clear if the toxin remains receptor-bound at low pH.

III. THE HYDROPHILIC-TO-HYDROPHOBIC SWITCH: CONFORMATIONAL CHANGES

Below the transition pH the toxin undergoes several distinct changes in physical properties which we have tried to characterize. Most important of these is the dramatic increase in hydrophobicity. We have developed new fluorescence quenching methods that measure this hydrophobicity through binding to micelles of mild nonionic detergents. In this assay the intrinsic protein (Trp) fluorescence intensity is measured in two samples, one of which contains ordinary micelles and the second contains micelles with fluorescence quenchers. If the protein is hydrophobic it will bind to the micelles and weaker fluorescence will be observed in the sample with quencher. This method considerably simplifies the measurement of hydrophobicity. As seen in Fig. 3, there is considerable detergent binding by toxin only below the transition pH. Binding studies show that binding of 10^{-7} *M* toxin to micelles of the detergent Brij 96 (critical micelle concentration, 3 μM), saturates at 15 μM detergent. An important question is whether this binding is tight enough to explain spontaneous insertion of organelle-trapped toxin molecules into the organellar membrane. Calculations suggest that this is so. If we assume again that toxin is trapped in an organelle of diameter 0.1–1.0 μm, then the apparent lipid concentration will be given by the moles of lipid facing the interior divided by the internal aqueous lumen volume. The moles of lipid can be calculated in turn from the internal surface area of the organelle membrane divided by the number of lipids per unit area. Assuming 70 Å^2 per lipid molecule, and that 50% of the surface is occupied by protein and 50% by lipid, gives a 7–70 m*M* apparent lipid concentration for a 1.0- or 0.1-μm-diameter organelle, respectively. This far exceeds the concentration necessary to observe tight binding *in vitro*.

Other changes in the structure of the toxin accompany the increase in hydrophobicity. There is an increase in average exposure of Trp residues to the solution, as judged both from a red shift in λ_{max} and an increase in acrylamide quenching at low pH. Circular dichroism (CD) shows a change in secondary structure at low pH. At neutral pH the protein appears to be rich in β sheets by CD although considerable α-helix may also be present (Blewitt *et al.*, 1985; Collins and Collier, 1985). The changes observed in CD at low pH must likely involve a slight increase in random coil.

These changes in secondary and tertiary structure could be interpreted as a partial denaturation at low pH. Support for this concept comes from the effect of high temperature on the conformation of the toxin and its transition pH (Zhao and London, 1985). In studies of the thermal denaturation transition we found that the thermally denatured conformation resembles the low-pH conformation in several respects, although they are not identical. In particular, both conformations show increased Trp exposure and hydrophobicity. Furthermore, as pH is decreased the thermal transition temperature is decreased, and as temperature is increased the transition pH increases. This implies that at least some of the interactions in the toxin that can be disrupted by thermal denaturation at high temperature are similarly disrupted at low pH. Therefore, the conformational

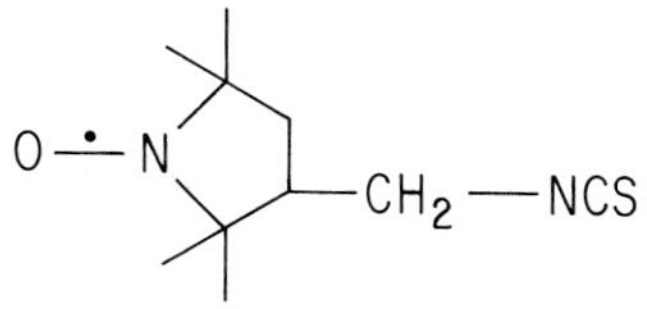

(I)

change at low pH can indeed by thought of in terms of partial denaturation. On the other hand, it must be emphasized that extensive unfolding of the polypeptide does not occur at low pH. The conformation of the toxin in guanidinium Cl or urea is clearly distinct from that at low pH or at high temperature. Unfolding may involve only a very limited domain at low pH or reflect the loss of interactions between domains.

These conclusions are supported by electron spin resonance (ESR) studies of dipththeria toxin spin-labeled at a 2 : 1 (mole/mole), ratio with the amino specific isothiocyanate probe (**I**).

At pH 7 the ESR signal of the labeled toxin indicates that there are two different spin-label environments that fall in the weakly to moderately "immobilized" range. A single isotropic mobile signal is found under completely unfolded conditions, such as in urea or guanidinium Cl, but not at low pH or after denaturation by high temperature. Unfortunately, the ESR signal appears to be sensitive to aggregation, which is extensive both at low pH and high temperature (Blewitt *et al.*, 1985; Zhao and London, 1986), and thus a more precise analysis of conformation from ESR spectra will be complex.

The finding of aggregation is important because it can complicate experimental interpretation in several types of experiments. For example, aggregation via contacts between hydrophobic sites can compete with detergent binding, reducing the amount of bound detergent and thus the apparent hydrophobicity judged from detergent binding. Aggregation of the low pH conformation could also affect the balance between the neutral and low pH conformations (and thus the apparent transition pH) by driving the reaction toward the low pH conformation. It could affect the degree of the cooperative dependence upon proton concentration as well. Also, CD spectra can be distorted by aggregation. Perhaps most important, if nonphysiological aggregates occur when toxin is inserted into model membranes, then serious artifacts may result. On the other hand, one cannot rule out a functional role for aggregation. Studies showing that a single A subunit artifically introduced into the cytoplasm can eventually kill a cell (Yamaizumi *et al.*, 1978) do not prove that toxin penetrates membranes in monomeric form. In any case, the influence of aggregation must be a careful consideration in future studies.

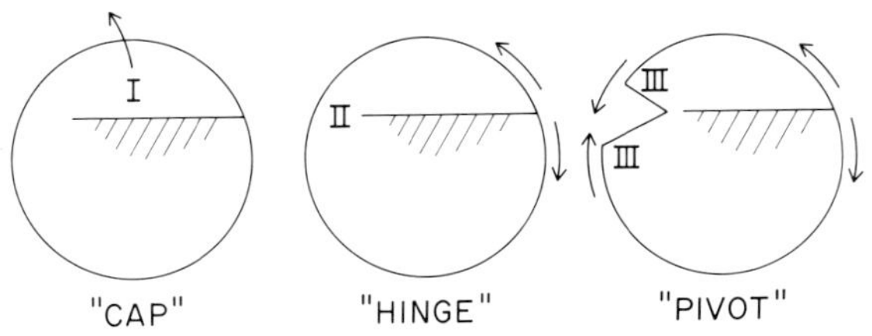

Fig. 4. Possible nature of conformational changes at low pH. Hatched area represents hydrophobic sites. Numerals show region of conformational change.

IV. MECHANISM OF THE CONFORMATIONAL CHANGES

On the basis of these results one can ask, What specific changes in toxin structure occur at low pH? Some possible models are shown in Fig. 4. First, a domain covering the hydrophobic site and buried Trp residues might unfold, thereby exposing buried sites. Second, a hinge holding two domains together could unfold, breaking interdomain interactions, causing domains to come apart, and thereby exposing buried sites. Such a mechanism has been proposed for the change undergone by staphylococcal α-toxin in the presence of lipids or detergents (Tobles *et al.,* 1985). Third, one could imagine a ''pivoting'' mechanism in which the change in conformation involves formation of intrapolypeptide interactions. For example, in the Bohr effect on hemoglobin conformation, protonation stabilizes certain bonding interactions while weakening the interaction with oxygen. However, the denaturation-like effects of low pH are more consistent with the first two models. Fourth, low pH could create a hydrophobic surface directly without a conformational change by protonation of surface Asp and Glu residues, as proposed recently for colicin E_1 (Davidson *et al.,* 1985). This must be ruled out in view of the evidence for a preformed buried hydrophobic site in the sequence data and in experiments with mutant toxins (Boquet *et al.,* 1976), evidence for large conformational changes and, as explained below, because of the effect of salt on the transition pH. As noted earlier, increasing ionic strength by addition of 150 m*M* salt increases the pH of the hydrophilic-to-hydrophobic transition by one unit, but salt shifts the pK_a values of Asp and Glu to lower pH, which means that this fourth mechanism predicts a decrease of the transition pH, opposite to the observed effect. Nevertheless, even though it cannot be the only change occurring at low pH, an increase in hydrophobicity due to charge neutralization could still be one component of the transition.

A second question is, What causes the change in conformation at low pH? Clearly, protonation of amino acid residues must be of central importance. If we let N represent the native conformation, which predominates at neutral pH, and L the conformation which predominates at low pH, then the binding of protons must affect the equilibrium between N and L states such that in the protonated state L predominates. The question can be restated as: How does protonation raise the energy of the N conformation relative to the L conformation?* To answer this question we assume, based on the conclusions above, that the L conformation shows increased exposure of buried groups and has unfolded, at least to the degree that its structure is less compact than in the N state. In this case, the changes in electrostatic interactions induced by protonation will affect the energy of the two states differently.† There are a number of ways in which stabilizing electrostatic interactions might be lost, and destabilizing ones formed, on protonation of acidic or basic amino acid residues. For example, protonation of buried acidic residues could break "stabilizing" internal salt bridges. Protonation of basic residues near other positively charged residues could give rise to electrostatic repulsions. Alternatively, if a basic residue is deeply buried in a hydrophobic site, an energetically unfavorable isolated charge could form on protonation. Such energetically unfavorable changes would tend to destabilize the N state more than the L state because the L state would tend to have an increased distance between charges and would tend to have more charges exposed to solvent and thus in a more polar environment, both of which would diminish the influence of electrostatic effects. Therefore, the equilibrium would shift toward the L state upon protonation. It should be noted that it is difficult to precisely identify the types of ionizable groups involved directly from the apparent pK_a of the transition, because it is a complex function of the pK_a values of all involved residues, both in the N and L states (Tanford, 1970). For example, although the apparent pK_a of the transition is below the normal pK_a for isolated, exposed basic residues, the participation of basic groups is *not* at all ruled out because they would have lower pK_a values when buried in the N state as described above.

The sensitivity of the transition to ionic strength suggests the importance of electrostatic interactions. As noted earlier, the transition pH shifts from pH 4 to 5 in 150 m*M* NaCl. Ionic strength has an effect on pK_a, decreasing the pK_a of acids and increasing the pK_a of bases. Therefore, one explanation for the salt-induced shift to increasing pH could be that

* Alternatively, because the conformational and ionization equilibria are coupled, it is equally valid to formulate the question as: Why does L bind protons more strongly than N?

† Of course, upon protonation there can also be some changes in hydrogen bonding, hydrophobicity, and Van der Waals interactions.

protonation of basic residues triggers the transition. However, the pH shift observed is much larger than the expected shift in pK_a (Perrin and Dempsey, 1974), and therefore other factors may be involved. Indeed, a further effect would arise from electrostatic repulsions between like-charged sites. The presence of such sites would also shift the equilibrium toward the L state in salt, because the stabilizing shielding of electrostatic repulsions by counterions would be more effective in the less compact L conformation, allowing more effective penetration of counterions from solution (Tanford, 1961). Another effect of increased ionic strength would be to weaken intramolecular salt bridges by competition from ions in solution, so again salt bridges could be important.

Examination of the amino acid sequence of the B subunit hints at the sites of possible electrostatic interactions. Figure 5 shows a plot of a parameter we have defined as the excess charge fraction (ECF). The ECF is simply the number of net formal excess positive or negative charges in a segment divided by the total number of charges in the segment. For this calculation we assign a charge of −1 for Glu and Asp and +1 for Arg, Lys, and His. The ECF shows that, as previously known, the C-terminal region of the B fragment has excess positive charge. Intriguingly, the N-terminal half of the B fragment has a large negative ECF. This represents not a large number of negative charges, but rather the fact that the few charged residues among the hydrophobic stretches tend to be negatively charged residues. One possibility suggested by this is that the C-terminal region folds over the N-terminal region so that critically important salt bridges form between the two. Indeed, studies with a mutant toxin have demonstrated that deletion of the C-terminal does expose hydrophobic sites (Boquet *et al.*, 1976). On the other hand, one can easily propose alternative interpretations. One possibility is that the acidic residues are not involved in salt bridges, but instead function in pH-dependent regulation of the degree of hydrophobicity of the N-terminal half. As for the C-terminal region, perhaps it unfolds due to increased electrostatic repulsions as its basic residues become protonated. Another possibility is that the excess positive charge in the C-terminal region is involved only in the previously identified receptor and/or polyanion binding sites and has no role in the transition. Even explanations of the transition mechanism in which electrostatic interactions have no role can be conceived. For example, it has been suggested that proline isomerization catalyzed by low pH could be important (Deleers *et al.*, 1983). Clearly, all such hypotheses are pure speculation at this point and will remain so until the crystal structure, now being determined by two groups of investigators, is available (Collier *et al.*, 1982; McKeever and Sarma, 1982). We must also caution that at present it is not really known whether the conformational changes

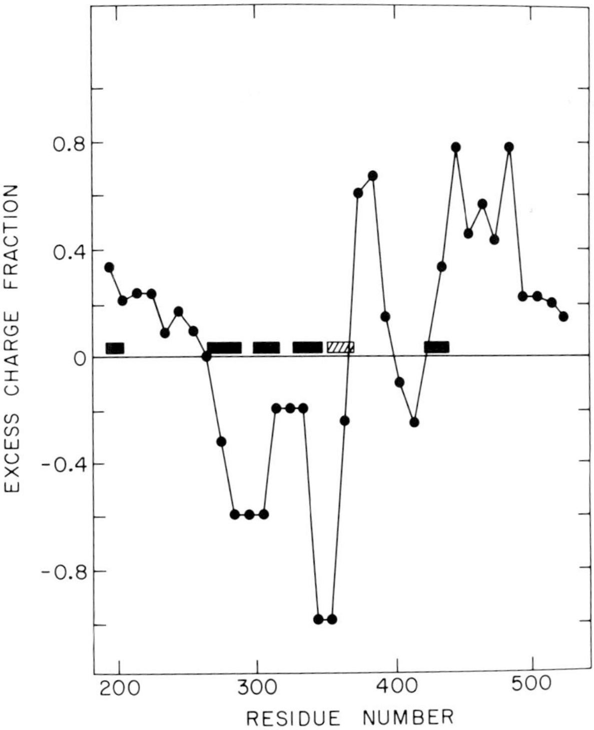

Fig. 5. Charge profile of B subunit. Stretches of 10 or more uncharged amino acid residues are shown as solid blocks. Lined block has only one charged residue. Data points are excess charge fraction (see text) calculated for a 30-residue block; point is placed at center residue. Residue number corresponds to position in overall polypeptide sequence (Greenfield *et al.*, 1983).

occur only within the B subunit, A subunit, the A–B interface, or a combination of these sites. It is very likely that the cooperative transition involves a number of different changes.

V. IMPLICATIONS FOR THE CONFORMATION OF OTHER PROTEINS AND DESIGN OF MODIFIED TOXINS

An important implication of these studies is the possibility that the design of diphtheria toxin conformation may be very similar to that of other proteins believed to undergo a conformational change triggered by low pH in acidic organelles. One of the best characterized cases is that of the influenza virus hemagglutinin, which is believed to play a critical role in low-pH fusion of viral envelope with the membrane of an acidic organelle (White *et al.,* 1983). The pH needed for influenza virus fusion is very similar to the transition pH for the toxin. Furthermore, like the toxin's transition, fusion occurs rapidly on exposure to low pH, and the switch between fusogenic and nonfusogenic states occurs over a narrow pH range. In addition, salt bridges in the hemagglutinin have already been implicated in the pH-induced conformational change (Wilson *et al.,* 1981; Daniels *et al.,* 1985), and a possible relationship between the effects of high temperature and low pH has been hinted at (Daniels *et al.,* 1985). Similar behavior at low pH has also been proposed for a number of other viruses (White *et al.,* 1983) and toxins (Hoch *et al.,* 1985). Furthermore, it is likely that low pH regulates the dissociation of some receptor–ligand complexes at low pH (Ciechanover *et al.,* 1983). So understanding the behavior of the toxin may help us understand the behavior of a whole class of proteins and vice versa. For example, the similarity between influenza hemagglutinin and toxins may help in the design of therapeutic agents. Amantadine, which is used prophylactically and therapeutically for influenza, probably acts as a lysosomotropic agent, increasing the pH in acidic organelles. It might be useful in treating diphtheria as well by preventing membrane penetration by the toxin.

What lessons do these studies have for the design of modified toxins? First, addition or removal of salt bridges may be a key to altering the stability of the protein. Second, basic as well as acidic residues may participate in acid triggering of a conformational change. Finally, any changes that affect thermal stability can also affect pH stability, even if they do not involve ionizable residues. From these ideas it seems that use of oversimplified models of protein structure to design specific changes will give unsatisfactory results in some cases, because the critical residues to change will not always be apparent just from the sequence. Of course, without the high-resolution X-ray structure and characterization of a series of single site mutants, these conclusions can only be a list of possibilities rather than a firm set of rules. Nevertheless, they serve as a starting point for the design of further experiments and suggest that when a combined approach to the structure of these molecules is pursued we will be able to use the understanding gained to design useful modified toxin proteins.

REFERENCES

Alving, C. R., Iglewski, B. H., Urban, K. A., Moss, J., Richards, R. L., and Sadoff, J. C. (1980). *Proc. Natl. Acad. Sci. U.S.A.* **77,** 1986–1990.

Barbieri, J. T., Carroll, S. F., Collier, R. J., and McCloskey, J. A. (1981). *J. Biol. Chem.* **256,** 12247–12251.

Blewitt, M. G., Zhao, J.-M., McKeever, B., Sarma, R., and London, E. (1984). *Biochem. Biophys, Res. Commun.* **120,** 286–290.

Blewitt, M. G., Chung, L. A., and London E. (1985). *Biochemistry* **24,** 5458–5464.

Boquet, P., Silverman, M. S., Pappenheimer, A. M., Jr., and Vernon, W. B. (1976). *Proc. Natl. Acad. Sci. U.S.A.* **73,** 4449–4453.

Cabiaux, V., Vandenbranden, M., Falmagne, P., and Ruysschaert, J. M. (1984). *Biochim. Biophys, Acta* **775,** 31–36.

Carroll, S. F., and Collier, R. J. (1984). *Proc. Natl. Acad. Sci. U.S.A.* **81,** 3307–3311.

Ciechanover, A., Schwartz, A. L., and Lodish, H. F. (1983). *J. Cell Biochem.* **23,** 107–130.

Collier, R. J. (1982). *In* "ADP Ribosylation Reactions: Biology and Medicine" (O. Hayashi and K. Ueda, eds.), pp. 575–592. Academic Press, New York.

Collier, R. J., Westbrook, E. M., McKay, D. B., and Eisenberg, D. (1982). *J. Biol. Chem.* **257,** 5283–5285.

Collins, C. M., and Collier, R. J. (1984). *J. Biol. Chem.* **259,** 15159–15162.

Collins, C. M., and Collier, R. J. (1985). *Biochim. Biophys, Acta* **828,** 138–143.

Collins, C. M., Barbieri, J. T., and Collier, R. J. (1984). *J. Biol. Chem.* **259,** 15154–15158.

Daniels, R. S., Downie, J. C., Hay, A. J., Knossow, M., Skehel, J. J., Wang, M. L., and Wiley, D. C. (1985). *Cell* **40,** 431–439.

Davidson, V. T., Brundsen, K. R., and Cramer, W. A. (1985). *Proc. Natl. Acad. Sci. U.S.A.* **82,** 1386–1390.

Deleers, M., Beugnier, N., Falmagne, P., Cabiaux V., and Ruysshaert, J.-M. (1983). *FEBS Lett.* **160,** 82–86.

Donovan, J. J., Simon, M. I., Draper, R. K., and Montal, M. (1981). *Proc. Natl. Acad. Sci. U.S.A.* **78,** 172–176.

Draper, R. K., and Simon, M. I. (1980). *J. Cell Biol.* **87,** 849–854.

Eidels, L., Ross, L. L., and Hart, D. A. (1982). *Biochem. Biophys, Res. Commun.* **109,** 493–499.

Eidels, L., Proia, R. L., and Hart, D. A. (1983). *Microbiol. Rev.* **47,** 596–620.

Geisow, M. J. (1984). *Exp. Cell Res.* **150,** 29–35.

Geisow, M. J., and Evans, W. H. (1984). *Exp. Cell Res.* **150,** 36–46.

Gray, G. L., Smith, D. H., Baldridge, J. S., Harkins, R. N., Vasil, M. L., Chen, E. Y., and Heyneker, H. L. (1984). *Proc. Natl. Acad. Sci. U.S.A.* **81,** 2645–2649.

Greenfield, L., Bjorn, M. J., Horn, G., Fong, D., Buck, G. A., Collier, R. J., and Kaplan, D. A. (1983). *Proc. Natl. Acad. Sci. U.S.A.* **80,** 6853–6857.

Hoch, D. H., Romerio-Mira, M., Ehrlich, B. E., Finkelstein, A., DasGupta, B. R., and Simpson, L. L. (1985). *Proc. Natl. Acad. Sci. U.S.A.* **82,** 1692–1696.

Hu, V. W., and Holmes, R. K. (1984). *J. Biol. Chem.* **259,** 12226–12233.

Hudson, T. H., and Neville, D. M., Jr. (1985). *J. Biol. Chem.* **260,** 2675–2680.

Kaczorek, M., Delpeyroux, F., Chenciner, N., Streeck, R. E., Murphy, J. R., Boquet, P., and Tiollais, P. (1983). *Science* **221,** 855–858.

Kagan, B. L., Finkelstein, A., and Colombini, M., (1981). *Proc. Natl. Acad. Sci. U.S.A.* **78,** 4950–4954.

Lambotte, P., Falmagne, P., Capiau, C., Zanen, J., Ruysschaert, J.-M., and Dirkx, J. (1980). *J. Cell Biol.* **87,** 837–840.

McKeever, B., and Sarma, R. (1982). *J. Biol. Chem.* **257,** 6923–6925.

Marnell, M. H., Shia, S.-P., Stookey, M., and Draper, R. K. (1984). *Infect. Immun.* **44,** 145–150.

Maxfield, F. R. (1982). *J. Cell Biol.* **95,** 676–681.

Misler, S. (1984). *Biophys. J.* **45,** 107–109.
Perrin, D. D., and Dempsey, B. (1974). "Buffers for pH and Metal Ion Control," p. 8. Chapman & Hall, London.
Ratti, G., Rappuoli, R., and Giannini, G. (1983). *Nucleic Acids Res.* **11,** 6589–6595.
Ross, A. H., Radhakrishnan, R., Robson, R. J., and Khorana, H. G. (1982). *J. Biol. Chem.* **257,** 4152–4161.
Sandvig, K., and Olsnes, S. (1980). *J. Cell Biol.* **87,** 828–832.
Sandvig, K., and Olsnes, S. (1981). *J. Biol. Chem.* **256,** 9068–9076.
Tanford, C. (1961). "Physical Chemistry of Macromolecules," pp. 472–473. Wiley, New York.
Tanford, C. (1970). *Adv. Prot. Chem.* **24,** 1–95.
Tobles, N., Wallace, B. A., and Bayley, H. (1985). *Biochemistry* **24,** 1915–1920.
Torchinsky, Yu. M. (1981). "Sulfur in Proteins." Pergamon, Oxford.
Van Ness, B. G., Howard, J. B., and Bodley, J. W. (1980). *J. Biol. Chem.* **255,** 10710–10716.
White, J., Kielian, M., and Helenius, A. (1983). *Q. Rev. Biophys.* **16,** 151–195.
Wilson, I. A., Skehel, J. J., and Wiley, D. C. (1981). *Nature (London)* **289,** 366–373.
Wright, H. T., Marston, A. W., and Goldstein, D. J. (1984). *J. Biol. Chem.* **259,** 1649–1654.
Yamaizumi, M., Makeda, E., Uchida, T., and Okada, Y. (1978). *Cell* **15,** 245–250.
Zalman, L. S., and Wisnieski, B. J. (1984). *Proc. Natl. Acad. Sci. U.S.A.* **81,** 3341–3345.
Zhao, J.-M., and London, E. (1986). *Proc. Natl. Acad. Sci. U.S.A.* **83,** 2002–2006.

8

Design and Total Chemical Synthesis of a Gene for Bovine Rhodopsin

DANIEL D. OPRIAN, MICHAEL NASSAL, LUCA FERRETTI, SADASHIVA S. KARNIK, AND H. GOBIND KHORANA

Departments of Biology and Chemistry
Massachusetts Institute of Technology
Cambridge, Massachusetts 02139

I. INTRODUCTION

Rhodopsin is the photoreceptor protein of vertebrate retinal rod cells. The amino acid sequence of bovine rhodopsin, an integral membrane protein containing 348 amino acids, has been determined by both protein (Ovchinnikov *et al.*, 1982; Hargrave *et al.*, 1983) and gene sequencing (Nathans and Hogness, 1983). The polypeptide chain contains seven predominantly hydrophobic segments of 19 to 28 amino acids that are thought to traverse the rod cell disk membrane as α-helical rods (Ovchinnikov, 1982; Dratz and Hargrave, 1983). A molecule of 11-*cis*-retinal attached to the ε-amino group of Lys 296 serves as the chromophore and, on absorption of light, undergoes isomerization to the all-*trans* form. This results in a conformational change in the protein which leads to the activation of retinal GTPase and, ultimately, hyperpolarization of the rod cell.

In an effort to gain deeper insight into the process of visual excitation, we have undertaken studies of the structure–function relationships of the visual pigment rhodopsin, retinal GTPase, rhodopsin kinase, and related proteins by the technique of site-specific mutagenesis. As the anticipated use of mutagenesis is extensive, we elected to use the approach of restric-

PROTEIN ENGINEERING
Applications in Science,
Medicine, and Industry

tion fragment replacement (Lo *et al.,* 1984). To facilitate these studies, we chemically synthesized a gene for bovine rhodopsin. The synthetic gene contains 28 unique restriction sites distributed throughout the sequence that define target fragments for mutagenesis which are, in general, no more than 60 base pairs (bp) in length. Mutations are introduced into the gene by simply replacing specific fragments with synthetic counterparts which contain the desired codon changes (Lo *et al.,* 1984). We describe here the design, total chemical synthesis, and *in vitro* expression of this gene.

II. DESIGN OF THE GENE

Our overriding concern was the placement of unique restriction sites in the synthetic gene. As shown in Fig. 1, 26 unique restriction sites have been placed in the coding region of the gene so that the distance between neighboring sites is, in general, 60 base pairs or less. These restriction sites were created by (1) introducing unique sites into the synthetic gene, (2) leaving undisturbed a few of the unique sites present in the native gene, and (3), in the case of repeated sites, destroying all except one site. In addition, unique sites have been placed at each end of the gene for easy transfer from one vector to another. We chose an *Eco*RI site for the 5′end of the gene and a *Bam*HI site for the 3′end because these two sites are found commonly in plasmid and bacteriophage vectors. However, since not all vectors have *Eco*RI and *Bam*HI sites, we ensured that both sites are easily changed, at the 3′end by replacement of *Nar*I–*Bam*HI or *Bst*XI–*Bam*HI fragments and at the 5′end by replacement of an *Eco*RI–*Kpn*I fragment. For each change only two oligonucleotides must be synthesized.

The synthetic gene is 1057 nucleotides in length and codes for a sequence of 348 amino acids corresponding to bovine opsin. There are 52 nucleotide changes in the synthetic gene from the naturally occurring sequence (Nathans and Hogness, 1983; Koike *et al.,*1983). Thirty-six of these changes were introduced to create the 26 unique restriction sites. Twelve changes were made to eliminate stretches of four or more G's or C's in order to avoid possible difficulties in chemical synthesis of oligonucleotides or DNA sequence analysis of the gene. In one instance, however, a stretch of four C's was retained in order to create a unique *Sfi*I restriction site. Three changes were made from a consideration of codon usage in *Escherichia coli* (Grantham *et al.,* 1981), as this was a likely host organism for expression of the synthetic gene. Thus, the UUG codon appearing for both Leu 119 and Leu 128 was changed to CUG, and the

AUA codon for Ile 219 was changed to AUC. Other than these changes, no attempt was made to optimize the synthetic gene sequence for codon usage in *E. coli*. Finally, one change was introduced at position 33 of the nucleotide sequence to destroy an *Eco*K restriction site near the 5′end of the gene. This site was removed because it is recognized by the restriction–modification system of *E. coli* K12 strains.

The gene was synthesized in three fragments of roughly equal lengths. Each carried restriction sites at the ends as shown in Fig. 1, so that they could be cloned and amplified at an intermediate step in the overall synthesis. Fragment EX (so named because it is flanked by *Eco*RI and *Xho*I restriction sites) was 335 base pairs in length and was built up from 32 oligodeoxynucleotides as shown (Fig. 1). Fragments XP and PB were 364 and 345 base pairs in length, respectively, and were constructed with 20 oligonucleotides each (Fig. 1). After cloning, the fragments were ligated to give the entire gene.

The DNA duplexes were constructed from single-stranded oligonucleotides according to general principles developed earlier (Khorana, 1979). The length of protruding cohesive ends formed from the annealing of complementary oligomers was four to five nucleotides. This ensured that most of the nucleotides of each oligomer were in base-paired regions of the duplexes during the ligation reactions and hence were shielded from participating in unwanted side reactions. Except for the ends of fragments to be cloned, cohesive ends were not self-complementary. This was to prevent the self-annealing that would lead to unwanted ligation of duplexes. Oligonucleotides with similar or identical cohesive ends were never used in the same ligation mixture. This was to ensure fidelity in the joining reactions. We did not know ahead of time the order in which oligonucleotides were to be joined to construct the three large fragments. Therefore, each of the cohesive ends within any given fragment was unique. A critical feature of the synthetic strategy was that all of the sites used in the ligase-catalyzed joining reactions were different from the restriction sites created in the gene. The only restriction sites required during synthesis of the gene were at the ends of the three fragments EX, XP, and PB, to allow cloning of the synthetic duplexes and their subsequent ligation to form the full-length gene. To prevent self-annealing and ligation of these palindromic ends, the 5′-OH end groups in oligonucleotides which formed the 5′ termini of each fragment were not phosphorylated.

In addition to the considerations described above for division of the double-stranded DNAs into single-stranded oligonucleotides, another principle followed was related to the characterization of the 5′ termini of the synthetic oligonucleotides. The nucleotides at the 5′ terminal and pen-

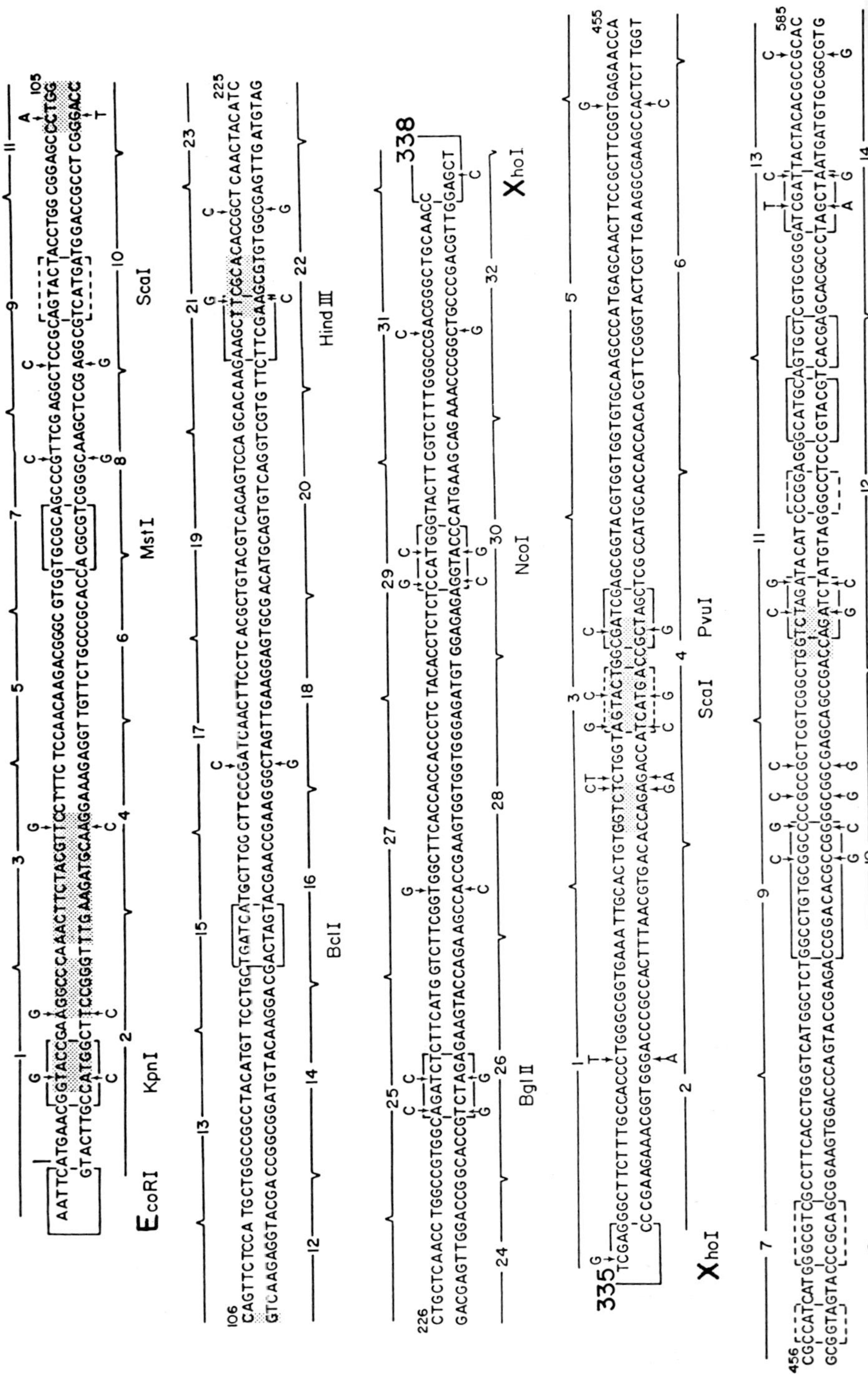
EcoRI
KpnI
MstI
ScaI
BclI
Hind III
Bgl II
NcoI
XhoI
PvuI
SfiI
XbaI
HpaII
SphI
HgiAI
ClaI
HgaI
AhaII

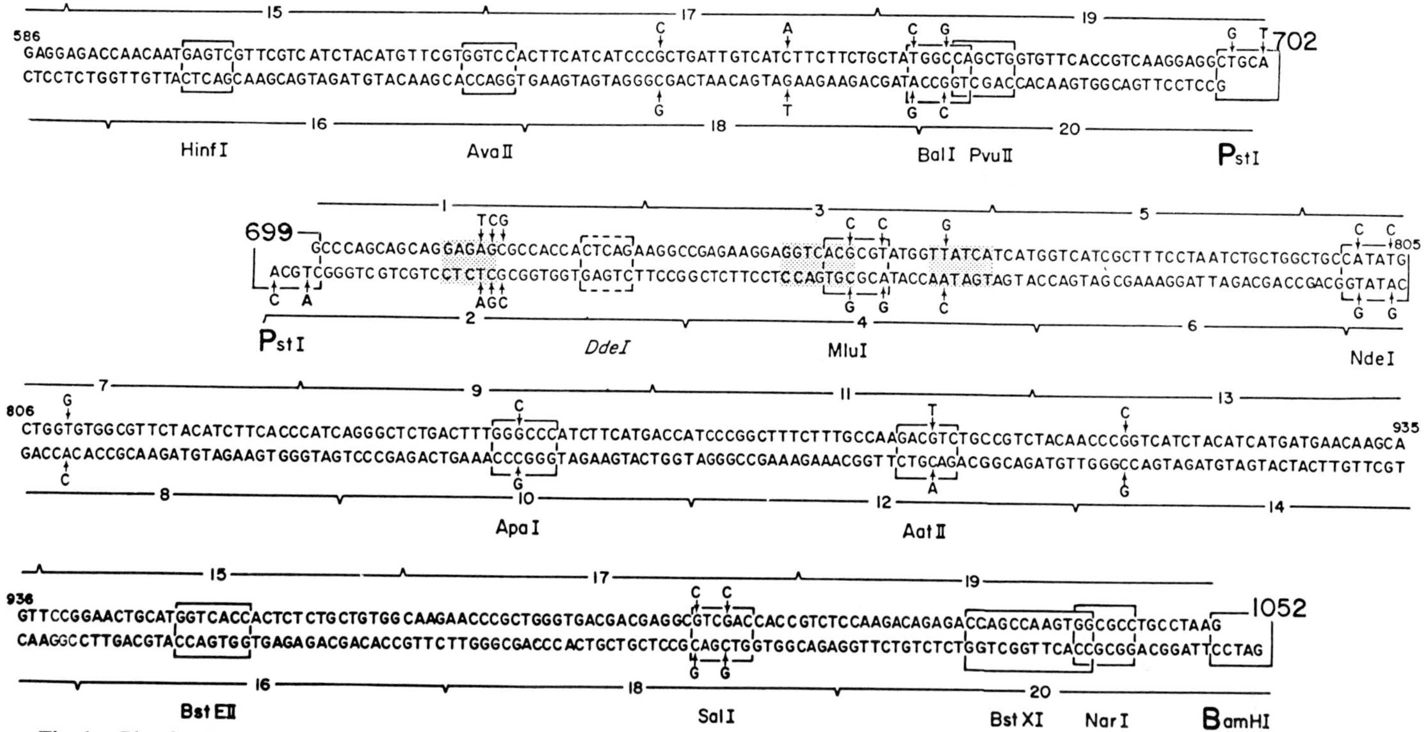

Fig. 1. Plan for the total synthesis of a gene for bovine rhodopsin. The gene is divided into three fragments; fragment EX, nucleotides -5 to 338; fragment XP, nucleotides 335 to 702; and fragment PB, nucleotides 699 to 1052. The oligonucleotides to be synthesized for each fragment are numbered and shown between carets, placed above and below the nucleotide sequences. Vertical arrows indicate nucleotides in the synthetic gene that have been changed from the native sequence. The unique restriction sites in the synthetic gene are shown boxed with a solid line. The restriction sites unique within a fragment but not unique in the gene are shown boxed with a dashed line. Restriction sites present in the native gene and removed in the synthetic gene are shown in shaded boxes. The *Dde*I site in fragment PB is distinguished by italics because there is a second *Dde*I site which overlaps the unique *Bam*HI site. However, the second *Dde*I site is not recognized if the gene fragment has been digested with *Bam*HI.

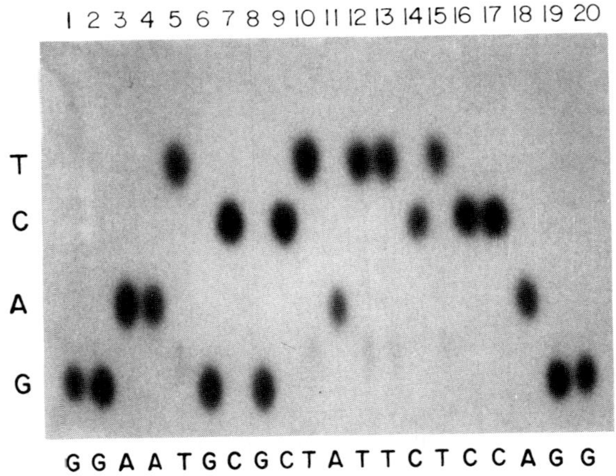

Fig. 2. The 5′-end nucleotide analysis of oligonucleotides corresponding to fragment PB. Each oligomer was labeled at the 5′end with [γ-^{32}P]ATP and T_4 polynucleotide kinase to a specific activity of 2.5 Ci/mmol. The labeled oligomers were then digested with snake venom phosphodiesterase. The resulting nucleoside 5′-monophosphates were separated by thin-layer chromatography on PEI-cellulose plates, developed with 1 *M* LiCl, and then visualized by autoradiography. Nucleoside 5′-monophosphate standards were also applied to the chromatogram and visualized with a UV lamp. The R_f values of the markers indicated to the left of the autoradiogram were as follows: dGMP, 0.30; dAMP, 0.40; dCMP, 0.51; dTMP, 0.59.

ultimate positions of each oligonucleotide were different. The rationale for this was as follows. Our purification procedure for the oligonucleotides was separation by size on preparative polyacrylamide gels, which was highly effective in separating the desired full-length product, *n*, from all of the lower homologs, $n-1$, $n-2$, etc., resulting from incomplete chain elongation. Since chain elongation in the method of chemical synthesis that we used proceeds in the 3′- to 5′-direction, the most likely impurity in the gel-purified oligomer preparations, the *n*-1 homolog, is missing one nucleotide at the 5′ end of the molecule. The presence of an *n*-1 contaminant in the full length oligomer preparation could be detected by 5′-end analysis (Fig. 2), provided the 5′ terminal and penultimate nucleotides were different.

III. SYNTHESIS AND ASSEMBLY OF THE GENE

A. Synthesis, Purification, and Characterization of Oligonucleotides

Oligonucleotides were synthesized on a solid support by the phosphite-triester approach, using the Applied Biosystems model 380A DNA synthesizer. Chain elongation was effected by condensation of the 5′-hydroxyl groups of the growing oligonucleotide chain bound to the solid support with a fully protected 2′-deoxynucleoside *N,N*-diisopropylphosphoramidite (McBride and Caruthers, 1983). Dimethoxytrityl (DMT) protecting groups, including that of the terminal 5′ nucleotide, were removed by treatment with 3% (w/v) dichloroacetic acid in dichloromethane. After the last cycle of synthesis, the terminal DMT group was removed, the oligomer was cleaved from the solid support, and the chemical blocking groups on the bases were removed according to standard protocols (Caruthers, 1982). The repetitive yield was calculated on the basis of final yield of purified product and was usually in the range of 94 to 96% per cycle.

Following complete deprotection, 10–15% (25–40 A_{260}) of the crude oligonucleotide mixture was applied directly to preparative polyacrylamide/8.3 *M* urea gels for purification. The acrylamide content of the gels was governed by the length of oligomer to be purified and ranged from 10 to 20% (w/v). Invariably, the major product was the full-length oligomer, with individual failed sequences representing only minor contaminants in the crude mixture. The full-length product was obtained by excision from the preparative gels. DNA was eluted from the gel slice (Maxam and Gilbert, 1980) and then desalted by chromatography on Sep-Pak C_{18} cartridges (Walters Associates) as described previously (Lo *et al.*, 1984). Following this treatment the oligomers were phosphorylated at their 5′ ends with [γ-^{32}P]ATP and T_4 polynucleotide kinase (Sekiya *et al.*, 1976). The oligomers were labeled with [γ-^{32}P]ATP of relatively low specific activity so that the same samples used for characterization of the purified products could be used for synthesis of the gene without concern for any damage caused by disintegration of the radionuclide.

Two different methods were used to assess the purity of the synthetic oligomers: (1) electrophoresis on denaturing polyacrylamide gels and (2) 5′-end analysis. Results of the 5′-end analysis of all 20 oligomers used in the synthesis of fragment PB are shown in Fig. 2. All oligonucleotides were judged to be at least 95% pure on the basis of electrophoretic homogeneity on denaturing polyacrylamide gels and the observation of a unique 5′-monophosphate from 5′-end analysis.

During characterization of the individual oligomers we found that in at least one case an oligonucleotide isolated as an *n*-1 band had the same 5′-end nucleotide as the full-length product. DNA sequence analysis by a modified Maxam and Gilbert procedure (Banaszuk *et al.*, 1983) showed the *n*-1 band to be identical to the full-length product except that it lacked the 3′-terminal nucleotide. This result suggests that the first incoming nucleoside phosphoramidite, designed as the second nucleotide in the growing chain, reacted directly with the solid support to initiate a new oligonucleotide chain. The functional groups on the solid support for attachment of the incoming phosphoramidite may have been groups that escaped capping by acetic anhydride. Alternatively, they may have arisen from fresh surfaces exposed by mechanical crushing of the glass beads during manipulation of the loaded support, or from groups exposed during deprotection of the first nucleoside with dichloroacetic acid. Although the nature of the linkage to the solid support is unknown at this time, its stability to acid and base must be similar to that of the first nucleoside. The presence of this unusual *n*-1 species in the crude oligonucleotide preparation was of no concern since the *n*-1 band was well separated from the full-length oligonucleotide by preparative gel electrophoresis.

In all, 72 oligodeoxynucleotides were chemically synthesized for assembly of the gene. The length of the oligomers increased as work progressed, largely as a consequence of increasing confidence in our ability to purify full-length oligomers from the *n*-1 impurities. Oligomers ranged from 15 to 27 nucleotides in length for the first fragment, EX, and 29 to 40 nucleotides in length for the last two, XP and PB.

B. Enzymatic Joining of Oligonucleotides: Synthesis of Fragment PB

The following procedures for the synthesis of fragment PB are also typical of those used for fragments EX and XP.

The optimal combination of oligomers for joining reactions and the order in which oligonucleotide duplexes were ligated to form larger fragments were determined empirically from the results of small-scale ligation reactions. Initially, the joining of each combination of four adjacent oligonucleotides at a time was tried. Usually these reactions went in very high yield (>90%); however, in a few cases, e.g., EX 9–12 (refer to Fig. 1 for numbering of oligonucleotides), the yields were low (<50%). In the case of PB 13–16, an unidentified higher molecular weight product, probably a hexamer, was formed. On the basis of these results and further test reactions with combinations of six or more oligonucleotides, two different procedures were used for the synthesis of fragment PB. In the first (Fig. 3I), oligonucleotides PB 1–6, PB 7–14, and PB 15–20 were joined separately and the three resulting duplexes were purified by preparative gel electrophoresis. The full-length fragment PB was obtained by ligation of the three gel-purified duplexes. Gel purification of the intermediate ligation products was necessary in order to avoid additional side products in subsequent ligation reactions. In the second method (Fig. 3II) duplexes

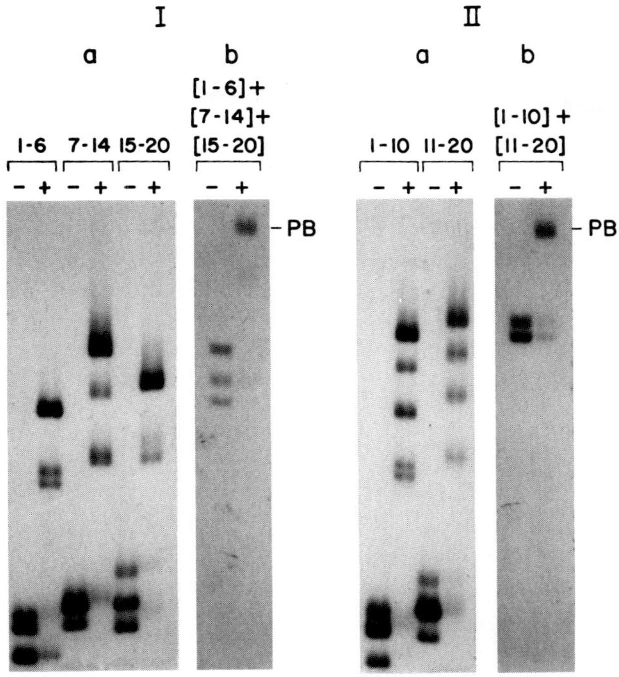

Fig. 3. Synthesis of fragment PB by T_4 ligase-catalyzed joining reactions. (Ia) The ligase-catalyzed joining reactions of oligonucleotides PB 1–6, PB 7–14, and PB 15–20 [without (−) and with (+) ligase]. The joined products were purified and joined in a three-component reaction (Ib) to form PB. (IIa) The ligase-catalyzed joining reactions of oligonucleotides PB 1–10 and PB 11–20 [without (−) and with (+) ligase].(IIb) Joining of purified products [1–10] and [11–20] from IIa to form PB. The ligation products were separated by electrophoresis in nondenaturing 8% polyacrylamide gels and visualized by autoradiography.

[1–10] and [11–20], purified from the first step ligation, were joined in a second step to form the full-length fragment PB (Fig. 3II).

The overall yield of the two different strategies was comparable. Slightly less product was formed in the first step from the more complex annealing mixtures 1–10 and 11–20. However, this was compensated for in the second step, which involved only a two-component ligation. Typically, the yield after each ligation step followed by gel purification was about 20%. Therefore, fragment PB, which was assembled in two stages, had an overall recovery of about 4%.

C. Cloning and DNA Sequence Analysis of Fragments EX, XP, and PB

Once all three synthetic fragments, EX, XP, and PB, had been assembled and purified by gel electrophoresis, each was cloned into M13 for amplification of the DNA. The nucleotide sequence of each fragment was determined at this stage by the dideoxy method, using the single-stranded M13 recombinant phage DNA as template and the synthetic oligomers used to assemble the fragments as primers. The sequence of each fragment corresponded exactly to that expected from the synthesis.

D. Assembly of the Complete Rhodopsin Gene from the Cloned Fragments

The fragments were excised at the flanking restriction sites from the RF form of the vector DNA and then purified by preparative agarose gel electrophoresis. The entire gene was assembled in a four-component ligation mixture, containing each of the three gel-purified fragments (EX, XP, and PB) and *Eco*RI-*Bam*HI doubly digested M13 vector, and then cloned into *E. coli* strain JM103. The nucleotide sequence of the entire gene was determined from single-stranded M13 templates using as primers the oligomers that were used to assemble the gene. In this manner, the 1-kbp sequence was determined to be exactly that predicted from the synthetic scheme shown in Fig. 1. The three fragments EX, XP, and PB and the entire gene EB, purified from M13 vector DNA, are shown in Fig. 4.

IV. *In Vitro* EXPRESSION OF THE SYNTHETIC RHODOPSIN GENE

We used a system for the *in vitro* synthesis of protein (Krieg and Melton, 1984) for expression of the synthetic rhodopsin gene. The gene was cloned into the plasmid vector pSP65 (Melton *et al.*, 1984) downstream of the SP6 promoter. Following runoff transcription with SP6 polymerase and capping of the nascent RNA chain with guanylyltransferase, the synthetic mRNA was translated in a wheat germ extract (Krieg and Melton, 1984). As is shown in Fig. 5, essentially a single protein band is labeled with [^{35}S]methionine. As expected, the ^{35}S-labeled band migrates faster than authentic rhodopsin due to the presence of oligosaccharide chains in the latter. That this band is indeed rhodopsin is suggested by the following: (1) it has the same electrophoretic mobility as rhodopsin synthesized from bovine retina mRNA template in the wheat germ system (not shown), and (2) it reacts with anti-rhodopsin antibodies, as shown in Fig. 5.

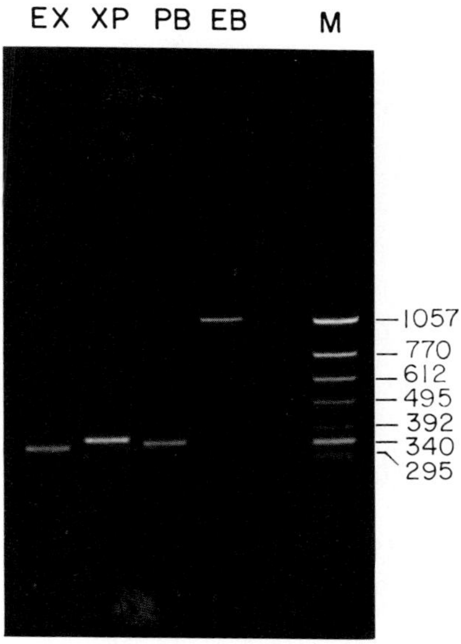

Fig. 4. Gel electrophoretic analysis of the cloned fragments EX, XP, and PB and the total synthetic gene EB. The DNA bands were visualized with ethidium bromide. The fragments were previously purified by gel electrophoresis after excision with the appropriate restriction enzymes from the M13 vector DNA. Molecular weight markers are *Hin*cII digestion products of ϕX174 DNA. The expected size of the fragments is 335, 364, 345, and 1049 bp for EX, XP, PB, and EB, respectively.

V. SUMMARY

We have chosen a method for oligonucleotide-directed, site-specific mutagenesis which involves the replacement of restriction fragments in a gene with synthetic DNA duplexes containing the desired codon changes (Lo *et al.,* 1984). The method is limited by the frequency and distribution of unique restriction sites in naturally occurring gene sequences. For this reason, we undertook the design and total chemical synthesis of a gene for bovine rhodopsin that is specifically tailored for use in site-specific mutagenesis. The synthetic rhodopsin gene has been engineered to contain 26 unique restriction sites within the coding region of the gene and *Eco*RI and *Bam*HI sites at the 5′ and 3′ ends of the gene, respectively. These 28 unique restriction sites define a set of 27 target fragments for site-specific mutagenesis. The recognition sites are relatively evenly distributed throughout the gene so that 90% of the sequence can be mutated with restriction fragments of 80 bp or less. Further, we have expressed the synthetic gene in a completely *in vitro* protein synthesis system.

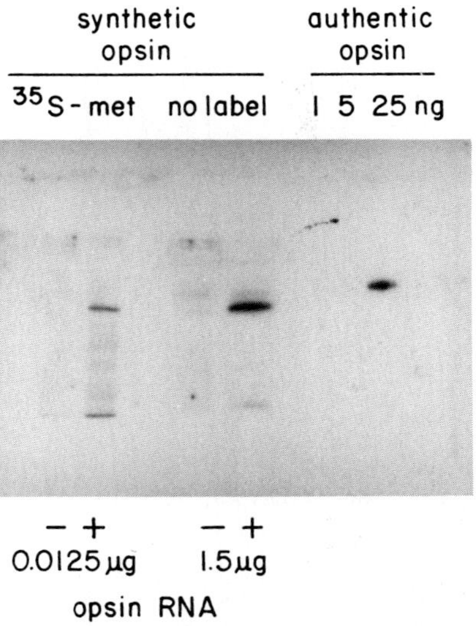

Fig. 5. *In vitro* expression of synthetic rhodopsin gene. mRNA was transcribed from the synthetic opsin gene by using SP6 polymerase as described in the text. Following translation in a wheat germ extract, aliquots of the reaction mixture were applied to a 15% acrylamide/SDS gel (Laemmli, 1970) and analyzed by Western blots (Burnette, 1981). The separated proteins, blotted onto nitrocellulose, were allowed to react with anti-rhodopsin antibodies and ^{125}I-labeled protein A (Burnette, 1981). Authentic opsin from bovine rod outer segments was applied to the gel as a control. The wheat germ translations were done with either unlabeled or ^{35}S-labeled methionine. That the signals from protein in the [^{35}S]methionine-labeled lanes were due to ^{35}S and not ^{125}I was demonstrated by the fact that the signal from this lane could be removed by autoradiography through a thin film of cellulose acetate, whereas the signals from the unlabeled translation mixture and from authentic rhodopsin were unaffected.

Restriction fragment replacement offers many advantages as a method for extensive site-specific mutagenesis of a gene. With the advent of automated methods of DNA synthesis combined with procedures for the rapid purification and characterization of synthetic oligonucleotides, gene design and total synthesis has become a practical method for tailoring a gene to the specific purpose of mutagenesis.

ACKNOWLEDGMENTS

This work was supported by grants GM28289 and AI11479 from the National Institutes of Health, grant PCM78-13713 from the National Science Foundation, and contract N00014-82-K-0668 from the Office of Naval Research. D.D.O. was supported by fellowship 5 F32 EY05616-02 from NEI, NIH. M.N. is a recipient of an EMBO long-term fellowship.

We thank Dr. K. Anne Kronis for bovine retina rod outer segments, Dr. Paul A. Hargrave for generously providing the anti-rhodopsin antibodies, and Dr. Uttam L. RajBhandary for invaluable discussions and encouragement throughout this work.

REFERENCES

Banaszuk, A.M., Deugau, K.V., Sherwood, J., Michalak, M., and Glick, B.R. (1983). *Anal. Biochem.* **128,** 281–286.

Burnette, W.N. (1981). *Anal. Biochem.* **112,** 195–203.

Caruthers, M.H. (1982). In "Chemical and Enzymatic Synthesis of Gene Fragments" H.G. Gasser and A. Lang eds, Verlag Chemie, Weinheim. pp. 71–79.

Dratz, E.A. and Hargrave, P.A. (1983). *Trends Biochem. Sci.* **8,** 128–131.

Grantham, R., Gautier, C., Gouy, M., Jacobzone, M., and Mercier, R. (1981). *Nucleic Acids Res.* **9,** r43–r74.

Hargrave, P.A., McDowell, J.H., Curtis, D.R., Wang, J.K., Juszczak, E., Fong, S.-L., Mohana Rao, J.K., and Argos, P. (1983). *Biophys. Struct. Mech.* **9,** 235–244.

Khorana, H.G. (1979). *Science* **203,** 614–625.

Koike, S., Nabeshima, Y.-I., Ogata, K., Fukui, T., Ohtsuka, E., Ikehara, M., and Tokunaga, F. (1983). *Biochem. Biophys. Res. Commun.* **116,** 563–567.

Krieg, P.A., and Melton, D.A. (1984). *Nucleic Acids Res.* **12,** 7057–7070.

Laemmli, U.K. (1970). *Nature* (*London*) **227,** 680–685.

Lo, K.-M., Jones, S.S., Hackett, N.R., and Khorana, H.G. (1984). *Proc. Natl. Acad. Sci. U.S.A.* **81,** 2285–2289.

McBride, L.J., and Caruthers, M.H. (1983). *Tetrahedron Lett.* **24,** 245–248.

Maxam, A.M., and Gilbert, W. (1980). In "Methods in Enzymology" (L. Grossman and K. Moldave, eds.), Vol. **65,** pp. 499–560. Academic Press, New York.

Melton, D.A., Krieg, P.A., Rebagliati, M.R., Maniatis, T., Zinn, K., and Green, M.R. (1984). *Nucleic Acids Res.* **12,** 7035–7056.

Nathans, J., and Hogness, D.S. (1983). *Cell* **34,** 807–814.

Ovchinnikov, Yu.A. (1982). *FEBS Lett.* **148,** 179–191.

Ovchinnikov, Yu. A., Abdulaev, N.G., Feigina, M.Yu., Artamonov, I.D., Zolotarev, A.S., Kostina, M.B., Bogachuk, A.S., Miroshnikov, A.I., Martinov, V.I., and Kudelin, A.B. (1982). *Bioorg. Khim.* **8,** 1011–1014.

Sekiya, T., Bresmer, P., Takeya, T., and Khorana, H.G. (1976). *J. Biol. Chem.* **251,** 634–641.

9

Surface-Simulation Synthesis and Its Applications in Protein Molecular Recognition

M. ZOUHAIR ATASSI

Marrs McLean Department of Biochemistry
Baylor College of Medicine
Houston, Texas 77030

I. INTRODUCTION

One of the major interests of this laboratory has been the understanding of the molecular immune recognition of proteins. Knowledge of the molecular features responsible for the antigenicity of native protein molecules lies at the basis of understanding in molecular terms the cellular events of the immune response and the mechanisms of immune disorders. Myglobin was selected as the first protein model for these studies, which started in 1963. The selection of myoglobin was dictated by necessity (Atassi, 1973) because it was the only protein for which the three-dimensional and covalent structures (Kendrew *et al.*, 1961; Edmundson, 1965) were known. Early in our studies we proposed (Atassi and Saplin, 1968) that protein antigenic sites can comprise residues which are distant in sequence but come into close proximity in the three-dimensional structure by virtue of the folding of the polypeptide chain. On completion of the determination of the entire antigenic structure of myoglobin (Atassi, 1975) it became evident that this recognition feature did not exist in myoglobin. It was found that the native protein has five antigenic sites, each comprising six or seven amino acid residues located in a continuous conformationally distinct surface region of the myoglobin molecule. Such

PROTEIN ENGINEERING
Applications in Science,
Medicine, and Industry

sites have been termed *continuous* antigenic sites (Atassi, 1978a; Atassi and Smith, 1978).

It is known that the "contact" residues in a protein binding site are not necessarily always in direct peptide bond linkage but are spatially adjacent in the molecule so as to form the binding surface. Many chemical and physiochemical approaches have been employed in probing protein active sites. The approaches, in general, have been able to describe and define the site and the environment near the site. In several proteins and enzymes, many of the residues involved in ligand or substrate binding have been identified by X-ray crystallography. However, this knowledge could not be translated into a useful experimental approach. Thus, a chemical approach was needed by which knowledge of the residues forming a protein binding site could be exploited to synthetically mimic a given protein activity. This was to come about from our studies on the antigenic structure of lysozyme.

Our studies on the antigenic structure of hen egg-white lysozyme relied on an intricate chemical strategy previously devised (Atassi, 1973) for the determination of the antigenic structure of myoglobin. The systematic application of this strategy to lysozyme revealed (Atassi *et al.,* 1976c; Atassi and Habeeb, 1977) that each of the antigenic sites of lysozyme is made up of residues which are generally not directly linked by peptide bonds but which, due to the folding of the polypeptide chain, come into close spatial proximity, describing a discrete area of the surface topography of the protein. In order to unequivocally verify the identity of such, then unknown, immunological recognition features, it was necessary to view protein interactions in a conceptually new light that would, in principle, lead to synthetic manipulation.

II. DEVELOPMENT OF SURFACE-SIMULATION SYNTHESIS

Protein interactions take part via specific area(s) comprising certain contact residues arranged on the surface of the binding site in a specific disposition and a specific shape. We reasoned (Atassi *et al.,* 1976b) that if the requirements of amino acid arrangement and shape of the binding site could be duplicated chemically on a matrix other than the parent protein, it should be feasible to duplicate the binding activity of the protein in a synthetic substance of appropriate design. This rationale led us to devise (Atassi *et al.,* 1976b) an unorthodox and novel concept by which the spatially adjacent residues constituting a protein binding site are directly linked via peptide bonds with appropriate spacing and directionality.

Thus, a peptide is generated which does not exist in the native protein but which attempts to mimic the construction of a specific surface area that forms the binding site (Atassi *et al.*, 1976b). Soon thereafter we proposed (Lee and Atassi, 1976) the term surface-simulation synthesis to describe this concept.

The idea of surface-simulation synthesis was accompanied by the realization that the success or failure of the concept of mimicking protein surfaces would be totally dependent on the ability of the design to meet certain structural and spatial requirements. When a portion of a polypeptide chain of known primary structure is to be synthesized, there is usually no ambiguity as to how to proceed. However, attempting to mimic an area of a protein surface was a totally different situation. On the protein surface, an area having a discontinuous architecture has no clear synthetic directionality. The first important task, therefore, was to determine the appropriate synthetic direction. The second important consideration was related to spacing. When two residues on the surface were separated by a distance which approximated the length of a peptide bond (about 4 Å) it might be obvious how to link them. But when two residues were separated by greater distances, the answer was not so clear. Other considerations were related to the shape of the surface to be mimicked and the physical properties of the surface-simulation synthetic peptide, such as its solubility and stability.

Studies designed to mimic the antigenic sites of lysozyme by surface-simulation synthesis revealed that the directionality of the synthesis and the spacing between the constituent residues of the site play critical roles in determining the binding activity of the synthetic peptide (Lee and Atassi, 1977a,b; Atassi and Lee, 1978). Before any synthetic studies can be done, it is necessary to know the identity of the contact residues of the site. Usually these are determined by chemical methods (for reviews, see Atassi, 1977a,b; Atassi and Habeeb, 1977) and/or X-ray crystallography of the protein crystals with the appropriate ligand. The C^{α}-to-C^{α} distances separating the contact residues of the site can be calculated from the X-ray coordinates.

When a peptide has been designed with appropriate spacing of the contact residues, another peptide can be made which has the same amino acids but in the reverse sequence (i.e., the *retro* sequence). It has been found that, generally, only one synthetic direction is appropriate for binding (Lee and Atassi, 1977a,b) (Fig. 1 and 2). The requirement for directionality is needed for appropriate side-chain orientation (Lee and Atassi, 1977a,b). Studies with computer graphics have shown (M. Z. Atassi and D. Barry, unpublished results) that, for a site mimicked by surface-simulation synthesis, only one synthetic direction can assume a conformation

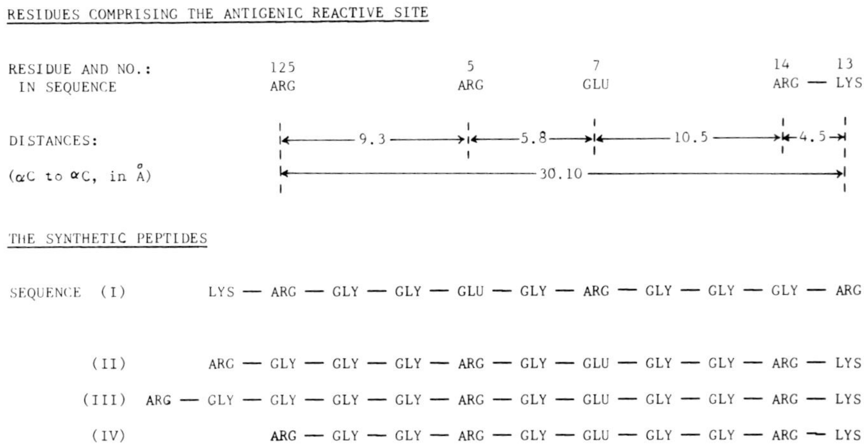

Fig. 1. (Top) Spatially adjacent surface residues constituting antigenic site 1 and their numerical positions in the primary structure of lysozyme. The distances (angstroms) separating the consecutive residues of the site are given as C^{α}-to-C^{α} distances together with the overall extended dimension of the site. (Bottom) Primary structures of the surface-simulation synthetic peptides that were designed to copy the site and investigate its directional and conformational requirements. Binding activity with anti-lysozyme antibodies was exhibited only by peptide IV. (From Atassi and Lee, 1978.)

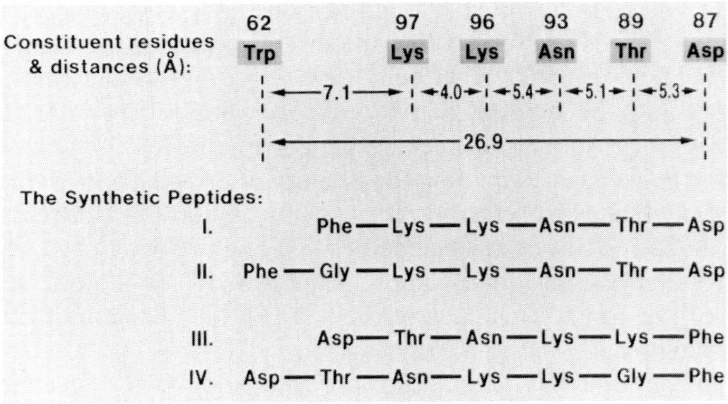

Fig. 2. (Top) Spatially adjacent surface residues constituting antigenic site 2 and their numerical position in the primary structure of lysozyme. The distances (angstroms) separating the consecutive residues of the site are given as C^{α}-to-C^{α} distances together with the overall extended dimension of the site. (Bottom) Structures of the synthetic peptides that were studied for effect of spacing and directionality on binding activity with anti-lysozyme antibodies. Note that peptides III and IV are the *retro* sequences of peptides I and II, respectively. Peptide II exhibited the maximum binding activity. Previously we showed (Lee *et al.*, 1976) that tryptophan can be replaced by phenylalanine with equal efficiency. (From Lee and Atassi, 1977a.)

in which the side-chain orientation is superimposable on that in the native protein. In the reverse synthetic direction the side-chain orientation will not fit.

To achieve the appropriate spacing of the contact residues, glycine spacers were initially used (Atassi, *et al.*, 1976b; Lee and Atassi, 1976, 1977a,b) when necessary. In subsequent studies we employed other amino acids with the expectation that they would impart certain structural characteristics to the synthetic peptide. However, it was found that glycine spacers give the best results, perhaps because of their flexibility and because they have no side chain that will interfere with the binding function of the contact residues.

III. APPLICATIONS OF SURFACE-SIMULATION SYNTHESIS

The introduction of the concept of surface-simulation synthesis (Atassi *et al.*, 1976b; Lee and Atassi, 1976, 1977a,b) led to the immediate realization (Atassi *et al.*, 1976b) that its application need not be confined to definition of protein regions that are recognized by antibodies and that it should be applicable to protein binding sites that discharge other functions. The following sections give examples of its application and demonstrate its versatility.

A. The Antigenic Sites of Lysozyme

Hen egg-white lysozyme is a "tight" protein (i.e., it is inaccessible to proteolytic attack and contains disulfide bonds). My interest in the antigenic structure of lysozyme was stimulated by the fact that such tight proteins (e.g., lysozyme, ribonuclease) have been difficult to study. Because of the inaccessibility of these tight proteins, it has not been possible to prepare a variety of overlapping peptides without rupturing the disulfide bonds. Investigators have therefore diverted their efforts to study the immunochemistry of preparations with broken disulfide bonds. Unfortunately, these unfolded preparations bear no immunochemical relationship to the native protein.

Through the systematic application of a comprehensive chemical strategy (see Atassi, 1973, 1975, 1978a), we found that lysozyme has three antigenic sites, each having spatially adjacent residues (that are otherwise distant in sequence) and describing a discrete area on the surface topography of the protein (for review, see Atassi, 1978a). Although we had proposed the existence of such protein antigenic sites quite early (Atassi and

Saplin, 1968), we subsequently found (Atassi, 1975; Twining and Atassi, 1979; Twining *et al.*, 1981) that they did not exist in myoglobin, and it was in lysozyme that we identified (Atassi *et al.*, 1976c) the first such sites. The need for chemical verification of such sites led us to introduce (Atassi *et al.*, 1976b; Lee and Atassi, 1976) the concept of surface-simulation synthesis, which enabled us to precisely delineate and chemically synthesize the three antigenic sites of lysozyme (for review, see Atassi, 1978a).

The identities of the lysozyme antigenic sites are shown in Fig. 3 and their locations in the three-dimensional structure of lysozyme can be seen in Fig. 4. The antigenic sites are briefly described below.

1. Site 1

This antigenic site is constructed (Atassi and Lee, 1978) by the side chains of the five spatially adjacent surface residues Arg 125, Arg 5, Glu 7, Arg 14, and Lys 13. The extended dimension of the site from Arg 125 to Lys 13 (C^{α}-to-C^{α}) is 30 Å. These residues, which describe part of the surface topography of the protein, bind antibody as if they are in direct peptide linkage (Atassi *et al.*, 1976b). In fact, the immunochemical reactivity of this site is fully satisfied by the surface-simulation synthesis peptide Arg-Gly-Gly-Arg-Gly-Glu-Gly-Gly-Arg-Lys, which does not exist in native lysozyme, but mimics the arrangement of residues in this surface region. The surface-simulation synthetic site exhibits a directional preference (Arg 125 to Lys 13), which appears to be independent of the species of the immunized animal, at least with the species tested so far (Atassi and Lee, 1978). The site has restricted conformational freedom, clearly evidenced by its sensitivity to variation of the spacing between its constituent residues in surface-simulation synthesis (Atassi and Lee, 1978). The intactness of the disulfide bond 6–127 in native lysozyme is critical for the integrity of this site (Atassi *et al.*, 1973, 1976a).

2. Site 2

This antigenic site also consists (Atassi *et al.*, 1976b; Lee and Atassi, 1977a) of spatially adjacent surface residues. These are Trp 62, Lys 97, Lys 96, Asn 93, Thr 89, and Asp 87. The extended dimension of the site

---

Fig. 3. The three antigenic sites of lysozyme. The diagram shows the spatially adjacent residues constituting each antigenic site and their numerical position within the primary structure. The distances (angstroms) separating the consecutive residues and the overall dimension of each site (in its extended form) are given, together with the dimension of each surface-simulation synthetic site. The latter assumes an ideal distance of 3.9Å between α carbons. The precise boundary and conformational and directional definitions of the sites are described in the text. The three sites account quantitatively for the entirety (96–100%) of the antigenic reactivity of lysozyme. (From Atassi, 1978a.)

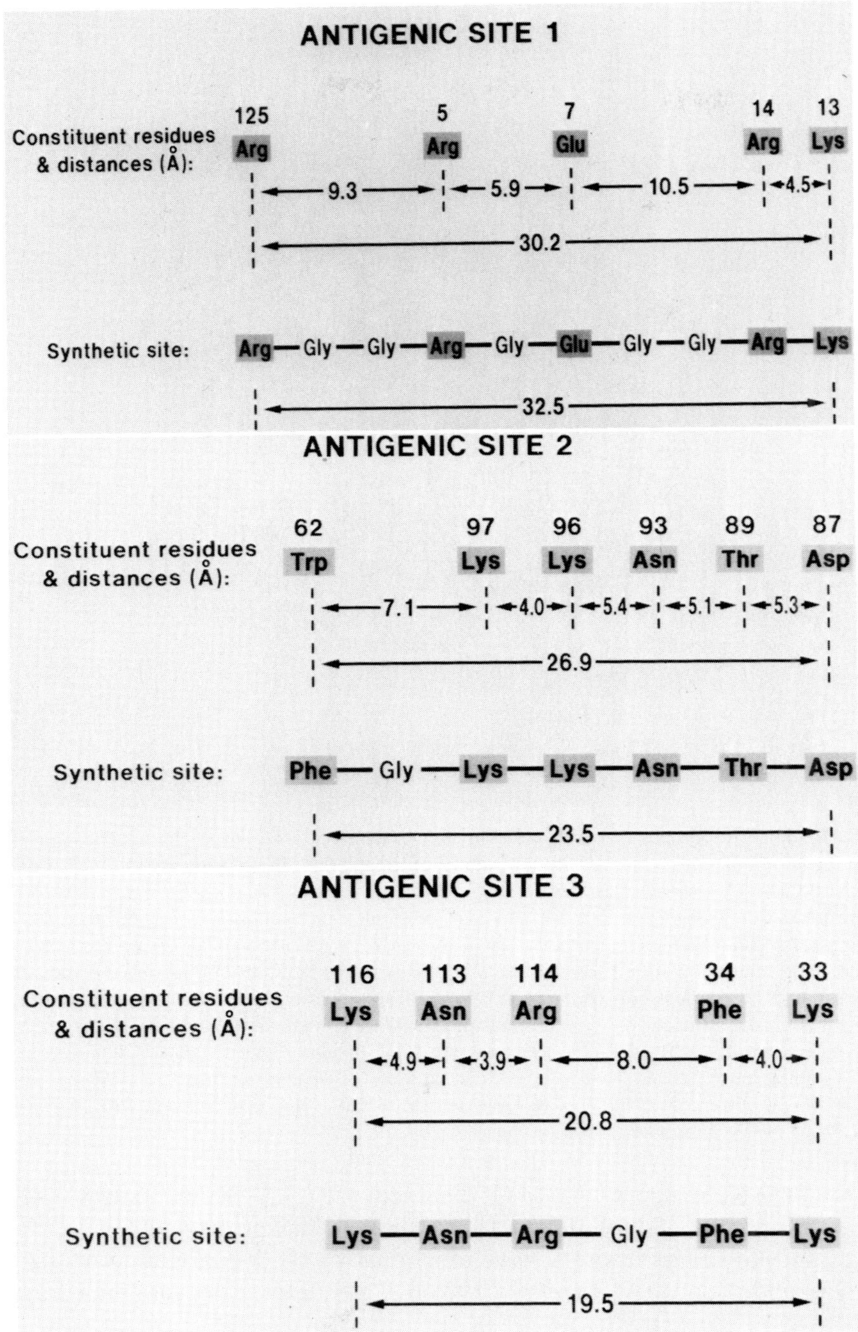
ANTIGENIC SITE 1
Constituent residues & distances (Å):
125 Arg
5 Arg
7 Glu
14 Arg
13 Lys
9.3
5.9
10.5
4.5
30.2
Synthetic site:
Arg—Gly—Gly—Arg—Gly—Glu—Gly—Gly—Arg—Lys
32.5
ANTIGENIC SITE 2
Constituent residues & distances (Å):
62 Trp
97 Lys
96 Lys
93 Asn
89 Thr
87 Asp
7.1
4.0
5.4
5.1
5.3
26.9
Synthetic site:
Phe—Gly—Lys—Lys—Asn—Thr—Asp
23.5
ANTIGENIC SITE 3
Constituent residues & distances (Å):
116 Lys
113 Asn
114 Arg
34 Phe
33 Lys
4.9
3.9
8.0
4.0
20.8
Synthetic site:
Lys—Asn—Arg—Gly—Phe—Lys
19.5

Fig. 4. Photographs of lysozyme models. (A) Relative positions of the residues constituting antigenic sites 1 and 2. The side chains of the residues in the sites are outlined, those making up site 1 with horizontal bars and those constituting site 2 with vertical bars. The preferred direction of site 1 (at least by surface-simulation synthesis) is Arg 125 to Lys 13. Site 2 also had a preferred direction (Trp 62 to Asp 87). (B) Position of antigenic site 3 on the molecule relative to sites 1 and 2. The side chains of the residues comprising the sites are outlined. The residues constituting site 3 have diagonal bars. This view is taken by rotating the model 125° anticlockwise on the vertical axis relative to the view shown in (A). From this perspective only parts of site 1 can be seen, i.e., residues Lys 13, Arg 5, and Arg 125 (horizontal bars). Of site 2, only Trp 62 can be seen (vertical bars). Site 3 showed the same directional preferences (Lys 116 to Lys 33) toward rabbit and goat antisera. (From Atassi and Lee, 1978, as modified from color in Atassi, 1978b.)

from Trp 62 to Asp 87 (C^α-to-C^α distance) is 27.3 Å. As with site 1, site 2 describes part of the surface topography of the protein molecule. The residues forming the site behave functionally as if they are directly linked by peptide bonds (Atassi *et al.*, 1976b). Thus, the surface-simulation syn-

Fig. 4. (*Continued.*)

thetic peptide Phe-Gly-Lys-Lys-Asn-Thr-Asp, which does not exist in lysozyme but simulates a surface region of it, carries the full reactivity of the site (Atassi *et al.*, 1976b; Lee and Atassi, 1977a). With the antisera studied so far, the antigenic site exhibits a preferred directionality in surface-simulation synthesis (Trp 62 to Asp 87). The site is subject to conformational restrictions indicated by a requirement for appropriate spacing between the constituent residues (Lee and Atassi, 1977a). Intactness of the disulfide bonds 64–80 and 76–94 is essential to bring together the various constituent residues of the site (Atassi *et al.*, 1973; Lee and Atassi, 1975). This antigenic site overlaps the enzymatic active site because they both share Trp 62 (Lee and Atassi, 1975, 1977a; Atassi *et al.*, 1976a,b).

3. Site 3

Like sites 1 and 2, this site is also constructed by spatially adjacent surface residues. The antigenic site comprises (Lee and Atassi, 1977b) the five residues Lys 116, Asn 113, Arg 114, Phe 34, and Lys 33 and, in its extended form, has a C^{α}-to-C^{α} dimension (from Lys 116 to Lys 33) of 21 Å. These residues describe part of the surface of the molecule and act functionally toward the antibody as if they are in direct peptide bond linkage. In fact, the surface-simulation synthetic peptide having the structure Lys-Asn-Arg-Gly-Phe-Lys (which does not exist in lysozyme) carries the full immunochemical reactivity of the site (Lee and Atassi, 1977b). With the antisera so far studied the synthetic antigenic site exhibited a preferred directionality (Lys 116 to Lys 33), since the reverse surface-simulation synthetic sequence was immunochemically inefficient (Lee and Atassi, 1977b). Intactness of the disulfide bond 30–115 is critical for the integrity of this antigenic site in lysozyme (Atassi *et al.*, 1973). Antigenic site 3 overlaps the hexasaccharide substrate binding site at the carbonyl group of Phe 34 and the side chain of Arg 114 (Lee and Atassi, 1977b).

It is important to note that the three surface-simulation synthetic peptides had the capacity to bind the bulk (97–99%) of the antibody response against native lysozyme (Table I). These three surface-simulation synthetic peptides were coupled to succinylated bovine serum albumin, and the conjugates were injected into rabbits (Atassi and Sakata, 1982). As expected, antibodies against each peptide reacted, only with that peptide, but it was also found that the antibodies could bind to lysozyme (Fig. 5 and Table II), and the complete specificity of this binding was rigorously established (Fig. 6) (Atassi and Sakata, 1982). This finding is extremely important in view of the fact that a surface-simulation synthetic peptide does not exist in the native protein but attempts to mimic a surface region of it. The concept provided a versatile experimental approach for synthetic mimicking of protein binding sites with diverse activities, and it was also proposed as a powerful tool for studying protein conformation (Atassi, 1978a; Atassi and Lee, 1978b). The latter would depend on making antibodies to surface-simulation peptides representing selected surface areas of a protein molecule and then studying the binding of these antibodies with the native protein. However, although it is a standard procedure in immunology to make antibodies to small peptides by immunizing with conjugates of the peptides on a high-molecular-weight carrier, this was the first demonstration that antibodies against surface-simulation synthetic peptides will, in fact, bind to the native protein.

In a more recent sudy, T-cell recognition of the surface-simulation

TABLE I

Binding of Radioiodinated Anti-Lysozyme Antibodies to Immunoadsorbents Carrying the Three Surface-Simulation Synthetic Sites[a,b]

	Binding of antibody from G9		Binding of antibody from G10	
Immunoadsorbent	(cpm)	(%)	(cpm)	(%)
Lysozyme[c]	50,180	100	73,330	100
Site 1[c]	15,760	31.4	20,335	27.7
Site 2[c]	18,015	35.9	27,600	37.6
Site 3[c]	14,950	29.8	22,300	30.0
Total of independent binding by three sites	48,425	97.1	20,232	95.8
Binding by passage[d] through sites serially	49,880	99.4	71,670	97.7

[a] From Atassi (1978a).

[b] The specific ^{125}I-labeled antibody fractions from antisera G9 and G10 were isolated on a lysozyme immunoadsorbent prior to use. The amounts of antibody applied were: G9, 5.21×10^4; G10, 7.56×10^4 counts per minute (cpm). Each value represents the average of four replicate analyses, which varied $\pm 1.3\%$ or less. Results have been corrected for the amount of antibody bound in control experiments with glycine–Sepharose, histidine–Sepharose, and myoglobin–Sepharose. Nonspecific background binding in the various controls ranged from 1 to 3% of the total label applied.

[c] Results for independent binding were obtained by passage of an aliquot of the antibody solution on only one of the immunoadsorbents indicated.

[d] Results obtained by serial passage of the same antibody sample on the immunoabsorbent of site 1, then site 2, then site 3.

synthetic antigenic sites has been explored (Bixler and Atassi, 1984) in mouse strain B10.BR which is a high responder to lysozyme (Okuda *et al.*, 1979). The discontinuous antigenic sites of lysozyme also had the capacity to stimulate proliferation of T cells driven by native lysozyme in long-term cultures (Bixler and Atassi, 1984). This constituted the first clear demonstration that, contrary to a long-held impression, T-cell recognition is not restricted only to sequence features, but can also be directed to protein discontinuous surface areas of high conformational dependence.

Thus, surface-simulation synthetic peptides (of the correct spacing and directionality) have the ability to bind antibodies raised against native lysozyme and to be recognized by T cells that are specific for the native protein. Furthermore, antibodies raised against the peptides bind to native protein. These findings clearly indicate that these peptides, even

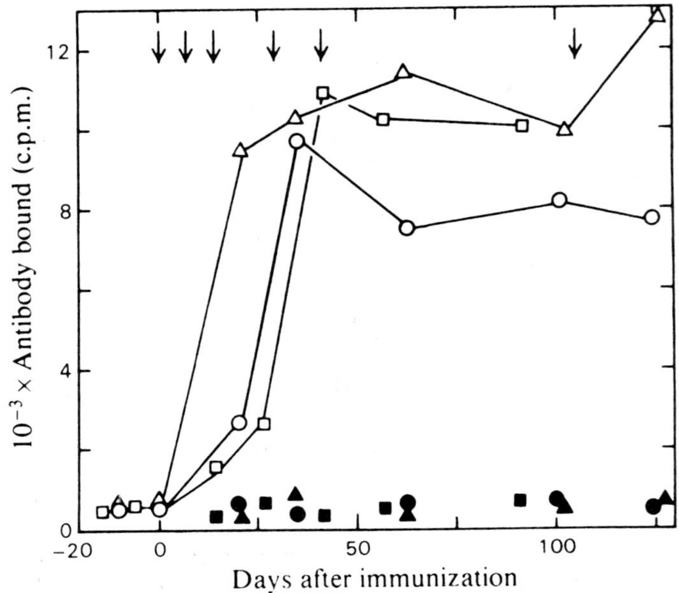

Fig. 5. Time-course study of the binding of rabbit anti-peptide antibodies with lysozyme and unrelated control proteins. Antibodies were determined by reaction with ^{125}I-labeled protein A by binding with lysozyme or control protein on polyvinyl chloride plate adsorbents. (○, ●) Serial antibodies against synthetic site 1 conjugate (rabbit L45); (△, ▲), serial antibodies against site 2 conjugate (rabbit L33); (□, ■), serial antibodies against site 3 conjugate (rabbit L43). Open symbols represent binding of the antibodies with lysozyme; filled symbols represent the average binding values of the respective antibodies with the control proteins, rabbit hemoglobin, bovine ribonuclease A, and bovine cytochrome *c*. Vertical arrows indicate days on which injections with each peptide conjugate were given. (From Atassi and Sakata, 1982.)

though they do not exist in lysozyme, must be true representatives of the respective discontinuous surface regions.

The potential application of surface-simulation synthesis in monitoring protein conformation depends on the preparation of antibodies to synthetic surface-stimulation peptides that are designed to mimic surface areas of a protein and on the ability of these antibodies to recognize and react with those surface regions in the native protein molecule. Such antibodies could then be used as conformational probes to double-check the three-dimensional structure of a protein in solution (Atassi, 1978a). This would also constitute an effective approach to monitoring the acquisition of correct residue arrangements of preselected regions of a protein (or fragment) on renaturation of a denatured protein or reformation of previously reduced disulfide bonds. Because this approach would provide

TABLE II

Immunoadsorbent Binding Studies of Anti-Site Antibodies to the Synthetic Sites, Lysozyme, and Control Proteins[a,b]

Immunoadsorbent	Anti-(site 1 peptide) (L45)	Anti-(site 2 peptide) (L33)	Anti-(site 3 peptide) (L35)
Site 1	4480	30	0
Site 2	10	5100	0
Site 3	0	0	3900
Lysozyme	4510	5180	3950
Ribonuclease	20	30	10
Cytochrome *c*	0	20	0
Site 1 of myoglobin	10	0	0

[a] From Atassi and Sakata (1982).

[b] Titrations were performed with a fixed amount of ^{125}I-labeled immunoglobulin G (5×10^4 cpm) from one antiserum against each peptide conjugate. The results shown here summarize the binding values obtained with 500 μ of each immunoadsorbent. They have been corrected for nonspecific binding (usually 3–5%) by an equal volume of uncoupled Sepharose. Each value is the average of triplicate analyses, which varied ±1.5% or less.

direct region-by-region scanning of the correct disposition of several residues on the surface, it might be useful to those interested in the mechanism of protein folding or in the prediction of protein conformation from its sequence. Thus, for example, the three surface-simulation synthetic sites of lysozyme show simultaneously the spatial interrelations of 16 surface residues in the native protein. The approach would therefore yield much more valuable and precise information than that derived from chemical (e.g., availability of side chains to chemical modification), spectral, or other physical measurements in solution. Of course, in principle, antibodies could be made to several surface-simulation peptides representing the surface areas that jointly encompass the entire surface of a protein molecule (Atassi and Sakata, 1982). These antibodies should recognize and bind to those surface areas even though the latter do not represent antigenic sites when the native protein is used as the immunogen. Such a battery of antibodies would provide the most powerful conformational probes available for monitoring the three-dimensional structure of a protein in solution. The ability to make antibodies to surface areas of a protein molecule that are preselected and mimicked by surface-simulation synthesis presents a versatile and powerful tool in exploiting therapeutic applications for helper, suppressive, tolerogenic, and allergenic sites on free proteins and those of other invasive agents (Atassi and Sakata, 1982).

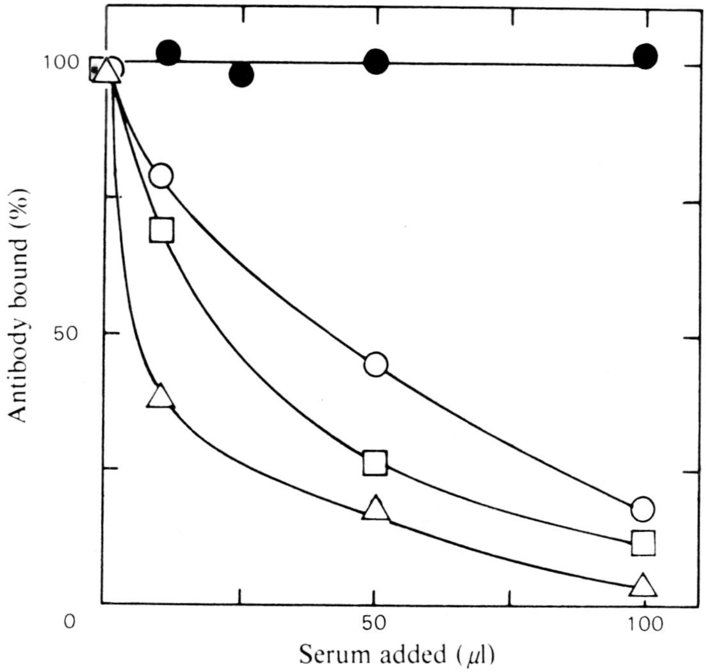

Fig. 6. Inhibition by various antisera of the binding to lysozyme–Sepharose of ^{125}I-labeled IgG against the synthetic sites. (○) Inhibition of binding of anti-(site 1 peptide) antibodies (L45, 125-day bleed) by goat anti-lysozyme antiserum G9; (△) inhibition of binding of anti-(site 2 peptide) antibodies (L34, 125-day bleed) by anti-lysozyme antiserum G9; (□) inhibition of binding of anti-(site 3 peptide) antibodies (L43, 93-day bleed) by anti-lysozyme antiserum G9. (●) Inhibitions (average values) by nonimmune goat serum, rabbit antisera against bovine serum albumin, and rabbit antisera against polymeric bovine cytochrome *c*. Inhibition with another anti-lysozyme antiserum (G10) gave results essentially similar to those shown here with antiserum G9. (From Atassi and Sakata, 1982.)

B. Surface-Simulation Synthesis of Antibody Combining Sites

From the foregoing short treatment, it is obvious that surface-simulation synthesis constituted a conceptual breakthrough without which the precise definition of the entire antigenic structure of lysozyme would have been unattainable. However, it was immediately realized (Atassi *et al.*, 1976b; Atassi, 1978a, 1979; Lee and Atassi, 1976) that the power of this concept should not in any way be confined to the determination of protein antigenic structures and that it should be possible to use it to synthetically mimic any type of protein binding sites involving the interaction of a protein with other proteins or with nonprotein molecules. One of the most

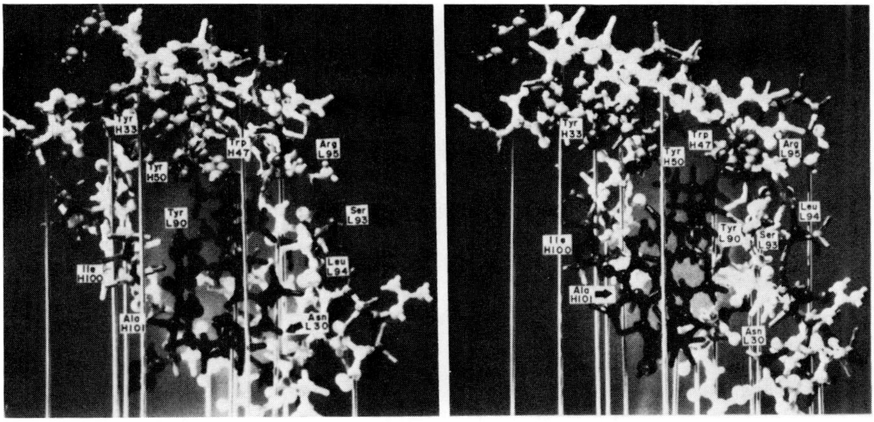

Fig. 7. Two views of a model of the IgG New combining site with bound Vit. K_1OH in place. Hypervariable light chain residues 27–31 and 90–95 and hypervariable heavy chain residues 31–33, 47–58, and 97–103 are shown. The labeled residues were linked, with appropriate spacing, by surface-simulation synthesis (see Fig. 8) into a single peptide with Ile 100H as its NH_2 terminus. (From Twining and Atassi, 1978.)

intriguing questions in immunology is whether an antibody combining site can be mimicked by surface-simulation synthesis. Ability to perform this task will obviously have far-reaching implications in immunology. Studies from our laboratory (described in the next section) indicate that this has indeed been accomplished. Thus, by mimicking the antibody combining site, at least in terms of binding function, the surface-simulation concept has made an important contribution to protein immunochemistry.

1. Surface-Simulation Synthesis of the Immunoglobulin New Combining Site to the γ-Hydroxyl Derivative of Vitamin K_1

Following the introduction and development of the concept of surface-simulation synthesis, we examined whether an antibody combining site can be reconstructed by surface-simulation synthesis. For our first attempt (Twining and Atassi, 1978) to test the feasibility of this idea, we chose the myeloma protein IgG New, which binds a hydroxyl derivative of vitamin K_1 (Vit. K_1OH) (Amzel *et al.*, 1974; Saul *et al.*, 1978). The combining site is shown in Fig. 7. The site residues (Ile 100H, Ala 101H, Asn 30L, Tyr 90L, Ser 93L, Leu 94L, Arg 95L, Trp 47H, Tyr 50H, Tyr 33H) were directly linked by peptide bonds, with appropriate intervening spacers (Twining and Atassi, 1978). Two peptides were synthesized (Fig. 8) which mimicked the combining site but differed in the absence (peptide

Contact residues & distances (Å):	V_H 100 **Ile**	V_H 101 **Ala**	7.2	V_L 30 **Asn**	6.3	V_L 90 **Tyr**	5.9	V_L 93 **Ser**	V_L 94 **Leu**	V_L 95 **Arg**	8.7	V_H 47 **Trp**	8.7	V_H 50 **Tyr**	6.4	V_H 33 **Tyr**

Synthetic site: **Ile**–**Ala**-Gly-**Asn**-Gly-**Tyr**-Gly-**Ser**-**Leu**-**Arg**-Gly-**Trp**-Gly-**Tyr**-Gly-**Tyr**

Control peptide: Trp-Arg-Asn-Gly-Leu-Gly-Trp-Tyr-Gly-Gly-Ser-Gly–Ile–Tyr-Ala-Gly

Fig. 8. Amino residues which make close contact with Vit. K_1OH in the hypervariable region of IgG New and structures of the two surface-simulation peptides (peptide B) designed to mimic the binding site. The distances (C^α to C^α) between the constituent residues of the combining site are also given. Also, a synthetic peptide A (not shown) was synthesized which did not have a spacer between Tyr 90L and Ser 93L. The control peptide has an amino acid composition identical to that of peptide B, except that its sequence is randomized. The diagram is not intended to imply that the site is an extended surface. It is, in fact, a cavity (see Fig. 7). (From Twining and Atassi, 1978.)

A) or presence (peptide B) of a spacer between Tyr 90L and Ser 93L. Also, a control peptide was synthesized having the same amino acids as peptide B but in a different random sequence. Peptides A and B showed remarkable binding activity toward Vit. K_1OH while the control peptide exhibited no binding activity (Fig. 9). Peptide B, approximating more closely the correct spatial separation between the side chains, had a higher binding activity than peptide A. Inhibition studies confirmed the specificity of the binding between Vit. K_1OH and peptide A or B (Fig. 9). The results showed that a complex binding site can be successfully mimicked by surface-simulation synthesis (Twining and Atassi, 1978).

2. *Surface-Simulation Synthesis of the Phosphorylcholine Combining Site of the Myeloma Protein M-603*

We also applied (Kazim and Atassi, 1980) the concept of surface-simulation synthesis to the combining site of the myeloma protein M-603. The X-ray coordinates of the binding site are known to a 3.1 Å resolution (Padlan *et al.*, 1976; Segal *et al.*, 1974). We have been able to synthetically mimic the phosphorylcholine (PC) binding characteristics of M-603 in the surface-simulation peptide Ser-Tyr-Gly-Gly-Arg-Tyr-Gly-Glu-Trp-Val (Fig. 10) (Kazim and Atassi, 1980). In a solid-phase radiometric assay, this peptide was found to bind to PC–Sepharose. In contrast, a control peptide which had the same amino acid composition but a different se-

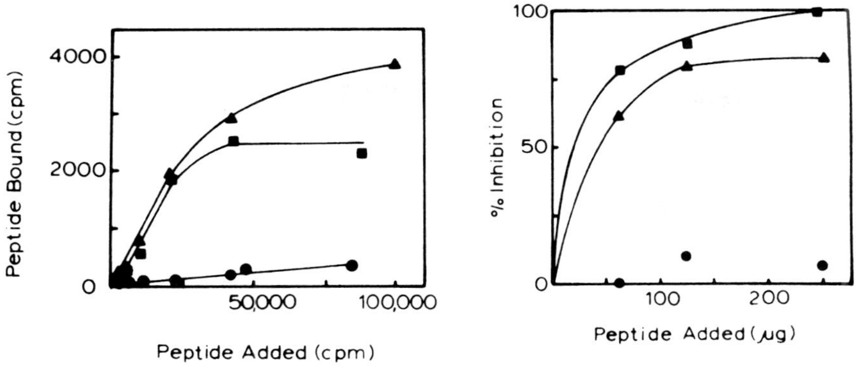

Fig. 9. (Left) Binding of increasing amounts of free ^{125}I-labeled peptides to constant amounts (2.5 μl) of Vit. K_1OH–Sepharose. (■) Surface-simulation peptide A; (▲) surface-simulation peptide B; (●) control peptide. (Right) Inhibition of binding of free ^{125}I-labeled peptide A to Vit. K_1OH-Sepharose by (■) peptide B; (▲) peptide A; (●) control peptide. (From Twining and Atassi, 1978.)

quence (see Fig. 10) showed little or no binding activity. The specificity of binding to PC–Sepharose by the surface-simulation peptide was further demonstrated by the inability of the control peptide to significantly inhibit the binding of M-603, which the surface-simulation peptide inhibited by

AMINO ACID RESIDUES IN PC COMBINING SITE	61H Ser	60H Tyr	52H **Arg**	33H **Tyr**	35H **Glu** –	36H **(Trp)** –	37H Val
DISTANCES IN M-603 (C^α-to-C^α, in Å)	←3.96→←13.04→←4.60→←6.83→←3.86→←3.73→						
SYNTHETIC PEPTIDES							
SURFACE-SIMULATION PEPTIDE A	Ser — Tyr — Gly — Gly — Arg — Tyr — Gly — Glu — Trp — Val						
DISTANCES IN PEPTIDE A (C^α-to-C^α, in Å)	←3.84→←11.52→←3.84→←7.68→←3.84→←3.84→						
CONTROL PEPTIDE B	Gly — Tyr — Glu — Arg — Trp — Gly — Ser — Val — Tyr — Gly						

Fig. 10. Amino acid residues in the phosphorylcholine binding site of M-603 and structure of the synthetic surface-simulation peptide A designed to mimic it. Synthetic control peptide B, having the same composition but a scrambled sequence relative to peptide A, is also shown. Amino acid residues in M-603 which have been implicated in phosphorylcholine binding are shown in bold type. Parentheses around Trp 35H indicate that it replaces Trp 104aH for reasons outlined elsewhere (Kazim and Atassi, 1980). The C^α-to-C^α distances in M-603 were calculated from the C^α atomic coordinates. Distances in peptide A were calculated assuming the C^α-to-C^α distance of each peptide bond to be 3.84 Å. (From Kazim and Atassi, 1980.)

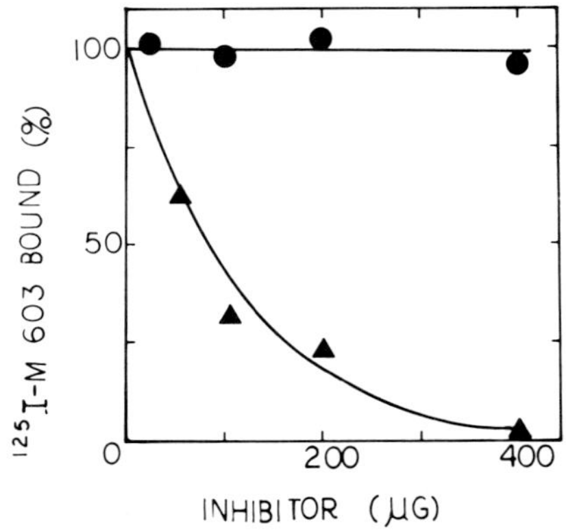

Fig. 11. Quantitative inhibition by unlabeled synthetic peptide A and control peptide B (see Fig. 10) of the binding of ^{125}I-labeled M-603 to phosphorylcholine–Sepharose. (▲) Peptide A; (●) peptide B. (From Kazim and Atassi, 1980.)

more than 90% (Fig. 11) in the concentration range examined (Kazim and Atassi, 1980).

These findings suggest that surface-simulation synthesis can be effectively employed to synthetically mimic antibody combining sites and may in the future be a valuable tool with which to examine idiotypic specificities of antibodies and to manipulate the immune response to clinically important antigens. These avenues of investigation are being pursued in this laboratory.

3. Possible Surface-Simulation Synthesis of Antibody Combining Sites to Lysozyme Antigenic Sites

The remarkable success of the surface-simulation synthesis concept in reconstructing antigenic sites of spatially adjacent surface residues and combining sites of myeloma proteins suggested to us (Atassi and Zablocki, 1977) its usefulness in investigating the feasibility of mimicking the antibody combining sites to protein antigenic sites. It is evident from the preceding sections that proper application of surface-simulation synthesis requires a detailed knowledge of the three-dimensional structure of the protein under study and unequivocal chemical identification of the residues constituting a binding site as well as their conformational spacing and directional requirements. Even though none of these parameters was known for the antibody combining sites against lysozyme, we reasoned (Atassi and Zablocki, 1977) that the situation was not entirely hopeless since, in all likelihood, the antibody combining site would comprise residues that are not necessarily directly linked by peptide bonds and that are complementary to those in the corresponding antigenic sites. Futhermore, the directionality of the antigenic site and the spacings between its constituent residues and the corresponding parameters of the antibody combining site must be equivalent or comparable for appropriate binding to take place. Thus, with precise knowledge of all the parameters of an antigenic site, it should be possible to create a reasonable design for its complementary antibody combining site. This has been done (Atassi and Zablocki, 1977) for two antigenic sites in native lysozyme.

Two peptides CS-2 and CS-3 (Fig. 12) were designed (Atassi and Zablocki, 1977) on the basis of complementarity to lysozyme antigenic sites 2 and 3, respectively, in ionic, hydrophobic, hydrophilic, and side-chain length of the constituent amino acids. Each of the two complementary peptides exhibited appreciable inhibitory activity toward the reaction of lysozyme with its antisera which compared favorably with the activity of the respective antigenic sites, and the inhibitory activities of the two peptides were additive. Peptide immunoadsorbents bound only lysozyme and not antibody or myoglobin. Neither of the two peptides had any immunochemical activity in the myoglobin or bovine serum albumin immune systems. Furthermore, control synthetic peptides of similar charge but different sequence had no inhibitory effect on the lysozyme immune reaction (Atassi and Zablocki, 1977). The evidence indicated that the antibody combining sites against antigenic sites 2 and 3 of native lysozyme were successfully mimicked synthetically, at least in terms of binding function.

We have stressed (Atassi and Zablocki, 1977) that the residues deduced in peptides CS-2 and CS-3 are in no way implied to be the actual residues brought together in the binding sites of the antibodies by the three-dimensional folding of the latter. This is difficult to know. Also, it was emphasized that the functional success of these two peptides does not imply a unique antibody site to a given lysozyme antigenic site. Other complementary amino acids may serve equally well in the antibody-combining site. For example, the role of leucine may be satisfied by isoleucine, valine, phenylalanine, etc., but we have not tested that yet. By employing related alternatives to each residue, it will not be difficult to rationalize heterogeneity and differences in affinity of the antibody combining site (Atassi and Zablocki, 1977).

The finding that two synthetic peptides (CS-2 and CS-3) which were designed to be complementary to antigenic sites 2 and 3 of lysozyme possessed antibody-like binding activity against their respective antigenic sites (Atassi and Zablocki, 1977) made it possible to investigate an important question. Since each of these peptides mimics an antibody combining site in terms of binding function to a lysozyme antigenic site, would

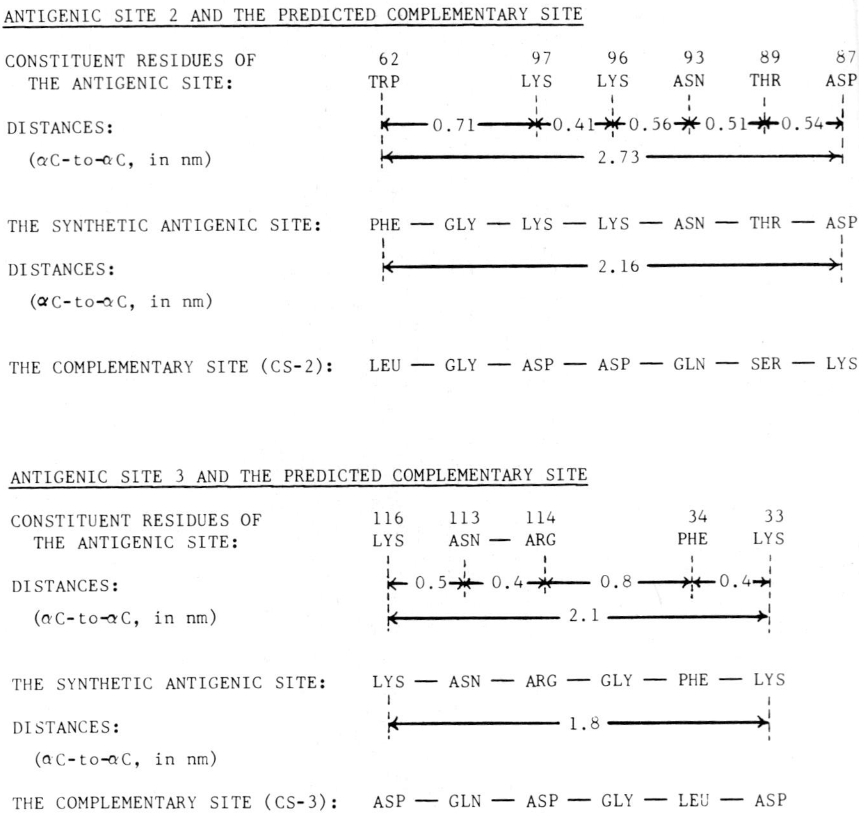

Fig. 12. Antigenic sites 2 and 3 of hen egg-white lysozyme and their predicted complementary antibody combining sites. The spatially adjacent surface residues constituting each antigenic site and the numerical positions of these residues in the primary structure of lysozyme are shown. The distances (nanometers) separating the consecutive residues and the overall dimension of each site are given, together with the dimension of the respective surface-simulation sythetic antigenic site. Below each antigenic site is given the structure of the respective complementary surface-simulation peptide which was predicted to mimic the antibody-combining site directed against the antigenic site. (From Atassi and Zablocki, 1977.)

antibodies to such a peptide react with antisera to lysozyme? If the antibody combining site is truly mimicked in peptides CS-2 and CS-3, it is extremely important to determine whether antibodies to these peptides would react with anti-lysozyme antibodies. Achievement of this would open up many hitherto untapped avenues in immunology, in particular those pertaining to regulation and manipulation of the immune response by anti-combining site antibodies.

TABLE III

Binding of Anti-Peptide CS-3 Antisera to Anti-Lysozyme Antibodies[a,b]

Adsorbent	Anti-peptide CS-3 antibodies bound (cpm) from antisera L42	L44
Anti-lysozyme G9	3200	3360
Anti-lysozyme G10	1750	1800
Controls		
Myoglobin	120	140
Nonimmune IgG	110	100
Anti-cytochrome *c*	130	120

[a] Results from Atassi and Sakata (1980).

[b] The results represent plateau binding values of immunoadsorbent titrations with fixed amounts of ^{125}I-labeled immune IgG of rabbit antisera (L42 and L44) against peptide CS-3 (designed to mimic an antibody site for lysozyme antigenic site 3; see Fig. 12) with increasing amounts of Sepharose adsorbents of immune IgG from two goat anti-lysozyme antisera (G9 and G10). Results represent the average of triplicate analyses, which varied ±1.2% or less.

In subsequent work (Atassi and Sakata, 1980) we raised antibodies against one of these peptides (CS-3), which is complementary to antigenic site 3 of lysozyme, to determine whether antibodies to this peptide will react with anti-lysozyme antibodies. Radioiodinated anti-peptide antibodies were bound by immunoadsorbents carrying immune IgG fractions of two goat anti-lysozyme antisera (Table III). The reaction was completely inhibited by lysozyme but not by serum albumin, myoglobin, hemoglobin, ribonuclease A, or cytochrome *c*. Thus, inhibition studies confirmed the specificity of the binding. The results clearly showed that antibodies to a peptide, which was designed to be complementary to a lysozyme antigenic site, react with anti-lysozyme antibodies. This indicated that the complementary peptide must have resembled closely the antibody combining sites in two lysozyme antisera with regard to the spacing, the directional requirements for surface-simulation synthesis, and the nature of the residues, presumably in the hypervariable regions on the heavy and light chains, of the anti-lysozyme antibodies. Thus, with precise knowledge of a protein antigenic site, a complementary peptide can potentially be designed (Atassi and Zablocki, 1977; Atassi and Sakata, 1980) to mimic the arrangement of residues in the combining sites of antibodies to that antigenic site. The success of the design can be tested by immunochemical studies on the complementary peptide and finally by the ability of its antibodies to react with anti-protein antisera.

C. Surface-Simulation Synthesis of the Substrate Binding Site of Trypsin

The three-dimensional structures of bovine trypsin, iPrP-trypsin, and complexes with trypsin inhibitor and benzamidine substrate analog are known in detail (Stroud *et al.,* 1974; Chambers and Stroud, 1977; Huber and Bode, 1978). It is therefore possible to design surface-simulation peptides that will mimic the binding site of a serine esterase. Recent studies (Atassi, 1985) were aimed at mimicking the substrate binding site of bovine trypsin in a surface-simulation peptide that expresses the binding activity of the enzyme molecule with substrates and inhibitors. Trypsin was chosen as the model for these studies because its X-ray structure is known in fine detail. The structure of iPrP-F-inhibited trypsin, initially determined by Stroud *et al.* (1971, 1974), was subsequently refined to 0.15 nm (1.5 Å) resolution (Chambers and Stroud, 1977). Also, the binding of trypsin to soybean trypsin inhibitor provided a convenient assay by which the surface-simulation peptides could be screened for binding activity.

The substrate binding site of trypsin in complex with substrate analog (benzamidine) and with trypsin inhibitor is known in detail at 0.19–0.15 nm (Huber and Bode, 1978; Chambers and Stroud, 1977; Bode and Schwager, 1975; Huber *et al.,* 1974, 1975). The following summarizes the amino acids that have been identified as the contact residues of the trypsin binding site (see also Fig. 13). Tyrosine 39 hydrogen-bonds with trypsin inhibitor. Histidine 40 is buried in the trypsin–inhibitor complex and participates in polar interaction with Arg 17 of trypsin inhibitor. Residues Phe 41, Ala 55, Ala 56, His 57, Tyr 94, and Leu 99 are all buried in the trypsin–inhibitor complex. Phenylalanine 41 participates in hydrogen bonding with trypsin inhibitor. Aspartic acid 189 participates in interaction with benzamidine and with ε-amino groups of lysine or guanidino groups of arginine in the substrate. Serine 190 is buried in the trypsin–inhibitor complex and participates in the binding to benzamidine. Cysteine 191 also resides in the pocket binding to benzamidine. Residues Gln 192, Gly 193, Ser 195, Ser 214, Trp 215, and Gly 216 all form hydrogen bonds with trypsin inhibitor and are buried in the trypsin–inhibitor complex. Also, Trp 215 forms the hydrophobic part of the benzamidine binding site. Finally, His 57, Asp 102, and Ser 195 are important residues in

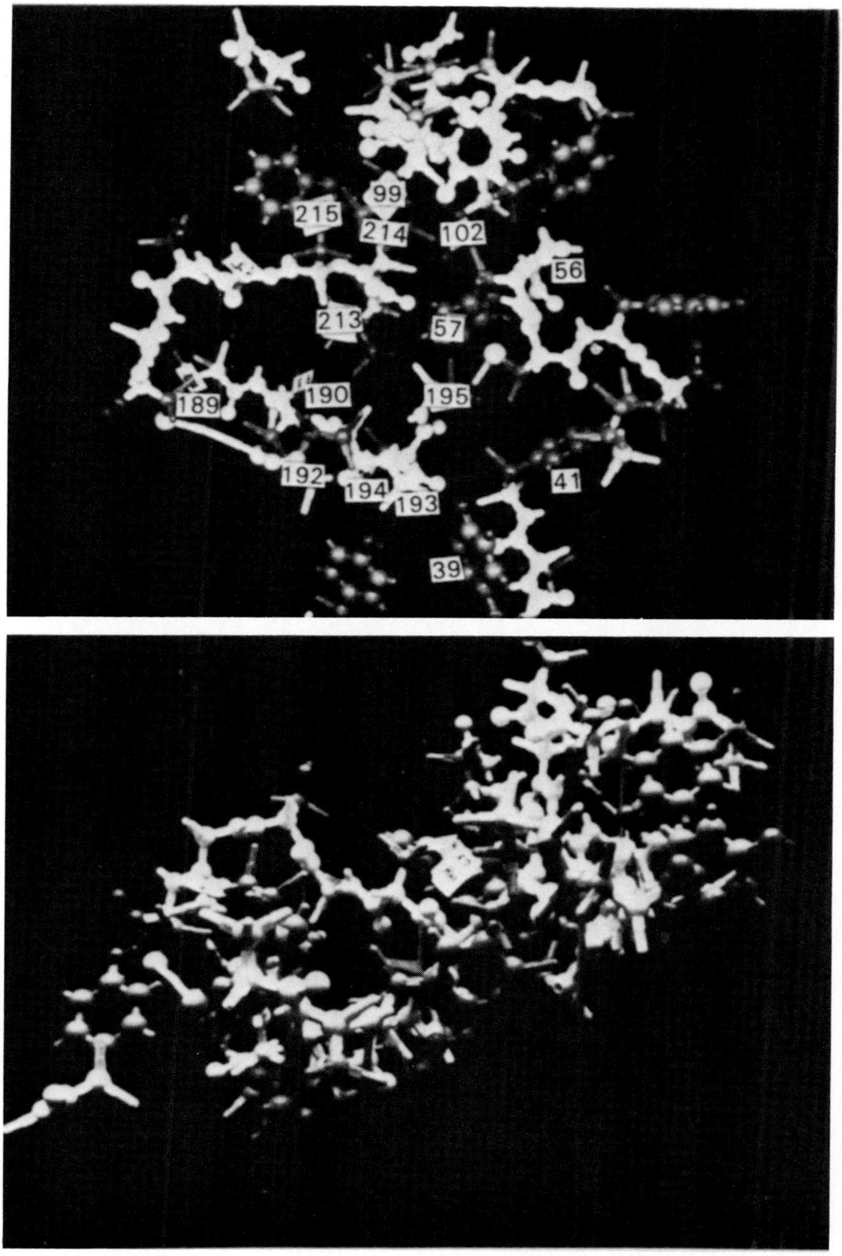

Fig. 13. Model showing the region of the substrate-binding site of trypsin (i.e., not the entire trypsin molecule). (Upper panel) Top view, looking down into the binding cavity. From this perspective, the following contact residues can be seen: Tyr-39, Phe-41, His-57, Ala-56, Asp-102, Leu-99, Trp-215, Ser-214, Val-213, Ser-195, Asp-194, Gly-193, Gln-192, Ser-190, and Asp-189. (Lower panel) Side view of the cavity, from the left side of the upper panel. (From Atassi, 1985.)

TABLE IV

Design of Surface-Simulation Peptide Mimicking Substrate Binding Site of Trypsin[a,b]

Contact residues and distances (A)	Surface-simulation synthetic peptide (TP) (B)	Control synthetic peptide (CTP) (C)
Asp 189	Asp (N terminus)	Trp (N terminus)
Ser 190	Ser	His
0.49 nm	Gly	Gly
Gln 192	Gln	Asp
Gly 193	Gly	Gly
Asp 194	Asp	Asp
Ser 195	Ser	Val
0.85 nm	Gly	Ala
	Gly	Gly
Val 213	Val	Gln
Ser 214	Ser	Phe
Trp 215	Trp	Ser
0.85 nm	Gly	Gly
	Gly	Gly
Leu 99	Leu	Asp
0.55 nm	Gly	His
Asp 102	Asp	Gly
0.53 nm	Gly	Ser
Ala 55	Ala	Gly
Ala 56	Ala	Ser
His 57	His	Gly
0.50 nm	Gly	Ala
Cys 42	Gly	Gly
Phe 41	Phe	Leu
His 40	His	Gly
Tyr 39	Tyr (C terminus)	Tyr (C terminus)

[a] From Atassi (1985).

[b] Column A identifies the contact residues in the binding cavity of the enzyme and the distances (C^{α} to C^{α}, in nanometers) separating appropriate residues. Column B indicates the surface-simulation synthetic peptide (TP) designed to mimic the binding site. Column C indicates control peptide (CTP) that has all the essential residues of the binding site except that they are randomized. The peptides are written with the N terminus on the top and the C terminus on the bottom.

the catalytic center of the enzyme. The distances (C^{α} to C^{α}) separating the residues of the trypsin binding site are given in Table IV.

From the X-ray coordinates and taking into account the participation of the residues listed above in the function of the enzyme, a peptide (TP) was designed (Atassi, 1985) by surface-simulation synthesis to mimic the binding site of trypsin. It should be emphasized that this peptide does not exist in trypsin, but rather mimics the topographic arrangement of the residues in its substrate-binding cavity. The elements of the design, which is based on spatial and distance considerations, are shown in Table IV. A control peptide was also prepared (Atassi, 1985) that incorporated all the contact and spacing residues but in random fashion (CTP). The control peptide was crucial for confirming that any binding activity found in the TP peptide is specific and not caused by some unusual charge or hydrophobic effects.

The radiolabeled TP peptide was able to bind to benzamidine and trypsin inhibitor adsorbents, whereas the control CTP peptide exhibited no binding activity (Atassi, 1985). This was encouraging and suggested that the peptide must have mimicked the architecture of the site. A relatively small amount of label was bound, which was of concern. This may have been due to the fact that the binding studies were carried out in the presence of 20% dimethylformamide because of the insolubility of both peptides in purely aqueous solvents. The presence of such an organic solvent would be expected to interfere in the binding of the peptide, which involves predominantly hydrophobic interactions on the peptide (Atassi, 1985).

To avoid the complications stemming from peptide insolubility, the peptide was coupled to a soluble carrier (Atassi, 1985). Fully succinylated lysozyme was used as the carrier because of its high solubility, but any soluble carrier should serve equally well. The coupling procedure (Atassi, *et al.*, 1981), which involves activation of the carboxyl side chains on the carrier by conversion into their ρ-nitrophenyl ester groups, avoids carrier–carrier cross-linking (due to prior succinylation of all the amino acid groups on the protein) and peptide polymerization side reactions, since the activated carboxyl groups reside only on the carrier and the reactive nucleophile is the α-amino group of the peptide. Having obtained soluble peptide conjugates, we investigated their ability to bind to trypsin inhibitor (Atassi, 1985). The binding of the TP peptide conjugates to both soybean trypsin inhibitor and α_1-antitrypsin (Table V) and the absence of any binding activity in the CTP peptide conjugates confirmed the correctness of the peptide design (Atassi, 1985).

We also investigated (Atassi, 1985) the action of other inhibitors known to react at the substrate binding site. The action of $PhMeSO_2F$ and iPrP-F on acetylcholinesterase, chymotrypsin, trypsin, and other serine proteinases has been well studied (e.g., Fahrney and Gold, 1963; Singer, 1967; Turini *et al.*, 1969; Morgan *et al.*, 1972). In each case, the binding between the enzyme and inhibitor results in the formation of a covalent bond via

TABLE V

Binding of Peptide–Suc-Lysozyme Conjugates to Soybean Trypsin Inhibitor and Human α_1-Antrypsin[a,b]

Protein or conjugate added	Anti-lysozyme bound (cpm) to	
	Soybean inhibitor	Human α_1-antitrypsin
TP peptide–Suc-lysozyme	19,110	18,540
CTP peptide–Suc-lysozyme	410	400
(Nonsense peptide)–Suc-lysozyme	460	420
Suc-lysozyme	430	410

[a] From Atassi (1985).

[b] Values represent the average of triplicate analyses, which varied ±1.3% or less. These values have not been corrected for nonspecific binding of the negative controls. The conjugates of nonsense peptide to Suc-lysozyme or Suc-lysozyme itself do not bind to trypsin inhibitor, and therefore they serve as negative controls for the binding of peptide–Suc-lysozyme conjugates. For a positive control, Suc-lysozyme was bound directly to the well (instead of trypsin inhibitor) and was followed (after appropriate washing) by 2% bovine serum albumin (instead of protein–peptide conjugate). The amount of anti-lysozyme antibodies bound to Suc-lysozyme was 19,760 cpm. This served as an appropriate positive control for the antisera and the reagents and gave an estimate of the maximum amount of label expected to be bound in the assay. The structure and synthesis of the nonsense peptide have been reported (Twining and Atassi, 1979).

the active serine residue of the enzyme. It was gratifying, and at the same time surprising, to find that the TP peptide reacted with both $PhMeSO_2F$ and iPrP-F, the reaction in each case taking place at a serine residue in the peptide. Again, the lack of reaction with the CTP peptide confirmed the specificity of reaction with the TP peptide. Although it is tempting to speculate that the reaction may have occurred with the serine residues at position 7 in the peptide (i.e., the residue representing Ser-195 in the mimicked structure), we have not yet determined which of the three serine residues on the peptide participate in the reaction with $PhMeSO_2F$ or iPrP-F.

The peptides exist in solution in a conformational equilibrium, the time average of which will be random, and at any given time there will exist conformational states that are energetically favored at appropriate energy minima and whose mode of folding approximates that required for binding (Atassi and Saplin, 1968). Those species are able to bind with the target molecule and are thus removed from the equilibrium, which should then shift in their favor (Atassi and Saplin, 1968). It should also be noted that, since peptides in the conjugates are attached only via their α-amino end, they would be able to fulfill the requirement of conformational flexibility for proper wrapping around the small ligands or folding that is favorable for binding with protein inhibitors. Appropriate peptide conformation can also be induced by the binding molecule (Atassi, 1975).

The TP peptide, either free or conjugated, exhibited no enzymatic activity with protein substrates (Atassi, 1985). Also, the soluble TP peptide–Suc-lysozyme conjugate had no hydrolytic activity (pH 8.1, room temperature) on N^{α}-*p*-tosyl-L-arginine methyl ester, a substrate commonly used to monitor trypsin enzymatic activity. Thus it was evident that the TP peptide, generated by surface simulation, expressed the binding activity of trypsin with substrates and inhibitors but not its catalytic activity (Atassi, 1985).

The absence of enzymatic activity with the TP peptide was disappointing but not entirely unexpected. The exact mechanisms of enzymatic catalysis are not fully understood, in spite of the voluminous literature on the subject. Catalysis is also subject to very stringent conformational and distance requirements, and since the distances and conformational relationships of the residues in a surface-simulation synthetic peptide do not precisely duplicate the corresponding parameters in the intact molecule, it is not surprising that only binding and not catalysis is duplicated (Atassi, 1985). Whether catalysis can be obtained by improvement in design and employment of other synthetic strategies remains to be determined.

IV. CONCLUSIONS

These studies have charted a strategy not only for determining protein antigenic sites but also for delineating and synthesizing other types of protein binding sites (Atassi *et al.*, 1976b; Atassi, 1978a, 1979, 1980). Furthermore, the concept of surface-simulation synthesis, which we introduced and developed during our determination of the antigenic structure of lysozyme (Atassi *et al.*, 1976b; Lee and Atassi, 1976, 1977a,b; Atassi and Lee, 1978a,b), has afforded great versatility in the synthetic mimicking of any type of protein binding site. The strategy and concepts developed during our determination of the antigenic structures of myoglobin and lysozyme have also enabled us to synthetically mimic sites related to other protein activities such as combining sites of antibodies (Atassi and Zablocki, 1977; Twining and Atassi, 1978; Kazim and Atassi, 1980; Atassi and Sakata, 1982) and substrate binding sites of enzymes (Atassi, 1985).

The strategy we developed initially with model proteins should be extremely useful in developing immunological therapeutic approaches (based on the use of free small synthetic peptides as immunogens) to antigens of clinical importance (e.g., allergens, viruses, toxins, tissue antigens). Furthermore, antibodies against surface-simulation peptides would provide powerful probes for the conformation of the parent protein. Another approach (Atassi, 1980) of yet untapped clinical potential would rely on the identification of the antibody combining site of selected monoclonal antibodies to the antigen. The antibody combining site can then be mimicked by the concept of surface-simulation synthesis. The synthetic peptides, suitably designed, can in principle be used to neutralize permanently the invasive antigen (e.g., a virus). The chemical strategies already worked out in this laboratory with selected model systems have removed any intellectual barriers to these approaches.

REFERENCES

Amzel, L. M., Poljak, R. M., Saul, F., Varga, J. M., and Richard, F. F. (1974) *Proc. Natl. Acad. Sci. U.S.A.* **71,** 1427–1430.

Atassi, M. Z. (1973). *Specific Recept. Antibodies, Antigens Cells, Int. Convoc. Immunol., 3rd 1972,* pp. 118–135.

Atassi, M. Z. (1975). *Immunochemistry* **12,** 423–438.

Atassi, M. Z. (1977a). *In* "Immunochemistry of Proteins" (M. Z. Atassi, ed.), Vol. 1, pp. 1–161. Plenum, New York.

Atassi, M. Z. (1977b). *In* "Immunochemistry of Proteins" (M. Z. Atassi, ed.), Vol. 2, pp. 77–176. Plenum, New York.

Atassi, M. Z. (1978a). *Immunochemistry* **15,** 909–936.

Atassi, M. Z. (1978b). *Adv. Exp. Biol. Med.* **98,** 41–99.

Atassi, M. Z. (1979). *CRC Crit. Rev. Biochem.* **6,** 371–400.

Atassi, M. Z. (1980). *Mol. Cell. Biochem.* **32,** 21–43.

Atassi, M. Z. (1985). *Biochem. J.* **226,** 477–485.

Atassi, M. Z., and Habeeb, A.F.S.A. (1977). *In* "Immunochemistry of Proteins" (M. Z. Atassi, ed.), Vol. 2, pp. 177–264. Plenum, New York.

Atassi, M. Z., and Lee, C. L. (1978a). *Biochem. J.* **171,** 419–427.

Atassi, M. Z., and Lee, C. L. (1978b). *Biochem. J.* **171,** 429–434.

Atassi, M. Z., and Sakata, S. (1980). *Biochim. Biophys. Acta* **624,** 573–578.

Atassi, M. Z., and Sakata, S. (1982). *Biochem. J.* **201,** 669–672.

Atassi, M. Z., and Saplin, B. J. (1968). *Biochemistry* **7,** 688–698.

Atassi, M. Z., and Smith, J. A. (1978). *Immunochemistry* **15,** 609–610.

Atassi, M. Z., and Zablocki, W. (1977). J. *Biol. Chem.* **252,** 8784–8787.

Atassi, M. Z., Habeeb, A.F.S.A., and Ando, K. (1973). *Biochim. Biophys. Acta* **303,** 203–209.

Atassi, M. Z., Koketsu, H., and Habeeb, A.F.S.A. (1976a). *Biochim. Biophys. Acta* **420,** 358–375.

Atassi, M. Z., Lee, C. L., and Pai, R. C. (1976b). *Biochim. Biophys. Acta* **427,** 745–751.

Atassi, M. Z., Lee, C. L., and Habeeb, A.F.S.A. (1976c). *Immunochemistry* **13,** 7–14.

Atassi, M. Z., Kazim, A. L., and Sakata, S. (1981). *Biochim. Biophys. Acta* **670,** 300–302.

Bixler, G. S., and Atassi, M. Z. (1984). *J. Immunogenet.* **11,** 245–250.
Bode, W., and Schwager, P. (1975). *J. Mol. Biol.* **98,** 693–717.
Chambers, J. L., and Stroud, R. M. (1977). *Acta Crystallogr., Sect. B* **B33,** 1824–1837.
Edmundson, A. B. (1965). *Nature (London)* **205,** 883–887.
Fahrney, D.E., and Gold, A. M. (1963). *J. Am. Chem. Soc.* **85,** 997–1000.
Huber, R., and Bode, W. (1978). *Acc. Chem. Res.* **11,** 114–122.
Huber, R., Dukla, D., Bode, W., Schwager, P., Bartels, K., Deisenhofer, J., and Steigmann, W. (1974). *J. Mol. Biol.* **89,** 73–101.
Huber, R., Bode, W., Kukla, K., Kohl, U., and Ryan, C. A. (1975). *Biophys. Struct. Mech.* **1,** 189–201.
Kazim, A. L., and Atassi, M. Z. (1980). *Biochem. J.* **187,** 661–666.
Kendrew, J. C., Watson, H. C., Strandberg, B. E., Dickerson, R. E., Phillips, D. C., and Shore, V. C. (1961). *Nature (London)* **190,** 666–670.
Lee, C. L., and Atassi, M. Z. (1975). *Biochim. Biophys. Acta* **405,** 464–474.
Lee, C. L., and Atassi, M. Z. (1976). *Biochem. J.* **159,** 89–93.
Lee, C. L., and Atassi, M. Z. (1977a). *Biochim. Biophys. Acta* **495,** 354–368.
Lee, C. L., and Atassi, M. Z. (1977b). *Biochem. J.* **167,** 571–581.
Lee, C. L., Pai, R. C., and Atassi, M. Z. (1976). *Immunochemistry* **13,** 681–687.
Morgan, P. H., Robinson, N. C., Walsh, K. A., and Neurath, H. (1972). *Proc. Natl. Acad. Sci. U.S.A.* **69,** 3312–3316.
Okuda, K., Sakata, S., Atassi, M. Z., and David, C. S. (1979). *J. Immunogenet.* **6,** 447–452.
Padlan, E. A., Davies, D. R., Rudikoff, S., and Porter, M. (1976). *Immunochemistry* **13,** 945–949.
Saul, F. A., Amzel, L. M., and Poljak, R. J. (1978). *J. Biol. Chem.* **253,** 585–597.
Segal, D. M., Padlan, E. A., Cohen, G. H., Rudikoff, S., Potter, M., and Davies, D. R. (1974). *Proc. Natl. Acad. Sci. U.S.A.* **71,** 4298–4302.
Singer, S. J. (1967). *Adv. Protein Chem.* **22,** 1–54.
Stroud, R. M., Kay, L. M., and Dickerson, R. E. (1971). *Cold Spring Harbor Symp. Quant. Biol.* **36,** 125–140.
Stroud, R. M., Kay, L. M., and Dickerson, R. E. (1974). *J. Mol. Biol.* **83,** 185–208.
Turini, P., Kurooka, S., Steer, M., Corbascio, A. N., and Singer, T. P. (1969). *J. Pharmacol. Exp. Ther.* **167,** 98–104.
Twining, S. S., and Atassi, M. Z. (1978). *J. Biol. Chem.* **253,** 5259–5262.
Twining, S. S., and Atassi, M. Z. (1979). *J. Immunol. Methods* **30,** 139–151.
Twining, S. S., David, C. S., and Atassi, M. Z. (1981). *Mol. Immunol.* **18,** 447–450.

II

Mutant Analysis

10

Functional Analysis of the Signal Peptide for Protein Secretion with Use of Oligonucleotide-Directed Site-Specific Mutagenesis

SUSAN LEHNHARDT, SUMIKO INOUYE, AND MASAYORI INOUYE

Department of Biochemistry
State University of New York at Stony Brook
Stony Brook, New York 11794

I. INTRODUCTION

Proteins which are exported from cells must be translocated across the hydrophobic membrane, their synthesis having been initiated in the hydrophilic cytoplasmic environment. This is true of both prokaryotic and eukaryotic systems, which are considered to have similar mechanisms for protein translocation. In all cases, secreted proteins are produced from precursors which contain an amino terminal extension of 15–30 amino acid residues. This region, termed the signal peptide, is cleaved from the protein on secretion. The information required for initiation of secretion is thought to reside within this short peptide extension. Therefore the signal peptide provides an ideal system for the study of protein structure–function relationships by oligonucleotide-directed site-specific mutagenesis. In this chapter we describe our approach to the study of the signal peptide by utilizing the signal peptides of bacterial secretory proteins.

PROTEIN ENGINEERING
Applications in Science,
Medicine, and Industry

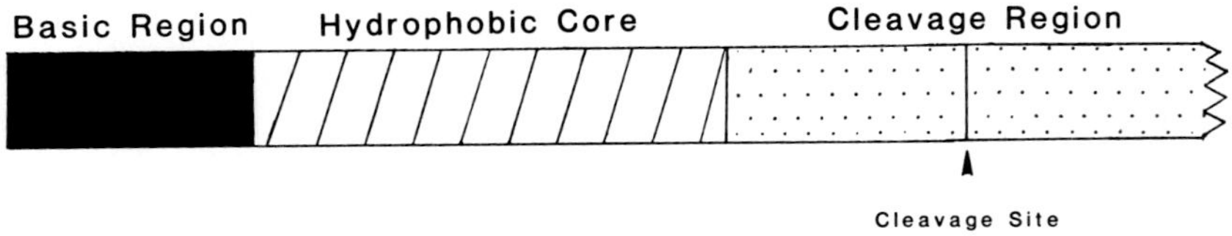

Fig. 1. Three consensus regions of the prokaryotic signal peptide. The amino terminal basic region is shown by the filled bar, the hydrophobic core region by hatching, and the cleavage region by stippling.

II. THE SIGNAL PEPTIDE

A. Common Features of the Prokaryotic Signal Peptide

When the amino acid sequence of the precursor of the major outer membrane prolipoprotein of *Escherichia coli* was first determined, it was found to have a unique amino terminal extension consisting of 20 amino acid residues (Inouye *et al.,* 1977). This was the first demonstration that prokaryotic secretory proteins have a signal peptide, and several unique features of the signal peptide were pointed out. Since then, the amino acid sequences of more than 50 prokaryotic signal peptides have been determined (Watson, 1984; Duffaud *et al.,* 1985). From the analysis of signal peptides one can find several common features in their structure (Inouye *et al.,* 1977; DiRienzo *et al.,* 1978; Inouye and Halegoua, 1980). These common features can be divided into the three regions shown in Fig. 1. The properties of each region are summarized as follows.

1. *Basic region:* one to three positively charged amino acid residues at the amino terminal region of the signal peptide. This region is termed the amino terminal basic region, which usually consists of four to seven amino acid residues.

2. *Hydrophobic core region:* a long hydrophobic sequence consisting of 14–20 amino acid residues following the amino terminal basic region. In many cases this region is punctuated by one or two proline and/or glycine residues. Often, serine and/or threonine residues are found at the end of the hydrophobic region.

3. *Cleavage region:* the region following the hydrophobic core region. This region encompasses the last few carboxy terminal amino acid residues of the signal peptide and the first few amino acid residues of the mature protein amino terminus. This region is usually less hydrophobic than the hydrophobic core region. The amino acid residue at the cleavage site (the carboxy terminal residue of the signal peptide) is a residue with a small side chain such as glycine or alanine.

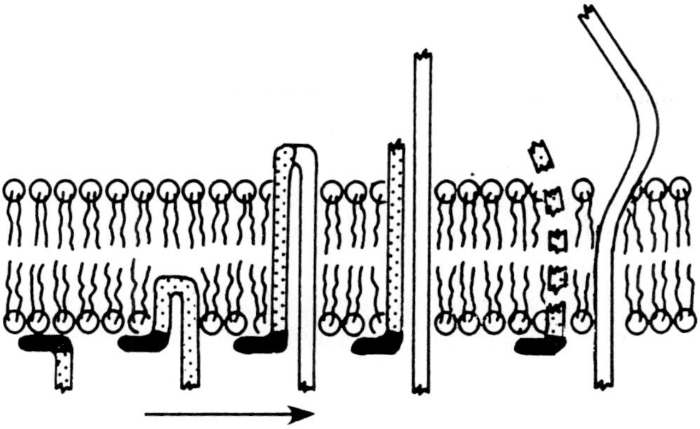

Fig. 2. Schematic diagram of the loop model. The signal peptide is shown as it sequentially associates and partitions into the phospholipid bilayer. The amino terminal basic region is black and the remainder of the signal peptide is stippled. The mature protein is depicted by an open bar.

B. Functional Model of the Signal Peptide

A model (loop model) has been proposed to take into account the function of the structural features presented above (Inouye *et al.*, 1977; Inouye and Halegoua, 1980; Fig. 2). In this model, the positively charged amino terminal region functions to stabilize the interaction between the signal peptide and the negative charges on the inner surface of the cytoplasmic membrane. The hydrophobic domain then forms a loop into the lipid bilayer of the membrane. Certain conserved features of the signal peptide support this model. For instance, the proline or glycine residue usually found within the hydrophobic region may play an important role in enabling the peptide to form a loop at the proper position. Once within the membrane, the loop would serve to anchor the protein in the lipid bilayer so that its translocation across the membrane may proceed. The exact mode of action for subsequent translocation of the protein across a membrane is not yet defined (for review, see Pollitt *et al.*, 1985a).

From genetic analyses, several protein factors are now considered to be involved in *E. coli* protein secretion as in the eukaryotic system (for review, see Walter *et al.*, 1984). However, the roles for each of the structural aspects of the signal peptide described above can still be explained by the loop model. The protein factors probably interact with the signal peptide to facilitate the roles of the signal peptide.

The most important aspect of the loop model is that the signal peptide changes its structure as protein secretion proceeds. When the signal peptide guides the protein to the cytoplasmic membrane and initiates its translocation across this membrane it does so in several discrete steps, each of which may require a different conformation and/or protein component. As the signal peptide emerges from the ribosome its conformation continually changes, and it may interact sequentially with different components of the secretory machinery.

The dynamic nature of the signal peptide offers the possibility of a specific conformational transition. Statistical secondary structural predictions of the hydrophobic region indicate that there is a high probability of forming either β conformation or α-helical structure (Chou and Fasman, 1978). Both conformations may be essential. A β structure would require only nine amino acids of the signal peptide to span the membrane bilayer, whereas an α-helical structure would require 20 residues. However, the signal peptide may first adopt a β structure to insert itself into the membrane and then convert the β structure to an α-helical conformation, pulling the secretory protein farther into the membrane so as to orient the protein properly for cleavage of the signal peptide. Such dynamic aspects of the signal peptide are discussed by Gierasch *et al.* (Chapter 11, this volume). Predictions such as these can be tested by analyzing functions of various signal peptide mutants, many of which have been generated by oligonucleotide-directed site-specific mutagenesis.

III. SITE-SPECIFIC MUTAGENESIS

To test the predictions described above, we engineered mutations in the signal peptide of the major lipoprotein of the *E. coli* outer membrane. The mature lipoprotein comprises 58 amino acid residues with a unique amino terminal modification of a glycerylcysteine to which three fatty acid residues are covalently attached. The lipoprotein is processed from a precursor prolipoprotein with an amino terminal extension consisting of 20 amino acid residues. The lipid modification of the cysteine residue at the cleavage site is absolutely required for cleavage of the signal peptide. Signal peptidase II, which cleaves the prolipoprotein signal peptide, is unique and specific for the general class of lipoproteins (for review, see Vlasuk *et al.*, 1985).

More recently, studies have been initiated on the signal peptide of another *E. coli* outer membrane protein, OmpA. The OmpA signal peptide is cleaved by another signal peptidase, signal peptidase I, which is considered to be responsible for the processing of all secretory proteins in *E. coli* except for the lipoproteins (Zimmerman and Wickner, 1983). Some aspects of these studies will also be discussed in this chapter.

A method of site-directed mutagenesis using double-stranded plasmid

DNA has been developed in our laboratory (Morinaga *et al.*, 1984; review and step-by-step protocol by Inouye and Inouye, 1986). In this method a synthetic oligodeoxyribonucleotide containing the desired DNA sequence change is used as the mutagen. For example, in the lipoprotein mutation I23I24 (Fig. 6 and Section IV) the DNA sequence was changed from the wild type by the 18-mer oligonucleotide TGCTCCATCATCGCTAAA. The oligonucleotide allows the wild-type sequence -Cys-Ser-Ser-Asn-Ala-Lys- (coded by TGCTCCAGCAACGCTAAA) to be changed to -Cys-Ser-Ile-Ile-Ala-Lys- in mutant I23I24. By choosing an appropriate sequence for the oligonucleotide, one is able to create not only substitutions as described above but also insertions and deletions of amino acid residues. As outlined in Fig. 3, a plasmid bearing the gene to be mutagenized ought to have three unique restriction sites. For instance, in the plasmid containing the lipoprotein gene, pYM141 (Fig. 3), sites for *Xba*I (X), *Eco*RI (E), and *Pst*I (P) are used. Sites X and E must flank the region of DNA which is to be mutagenized (the target site) and site P is preferably within a drug resistance marker gene. Equal amounts of the plasmid sample are subjected to restriction digestion either at the P site or at sites X and E to yield linear DNAs I and II, respectively. The P site is then treated with the Klenow fragment of DNA polymerase or bacterial alkaline phosphatase in order to prevent self-ligation and disrupt the gene for β-lactamase which confers ampicillin resistance to the cells. Also, sites X and E are preferably cleaved by two different restriction enzymes in order to prevent self-ligation. In this way, after denaturation and renaturation, the DNAs I and II do not result in plasmids which confer ampicillin resistance to the cells.

As depicted in Fig. 3, linear DNAs I and II are mixed in equimolar amounts in the presence of a 100- to 500-fold excess of the synthetic oligodeoxyribonucleotide mutagen, which has mismatch(es) with the target region according to the mutation to be isolated (step 1). The mixture is boiled to denature the linear DNAs (step 2) and cooled slowly to allow their renaturation (step 3). This step now allows the formation of not only the two original linear molecules c (II) and d (I) but also the new open-circular molecules a and b (Fig. 3). The target region of the newly formed open-circular molecule is single-stranded, and the mutagen hybridizes to form a heteroduplex as shown in molecule a at step 3 in Fig. 3. The occurrence of the open-circular molecules a and b in addition to the linear molecules c and d is monitored by subjecting the products of denaturation and renaturation to agarose gel electrophoresis (Fig. 4). Since formation of the open-circular molecules is absolutely essential for success of the mutagenesis, analysis by this method allows one to examine the products of the procedure at an early critical step. Subsequent addition of the

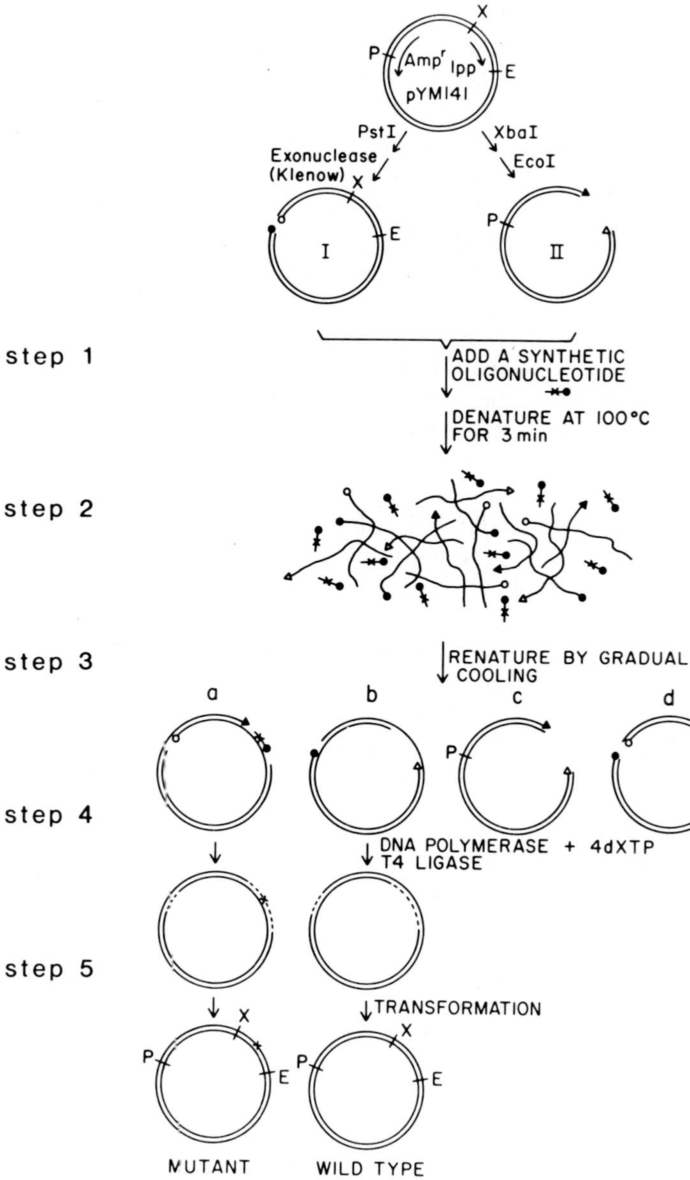

Fig. 3. Schematic diagram depicting the steps involved in oligonucleotide-directed site-specific mutagenesis with a circular double-stranded plasmid DNA. In this case the mutagenesis target site has been cloned between the restriction endonuclease sites *Xba*I (X) and *Eco*RI (E). Experimental details are outlined in the text.

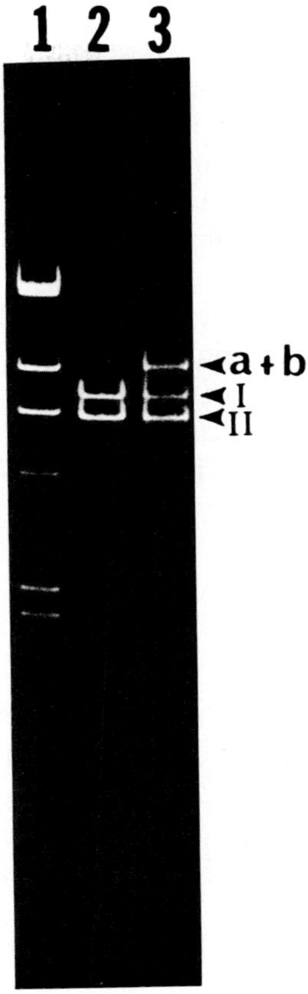

Fig. 4. Analysis of DNA after denaturation and subsequent renaturation. Lane 1, *Hin*dIII restriction endonuclease digest of bacteriophage λ DNA as a molecular weight standard; lane 2, the mixture of fragments I and II; lane 3, the same mixture after denaturation and renaturation by gradual cooling. Bands a + b, I, and II correspond to DNA a, DNA b, and fragments I and II, respectively, as shown in Fig. 3.

Klenow fragment of DNA polymerase I and ligase causes the single-stranded target region to be filled in and sealed (step 4). The mutagen serves as a primer and becomes integrated into the newly synthesized DNA. This plasmid thus contains the wild-type sequence on one strand and the mutated sequence on the other strand.

Transformation of competent *E. coli* cells with this mixture allows propagation of the mutant plasmid (step 5). The resultant colonies are screened for the presence of the mutant genotype. This is done by fixing the DNA from the colonies onto a solid support such as nitrocellulose or

Whatman 3-mm filter paper and probing with the ^{32}P-labeled oligonucleotide used as the mutagen (step 6). Mutants can be easily distinguished (Fig. 5A). Since the positive colonies from the primary transformation usually contain both the wild type and the mutant plasmid, secondary transformations should be performed. Screening these retransformants results in identification of cells which contain only the mutant plasmid (Fig. 5B). The plasmid DNA from these cells is then sequenced to verify the presence of the mutation.

This method yields desired mutations at a level of 3–25%. Oligonucleotide-directed site-specific mutagenesis can also be carried out with M13 single-stranded DNA vectors. Which method should be used for the mutagenesis depends on the structure of the clone of interest and the overall strategy of the project. These problems are also discussed in the review article cited above (Inouye and Inouye, 1986).

IV. MUTANT ANALYSIS

The results of mutagenesis to three areas of the lipoprotein signal peptide are briefly discussed below. The areas analyzed are the amino terminal basic region, the hydrophobic core region, and the cleavage site region (Fig. 1). In some cases results obtained with mutagenesis to the OmpA signal peptide will be included. The lipoprotein mutations so far synthesized are listed in Fig. 6.

A. The Amino Terminal Basic Region

Many alterations have been made in this region in order to investigate the role of the positively charged amino acid residues which are found in all prokaryotic signal peptides identified so far. To this end, positively charged residues were deleted and/or negatively charged residues were inserted in order to sequentially decrease the net charge (Fig. 6). Mutations in this region exert two effects. The reduction of the net charge from +2 (wild type) to +1 (D3, Δ2, N5) and to 0 (Δ2D3, Δ2N5) had little effect on the ability of the lipoprotein signal peptide to be cleaved. However, further reduction of the charge to −1 (E2D3) and −2 (E2D3N5) resulted in posttranslational translocation of the protein across the membrane. The wild-type lipoprotein was secreted as the protein was synthesized (cotranslational translocation), while the −2 mutation in particular had a severe effect on this process, resulting in the accumulation of large

A

B

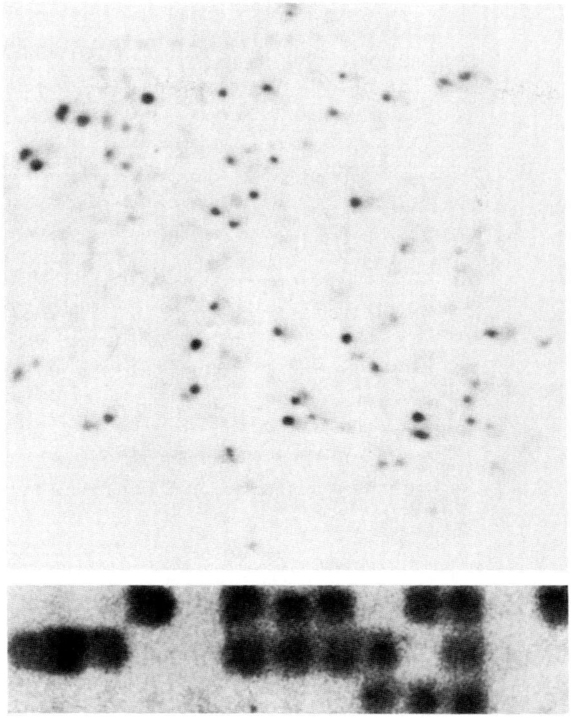

Fig. 5. Colony hybridization analysis with ^{32}P-labeled oligonucleotide mutagen as the probe. (A) Autoradiogram of primary transformants, which were blotted onto a nitrocellulose filter; (B) autoradiogram of secondary transformants, which were picked onto Whatman 3-mm filter paper and then hybridized to the ^{32}P probe.

amounts of prolipoprotein in the cytoplasm (Inouye *et al.*, 1982; Vlasuk *et al.*, 1983). Deleting the charged amino acid residues in the OmpA signal peptide also led to the accumulation of unprocessed precursor protein in the cytoplasm (S. K. Lehnhardt, S. Pollitt, and M. Inouye, unpublished data). These results support the prediction of the loop model that the amino terminal charges will faciliate the interaction between the signal peptide and the negatively charged phospholipid head groups of the cytoplasmic membrane.

The other effect of the charge mutations is a decrease in the rate of prolipoprotein synthesis, which correlates well with the decrease in the total positive charge of the signal peptide. As the charge decreases from

		1 5 10 15 20▼
Wild Type		Met-Lys-Ala-Thr-Lys-Leu-Val-Leu-Gly-Ala-Val-Ile-Leu-Gly-Ser-Thr-Leu-Leu-Ala-Gly-Cys-
Mutant	D3	Met-Lys-Asp-Thr-Lys-Leu-Val-Leu-Gly-Ala-Val-Ile-Leu-Gly-Ser-Thr-Leu-Leu-Ala-Gly-Cys
	Δ2	Met-○-Ala-Thr-Lys-Leu-Val-Leu-Gly-Ala-Val-Ile-Leu-Gly-Ser-Thr-Leu-Leu-Ala-Gly-Cys
	Δ2D3	Met-○-Asp-Thr-Lys-Leu-Val-Leu-Gly-Ala-Val-Ile-Leu-Gly-Ser-Thr-Leu-Leu-Ala-Gly-Cys
	E2D3	Met-Glu-Asp-Thr-Lys-Leu-Val-Leu-Gly-Ala-Val-Ile-Leu-Gly-Ser-Thr-Leu-Leu-Ala-Gly-Cys
	N5	Met-Lys-Ala-Thr-Asn-Leu-Val-Leu-Gly-Ala-Val-Ile-Leu-Gly-Ser-Thr-Leu-Leu-Ala-Gly-Cys
	Δ2N5	Met-○-Ala-Thr-Asn-Leu-Val-Leu-Gly-Ala-Val-Ile-Leu-Gly-Ser-Thr-Leu-Leu-Ala-Gly-Cys
	E2D3N5	Met-Glu-Asp-Thr-Asn-Leu-Val-Leu-Gly-Ala-Val-Ile-Leu-Gly-Ser-Thr-Leu-Leu-Ala-Gly-Cys
	V9	Met-Lys-Ala-Thr-Lys-Leu-Val-Leu-Val-Ala-Val-Ile-Leu-Gly-Ser-Thr-Leu-Leu-Ala-Gly-Cys
	Δ9	Met-Lys-Ala-Thr-Lys-Leu-Val-Leu-○-Ala-Val-Ile-Leu-Gly-Ser-Thr-Leu-Leu-Ala-Gly-Cys
	V14	Met-Lys-Ala-Thr-Lys-Leu-Val-Leu-Gly-Ala-Val-Ile-Leu-Val-Ser-Thr-Leu-Leu-Ala-Gly-Cys
	Δ14	Met-Lys-Ala-Thr-Lys-Leu-Val-Leu-Gly-Ala-Val-Ile-Leu-○-Ser-Thr-Leu-Leu-Ala-Gly-Cys
	V9V14	Met-Lys-Ala-Thr-Lys-Leu-Val-Leu-Val-Ala-Val-Ile-Leu-Val-Ser-Thr-Leu-Leu-Ala-Gly-Cys
	V9Δ14	Met-Lys-Ala-Thr-Lys-Leu-Val-Leu-Val-Ala-Val-Ile-Leu-○-Ser-Thr-Leu-Leu-Ala-Gly-Cys
	Δ9V14	Met-Lys-Ala-Thr-Lys-Leu-Val-Leu-○-Ala-Val-Ile-Leu-Val-Ser-Thr-Leu-Leu-Ala-Gly-Cys
	Δ9Δ14	Met-Lys-Ala-Thr-Lys-Leu-Val-Leu-○-Ala-Val-Ile-Leu-○-Ser-Thr-Leu-Leu-Ala-Gly-Cys
	Δ7Δ9Δ14	Met-Lys-Ala-Thr-Lys-Leu-○-Leu-○-Ala-Val-Ile-Leu-○-Ser-Thr-Leu-Leu-Ala-Gly-Cys
	Δ9Δ13Δ14	Met-Lys-Ala-Thr-Lys-Leu-Val-Leu-○-Ala-Val-Ile-○-○-Ser-Thr-Leu-Leu-Ala-Gly-Cys
	Δ7Δ9Δ13Δ14	Met-Lys-Ala-Thr-Lys-Leu-○-Leu-○-Ala-Val-Ile-○-○-Ser-Thr-Leu-Leu-Ala-Gly-Cys
	D9	Met-Lys-Ala-Thr-Lys-Leu-Val-Leu-Asp-Ala-Val-Ile-Leu-Gly-Ser-Thr-Leu-Leu-Ala-Gly-Cys
	D14	Met-Lys-Ala-Thr-Lys-Leu-Val-Leu-Gly-Ala-Val-Ile-Leu-Asp-Ser-Thr-Leu-Leu-Ala-Gly-Cys
	R9	Met-Lys-Ala-Thr-Lys-Leu-Val-Leu-Arg-Ala-Val-Ile-Leu-Gly-Ser-Thr-Leu-Leu-Ala-Gly-Cys
	A15	Met-Lys-Ala-Thr-Lys-Leu-Val-Leu-Gly-Ala-Val-Ile-Leu-Gly-Ala-Thr-Leu-Leu-Ala-Gly-Cys
	A16	Met-Lys-Ala-Thr-Lys-Leu-Val-Leu-Gly-Ala-Val-Ile-Leu-Gly-Ser-Ala-Leu-Leu-Ala-Gly-Cys
	A15A16	Met-Lys-Ala-Thr-Lys-Leu-Val-Leu-Gly-Ala-Val-Ile-Leu-Gly-Ala-Ala-Leu-Leu-Ala-Gly-Cys
	Δ20	Met-Lys-Ala-Thr-Lys-Leu-Val-Leu-Gly-Ala-Val-Ile-Leu-Gly-Ser-Thr-Leu-Leu-Ala-○-Cys
	Δ18	Met-Lys-Ala-Thr-Lys-Leu-Val-Leu-Gly-Ala-Val-Ile-Leu-Gly-Ser-Thr-Leu-○-Ala-Gly-Cys
	A20	Met-Lys-Ala-Thr-Lys-Leu-Val-Leu-Gly-Ala-Val-Ile-Leu-Gly-Ser-Thr-Leu-Leu-Ala-Ala-Cys
	S20	Met-Lys-Ala-Thr-Lys-Leu-Val-Leu-Gly-Ala-Val-Ile-Leu-Gly-Ser-Thr-Leu-Leu-Ala-Ser-Cys
	V20	Met-Lys-Ala-Thr-Lys-Leu-Val-Leu-Gly-Ala-Val-Ile-Leu-Gly-Ser-Thr-Leu-Leu-Ala-Val-Cys
	T20	Met-Lys-Ala-Thr-Lys-Leu-Val-Leu-Gly-Ala-Val-Ile-Leu-Gly-Ser-Thr-Leu-Leu-Ala-Thr-Cys
	L20	Met-Lys-Ala-Thr-Lys-Leu-Val-Leu-Gly-Ala-Val-Ile-Leu-Gly-Ser-Thr-Leu-Leu-Ala-Leu-Cys
	G21	Met-Lys-Ala-Thr-Lys-Leu-Val-Leu-Gly-Ala-Val-Ile-Leu-Gly-Ser-Thr-Leu-Leu-Ala-Gly-Gly
	20G21	Met-Lys-Ala-Thr-Lys-Leu-Val-Leu-Gly-Ala-Val-Ile-Leu-Gly-Ser-Thr-Leu-Leu-Ala-Gly-Cys (Gly inserted)

Mature Region		
		20▼
Wild Type		Gly-Cys-Ser-Ser-Asn-Ala-Lys
	I23I24	Gly-Cys-Ser-Ile-Ile-Ala-Lys
	A20I23I24	Ala-Cys-Ser-Ile-Ile-Ala-Lys
	I23K24	Gly-Cys-Ser-Ile-Lys-Ala-Lys
	A20I23K24	Ala-Cys-Ser-Ile-Lys-Ala-Lys

+2 to +1, 0, −1, and −2 the rate of synthesis decreases to 75, 68, 45, and 23%, respectively. This indicates that the positive charge of the signal peptide plays at least two roles in the secretory pathway, one involving translation and the other involving interaction of the protein with the membrane (Vlasuk *et al.*, 1983). It has been suggested that the decrease in translation as a result of charge changes to the signal peptide may be a direct result of poor signal peptide–membrane interaction (Hall and Schwartz, 1982).

B. The Hydrophobic Core Region

Many secretion-defective mutants have been obtained in this region by *in vivo* selection (Lin *et al.*, 1978; Bedouelle *et al.*, 1980; Emr and Silhavy, 1983; Emr *et al.*, 1980; Michealis *et al.*, 1983). The fact that these mutations were substitutions by charged amino acid residues or large deletions within the hydrophobic core region supports models which propose that the hydrophobic core region is important in membrane partitioning (Engelman and Steitz, 1981; von Heijne and Blomberg, 1979).

According to the loop model, the two glycine residues at positions 9 and 14 of the prolipoprotein signal peptide may be important in the flexibility of the signal peptide. Deletion of Gly 9 (Δ9) has no effect, while deletion of Gly 14 (Δ14) results in the accumulation of unmodified precursor. However, when these two deletion mutations are combined (mutant Δ9 Δ14), the Gly 9 deletion appears to suppress the effect of the Gly 14 deletion so that normal modification and processing are restored (Inouye *et al.*, 1984). Further deletions at two other positions on the Gly 9–Gly 14 double deletion have a similar effect. Deletion of Leu 13 (Δ9 Δ13 Δ14) results in accumulation of the unprocessed lipoprotein precursor, while deletion of Val 7 (Δ7 Δ9 Δ13 Δ14) acts to suppress the effect of the Leu 13 deletion (Pollitt *et al.*, 1985b).

Like the amino terminal basic region mutants, mutations in the hydro-

---

Fig. 6. Mutants of the major prolipoprotein of the *Escherichia coli* outer membrane. All mutants were engineered by oligonucleotide-directed site-specific mutagenesis with double-stranded plasmid DNA. The amino acid residues mutated are circled. Deletion mutants are shown by empty circles. Arrowheads indicate the position of the signal peptide cleavage site. A systematic designation is given for each mutation based on four rules (Duffaud *et al.*, 1985): (1) numbers always refer to the wild-type amino acid positions, (2) substitution is indicated by the single-letter code of the new amino acid followed by the position at which that substitution takes place, (3) deletion mutants are designated by a Δ preceding the number of the amino acid deleted, and (4) insertions are indicated by the single-letter code for the inserted amino acid residue flanked by the numbers of the amino acid residues between which it was inserted.

phobic region also affect the rate of synthesis of lipoprotein. The Gly 14 deletion increases synthesis 5-fold over the wild-type rate and the Val 7 deletion lowers synthesis 5- to 10-fold compared to the wild type (Pollitt *et al.*, 1985b). It is important to note that the function of the hydrophobic region depends not only on its hydrophobicity but also on the nature of individual amino acid residues. This indicates a strict structural requirement of this region for proper function of the signal peptide. It should also be noted that this region is implicated in more than one step in the secretion process. This is particularly evident in the case of mutant Δ7 Δ9 Δ13; Δ14, where synthesis is reduced without having any effect on modification and processing.

When residues 6–9 (Δ6–9) were deleted in the hydrophobic core of the OmpA signal peptide, processing was impaired (S. K. Lehnhardt, S. Pollitt, and M. Inouye, unpublished data). The OmpA signal peptide may be more sensitive to deletions within its hydrophobic core than lipoprotein and thus more closely resembles the outer membrane protein LamB, in which the loss of four amino acid residues within the hydrophobic core resulted in loss of processing and secretion (Emr and Silhavy, 1983). In both instances the loss of function can be correlated with a loss of predicted α-helical conformation.

In many signal peptides there are serine and/or threonine residues near the end of the hydrophobic core region. According to the rules of Chou and Fasman (1978), serine and to a lesser extent threonine could be expected to break a possible α helix. Mutations were engineered in the lipoprotein signal peptide which systematically substituted alanine residues for the serine–threonine dipeptide terminating the hydrophobic core region. Alanine is a helix-forming residue and thus could counter the effect of serine or threonine residues, which are more polar. Replacement of the serine residue with alanine (A15) but not the threonine residue with alanine (A16) resulted in the accumulation of membrane-bound unmodified precursor, which is rapidly modified and processed. Replacing both residues with alanine (A15A16) resulted in greater accumulation and slower processing than were found in mutant A15. Therefore the function of this area of the signal peptide appears to be well associated with the predicted ability of the amino acid residues to break an α-helix (Vlasuk *et al.*, 1984).

C. The Cleavage Region

The carboxy terminal residue of the signal peptide often has a small side chain such as alanine, glycine, or occasionally serine. In the case of lipoprotein, a glycine residue (Gly 20) is at this position (Fig. 6). Two types of mutations have been engineered in this region. The first varied the number of amino acid residues which separate the hydrophobic region from the cleavage site. The second used substitutions of the glycine residue at the cleavage site with other residues to test the specificity of the prolipoprotein signal peptide endopeptidase, signal peptidase II. Both types of alterations may also effect the conformation of this region and thus interfere with processing and/or modification.

It was found that deletion of the glycine residue at position 20 (Δ20) resulted in completely defective processing of prolipoprotein (Inouye *et al.*, 1984). This deletion mutation resulted in the sequence -Ala-Cys at the cleavage site. Since substitution of the glycine residue at position 20 with alanine (A20) had no effect on processing and modification, the defect caused by the glycine deletion is not due to the primary sequence at the cleavage site. Thus, the glycine deletion mutation is considered to have a serious effect on the conformation at this region, which is not well understood at present. In contrast to the glycine deletion mutation, insertion of a glycine after Gly 20 (20G21) had no effect (S. Pollitt, S. Inouye, and M. Inouye, unpublished data). The spacing between the hydrophobic region and the cleavage site is therefore flexible.

The ability of signal peptidase II to cleave after the carboxy terminal residue of the signal peptide correlates directly with the size of the amino acid side chain of that residue (Pollitt *et al.*, 1986). For example, processing and modification are prevented by replacement of the cleavage site glycine by valine (V20) and/or leucine (L20). Interestingly, although a serine is rarely found in this position, it is tolerated in the mutant S20. Further experiments utilizing the OmpA signal peptide should indicate whether the structural requirements of signal peptidase I are similar to those of signal peptidase II.

It is interesting to note that mutations at the amino terminal region of the mature protein affected processing of the prolipoprotein and that the extent of the effect depended on the compatibility between the signal peptide and the mature protein. When the sequence -Ser-Asn- (positions 23 and 24 of the prolipoprotein or positions 3 and 4 of the mature lipoprotein, respectively) was substituted with -Ile-Ile (I23I24) or -Ile-Lys- (I23K24), these mutant prolipoproteins could be modified and processed at least at 30°C. However, combining these mutations with a silent mutation in the signal peptide (A20) resulted in complete lack of processing of prolipoprotein. This deficiency was found to be well correlated with the disruption of predicted β-turn structure at the cleavage site. Using the rules of Chou and Fasman (1978), the wild-type prolipoprotein is predicted to have a high probability of β-turn structure at the cleavage site region. The mutants I23K24 and I23I24 are partially functional and are predicted to retain β-turn structure. However, when either one was combined with the otherwise nondefective mutant A20 (A20I23I24, A20I23K24) the predicted β-turn was abolished and cleavage of the signal peptide was totally impaired (S. Inouye, G. D. Duffaud, and M. Inouye, unpublished results). These results indicate that there is a strict structural requirement at the cleavage site, contributed to by the primary sequences of both the signal peptide and the mature protein near the cleavage site.

V. CONCLUSION

Mutagenesis to the signal peptide has been effective in elucidating the relationship between its structure and function. The charged amino acid residues in the amino terminal basic region are seen to be important not only in proper processing of the signal peptide but also in the levels of translation of lipoprotein. These observations support the hypothesis that this region is necessary in the initial interaction of the signal peptide with the cytoplasmic membrane. The hydrophobic core of the signal peptide is seen to exert its effect not only by its hydrophobicity but also by the interaction of specific amino acid residues, indicating a certain structural requirement. Studies on the cleavage region reveal a specific structural requirement at this region, which is fulfilled by not only the signal peptide but also the mature protein. Oligonucleotide-directed site-specific mutagenesis has allowed us to look at these areas individually to determine their contribution to the function of the signal peptide and to correlate their effect with their structural alterations.

Future studies will include analysis by an approach complementary to the functional assay which we have undertaken. Solid-phase synthesis of signal peptides, both wild type and mutants, and subsequent analysis of their structural features have already proved informative with the LamB signal peptide (Briggs and Gierasch, 1984) and are being undertaken with the OmpA signal peptide. The existence of certain structural features can be detected by *in vitro* methods such as circular dichroism and nuclear magnetic resonance techniques and function can be assayed by monitoring the ability of a mutant to function in secretion. This twofold approach will shed more light on the structure–function relationship of the signal peptide.

REFERENCES

Bedouelle, H., Bassford, P. F., Fowler, A. V., Zabin, I., Beckwith, J., and Hofnung, M. (1980). *Nature (London)* **285,** 78–81.

Briggs, M. S., and Gierasch, L. M. (1984). *Biochemistry* **23,** 3111–3114.

Chou, P. Y., and Fasman, G. D. (1978). *Annu. Rev. Biochem.* **47,** 251–276.

DiRienzo, J. M., Nakamura, K., and Inouye, M. (1978). *Annu. Rev. Biochem.* **47,** 481–532.

Duffaud, G. D., Lehnhardt, S. K., March, P. E., and Inouye, M. (1985). *Curr. Top. Membr. Transp.* **24,** 65–104.

Emr, S. D., and Silhavy, T. J. (1983). *Proc. Natl. Acad. Sci. U.S.A.* **80,** 4599–4603.

Emr, S. D., Hedgpeth, J., Clement, J., Silhavy, T. J., and Hofnung, M. (1980). *Nature (London)* **285,** 82–85.

Engelman, D. M., and Steitz, T. A. (1981). *Cell* **23,** 411–422.

Hall, M. N., and Schwartz, M. (1982). *Ann. Microbiol. (Paris)* **133A,** 123–127.

Inouye, M., and Halegoua, S. (1980). *CRC Crit. Rev. Biochem.* **7,** 339–371.

Inouye, S., and Inouye, M. (1986). *In* "DNA RNA Synthesis" (S. Narang, ed.), Academic Press, New York (in press).

Inouye, S., Wang, S., Sekizawa, J., Halegoua, S., and Inouye, M. (1977). *Proc. Natl. Acad. Sci. U.S.A.* **74,** 1004–1008.

Inouye, S., Soberon, S., Franceschini, T., Nakamura, K., Itakura, K., and Inouye, M. (1982). *Proc. Natl. Acad. Sci. U.S.A.* **79,** 3438–3441.

Inouye, S., Vlasuk, G. P., Hsiung, H., and Inouye, M. (1984). *J. Biol. Chem.* **259,** 3729–3733.

Lin, J. J. C., Kanazawa, H., Ozols, J., and Wu, H. C. (1978). *Proc. Natl. Acad. Sci. U.S.A.* **75,** 4890–4895.

Michealis, S., Inouye, H., Oliver, D., and Beckwith, J. (1983). *J. Bacteriol.* **154,** 366–374.

Morinaga, Y., Franceschini, T., Inouye, S., and Inouye, M. (1984). *Bio/Technology* **2,** 636–639.

Pollitt, S., Inouye, S., and Inouye, M. (1985a). *In* "Microbiology—1985" (D. Schlessinger, ed.), pp. 308–311. Am. Soc. Microbiol., Washington, D.C.

Pollitt, S., Inouye, S., and Inouye, M. (1985b). *J. Biol. Chem.* **260,** 7965–7969.

Pollitt, S., Inouye, S., and Inouye, M. (1986). *J. Biol. Chem.* **261,** 1835–1837.

Vlasuk, G. P., Inouye, S., Ito, H., Itakura, K., and Inouye, M. (1983). *J. Biol. Chem.* **258,** 7141–7148.

Vlasuk, G. P., Inouye, S., and Inouye, M. (1984). *J. Biol. Chem.* **259,** 6195–6200.

Vlasuk, G. P., Ghrayeb, J., and Inouye, M. (1985). *In* "Enzymes of Biological Membranes" (A. A. Martonosi, ed.), Vol. 2, pp. 309–330. Plenum, New York.

von Heijne, G., and Blomberg, C. (1979). *Eur. J. Biochem.* **97,** 175–181.

Walter, P., Gilmore, R., and Blobel, G. (1984). *Cell* **38,** 5–8.

Watson, M. E. E. (1984). *Nucleic Acids Res.* **12,** 5145–5164.

Zimmerman, R., and Wickner, W. (1983). *J. Biol. Chem.* **258,** 3920–3925.

11

Physical Properties of Genetically Defined Synthetic Signal Sequences Suggest Initial Steps in Protein Export

LILA M. GIERASCH,* MARTHA S. BRIGGS*,† AND DONALD G. CORNELL‡

** Department of Chemistry*
University of Delaware
Newark, Delaware 19716
and
† Department of Molecular Biophysics and Biochemistry
Yale University
New Haven, Connecticut 06511
and
‡Eastern Regional Research Center
United States Department of Agriculture
Philadelphia, Pennsylvania 19118

I. INTRODUCTION

This chapter describes our recent studies applying biophysical methods to the question of how a signal sequence mediates the process of protein translocation across a membrane. As discussed in this section, the roles of the signal sequence are not yet clearly defined, but its importance is well established. Is it possible to correlate the properties of various signal sequences with their functions *in vivo?* Our results suggest that the answer to this question is yes, and imply an active role for the signal se-

PROTEIN ENGINEERING
Applications in Science,
Medicine, and Industry

quence in the export process. Circular dichroism studies of isolated synthetic signal peptides, presented in Section II, reveal a correlation of conformation with function. Interactions of signal peptides with a lipid monolayer, as monitored by surface tension measurements, also correlate with export function *in vivo* (Section III). Analysis of the conformations taken up by the signal peptides upon interaction with the lipid monolayer, summarized in Section IV, suggests that both the conformational and lipid interaction properties of the signal sequences derive from intrinsic related characteristics of the functional sequences. Furthermore, although our results for isolated signal sequences in no way imply that the secretion process takes place without the participation of additional components, we can use them to propose a working model for the initial encounter of the signal sequence with a membrane (Section V).

A. Models of Protein Secretion

The problem of protein export is universal, and certain features of the mechanism by which it takes place appear to be general from prokaryotes to higher eukaryotes. Nearly every known protein that is destined to be translocated across a membrane is synthesized with a 15- to 30-residue-long N-terminal extension, termed a signal sequence, that is subsequently cleaved by a signal peptidase. Although the detailed mechanism of the translocation process and the roles of the molecular species taking part in it are not yet firmly established, a great deal of recent research activity on this problem has clarified many aspects. (Protein secretion and the role of signal sequences have been reviewed by Silhavy *et al.*, 1983; Kreil, 1981; Austen and Ridd, 1981; Inouye and Halegoua, 1980; Walter *et al.*, 1984; Briggs and Gierasch, 1986). Examination of the current models for protein secretion in prokaryotes and eukaryotes suggests several possible functions of the signal peptide.

The signal hypothesis, first put forth by Blobel and Dobberstein (1975a,b), is supported by many results for eukaryotic systems. Several observations on *in vitro* protein translocation systems have enabled components of the secretory apparatus to be isolated and characterized (Walter *et al.*, 1984; Gilmore *et al.*, 1984; Meyer, 1982). The essential features of this model (Fig. 1) include: (1) formation of a translation-arrested complex between the ribosome with its nascent protein chain (up to 40 residues of the mature protein plus the signal sequence) and the signal recognition particle (SRP), an assembly of six polypeptide chains and one piece of RNA; (2) binding of this translation-arrested complex to a membrane-bound receptor, termed the docking protein (Meyer *et al.*, 1982) or SRP receptor (Gilmore *et al.*, 1982), which releases the translation arrest and

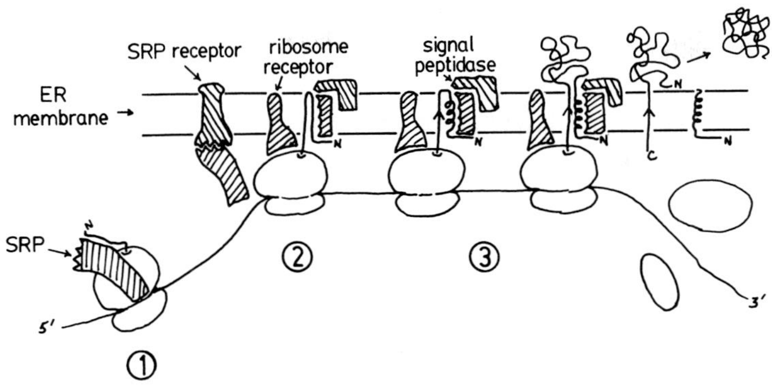

Fig. 1. Schematic diagram of the signal hypothesis as it is currently formulated.

leads to liberation of SRP from the export complex; (3) translocation of the protein chain concomitant with synthesis, mediated by a coupling device as yet uncharacterized (postulated to be a proteinaceous export site). The components of the translation/translocation complex would presumably disassemble on completion of the synthesis.

Export of proteins from prokaryotes (to outer membrane locations or the periplasmic space in gram-negative bacteria or to the medium in gram-positive bacteria) is viewed as being less tightly coupled to their synthesis. Some results have suggested that much of the protein chain is synthesized before translocation commences (Randall, 1983). Other studies, primarily on bacteriophage coat proteins in Wickner's laboratory, demonstrate that nascent exported proteins can cross a membrane posttranslationally (Wickner, 1979, 1980; Mueller and Blobel, 1984; Chen *et al.*, 1985). These findings led Wickner to hypothesize that the signal sequence functions by stabilizing a nascent secreted protein in a conformation that partitions favorably into a membrane. Cleavage of the signal peptide would initiate a conformational change to a form of the protein that favors the aqueous phase (the native structure) and spontaneously emerges from the membrane (Wickner, 1979, 1980). While this model may describe the translocation of some prokaryotic exported proteins, in other cases genetic and biochemical evidence argues for a cotranslational process mediated by SRP-like species (Emr *et al.*, 1981; Horiuchi *et al.*, 1983). However, relatively little is known about such species because until recently (Mueller and Blobel, 1984; Chen *et al.*, 1985) *in vitro* experiments were not readily carried out. In contrast to the situation with eukaryotes, though, many genetic data are available for prokaryotes which help to define requirements for export, and genetic manipulations of exported and nonexported proteins have enabled the testing of these requirements (Silhavy *et al.*, 1983; Benson *et al.*, 1985).

B. Possible Roles of Signal Sequences

Consideration of the steps and components implicated in the secretion process in prokaryotes and eukaryotes suggests that the signal sequence may function in a variety of ways. Many of these possible roles have already been described; others are listed here for completeness. Lacking specific evidence to the contrary, they cannot be eliminated at present. First, the signal sequence may bind to the ribosome. Second, SRP binding by the signal sequence is a likely step in formation of the translation-arrested complex (Walter *et al.*, 1981; Walter and Blobel, 1981a,b). By contrast, there is evidence that any interaction of the signal sequence with the SRP receptor at the membrane is weak (Gilmore and Blobel, 1983). Third, the signal sequence is proposed to play a key role in initiating the translocation process at the membrane, either by direct interaction with lipids (Wickner, 1979, 1980; Engelman and Steitz, 1981) or by interaction with a proteinaceous export site (Gilmore and Blobel, 1985; Prehn *et al.*, 1981; Blobel and Dobberstein, 1975a,b), or both. Inouye's loop model (Inouye and Halegoua, 1980) postulates that the signal sequence anchors the N terminus of the nascent chain on the inside surface of the membrane (to lipid or protein), allowing the adjacent regions to insert. Posttranslational translocation requires that the signal sequence play an active role in initiating membrane interaction. A proteinaceous export site may also participate in this pathway. Wickner's conformation-based "membrane trigger" hypothesis (Wickner, 1979, 1980) implicates the signal sequence at the level of influencing the nascent protein's conformational behavior.

C. Comparisons of Signal Sequences

These multifaceted potential roles of signal sequences are a challenge to the seeker of structure–function correlations. This challenge is compounded by the immense variety found among the over 277 known signal sequences (Watson, 1984). Even between closely related proteins or species little homology is seen in the signal region. Yet signal sequences can be exchanged between proteins in the same organism (Takahara *et al.*, 1985), or between organisms (Mueller *et al.*, 1982; Talmadge *et al.*, 1980), and function is conserved. Close examination of these functionally homologous but highly disparate sequences reveals some patterns and common features (Inouye and Halegoua, 1980; von Heijne, 1985). While the generality of these features is particularly striking among prokaryotic

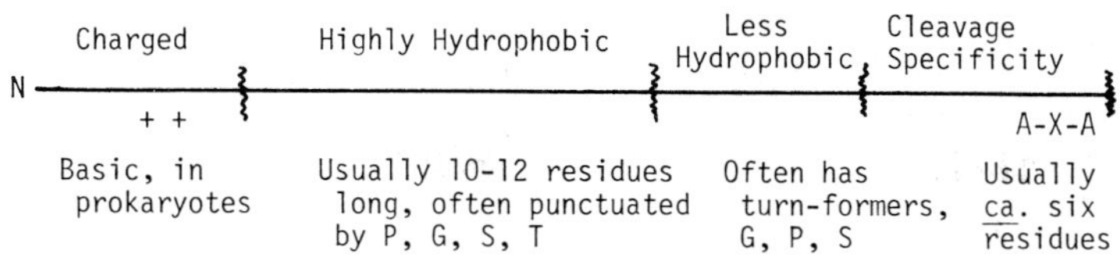

Fig. 2. General features of signal sequences.

signal sequences, close correspondences are even seen in comparisons of signal sequences from *Escherichia* coli and humans (Watson, 1984). The N-terminal region of signal sequences generally has one or more charged residues. In prokaryotes, these are positively charged. Following this region is a span of hydrophobic residues usually 10–12 long. A less hydrophobic sequence terminates this portion and leads into the signal peptidase recognition site (von Heijne, 1983; Perlman and Halvorson, 1983). Another general characteristic of signal sequences is that they are often punctuated by Pro, Gly, or Ser residues, even in their hydrophobic segments. These features are summarized in Fig. 2. Secondary structural predictions show that signal sequences have a high probability of adopting either an α-helical conformation or a β structure (Austen, 1979; Austen and Ridd, 1981). Also, a β turn has been suggested for the region following the hydrophobic stretch (Austen and Ridd, 1981; Inouye and Halegoua, 1980).

Genetic data for prokaryotes have established that certain mutations that alter these general characteristics of signal sequences indeed cause export defects. Among these are introduction of a charge in the hydrophobic region (Silhavy *et al.*, 1983) and removal of the positive loci near the N terminus (Inouye *et al.*, 1982). Furthermore, availability of such mutant strains with altered signal sequences provides the means to dissect the several potential roles.

It is surprising that throughout these efforts to understand the roles of the signal sequence in protein secretion, there have been very few examinations of its intrinsic properties (Rosenblatt *et al.*, 1980; Shinnar and Kaiser, 1984) and, before the studies discussed here, no attempts to correlate these properties with functions.

D. LamB Signal Sequence Mutants: A System to Test the Importance of Signal Sequence Intrinsic Properties

A set of *E. coli* mutants, isolated by Emr and Silhavy (1983), in which the signal sequence for the LamB outer membrane protein is altered provides an opportunity to seek correlations of intrinsic physical proper-

```
Wild Type
Met Met Ile Thr Leu Arg Lys Leu Pro Leu Ala Val Ala Val Ala Ala Gly Val Met Ser Ala Gln Ala Met Ala/
-25                 -20                 -15                 -10                 -5                 -1

Deletion Mutant
Met Met Ile Thr Leu Arg Lys Leu Pro ————— Val Ala Ala Gly Val Met Ser Ala Gln Ala Met Ala/

Gly → Cys Pseudorevertant
Met Met Ile Thr Leu Arg Lys Leu Pro ————— Val Ala Ala Cys Val Met Ser Ala Gln Ala Met Ala/

Pro → Leu Pseudorevertant
Met Met Ile Thr Leu Arg Lys Leu Leu ————— Val Ala Ala Gly Val Met Ser Ala Gln Ala Met Ala/
```

Fig. 3. Amino acid sequences of the LamB protein signal sequences. A slash indicates the site of cleavage of the signal sequence from the mature protein by signal peptidase. Arrows indicate the point mutations mentioned in the text. The fragments synthesized are underlined.

ties of signal sequences with their capacities to function in protein export. An export-defective phenotype was found to be associated with one strain, in which four residues were deleted from the hydrophobic region. Two pseudorevertants with restored export of the LamB protein were found to have secondary point mutations in the signal region. As shown in Fig. 3, in one of the pseudorevertants a Pro residue (at position −17, numbered from the cleavage site) is replaced by a Leu. In the other, a Gly at −8 is replaced by a Cys. Emr and Silhavy considered the effects of these changes on the predicted (Chou and Fasman, 1978) secondary structure of the signal sequences and hypothesized that adoption of an α-helical conformation was requisite to signal sequence function. Their argument was supported by the expectation that deletion of four residues disfavored the α-helical conformation in LamB because the deletion brings the helix-destabilizing residues, Pro and Gly, within three residues of each other. In the wild type these two are separated by seven residues. The two isolated pseudorevertants support this explanation, since one lacks the Pro and the other the Gly. This hypothesis implies that properties of the isolated signal sequence, in the absence of the mature protein, should be correlated with function, and, furthermore, this hypothesis is testable.

II. ARE CONFORMATIONAL PROPERTIES OF SIGNAL SEQUENCES CORRELATED WITH FUNCTION?

A. Synthesis of LamB Signal Peptides

Using solid-phase methods, we have synthesized fragments and entire signal sequences of the LamB signal region for the wild-type and mutant strains described in Section I,D (Fig. 3). The detailed synthetic procedures were reported previously (Briggs and Gierasch, 1984; Briggs, 1986) and are only summarized here. The solid support was a Merrifield resin [(chloromethyl)polystyrene/1% divinylbenzene], and couplings were carried out by the dicyclohexylcarbodiimide method, using hydroxybenzotriazole esters. Peptides were cleaved from the resin by treatment with anhydrous HF (90%) in the presence of anisole (10%). Extraction of the longer peptides from the resin usually required acetonitrile/acetic acid/water mixtures, while the shorter fragments could be extracted with just acetic acid/water. Purification by high-pressure liquid chromatography on either Hamilton PRP-1 or Vydac C4 semipreparative columns yielded signal peptides of the appropriate amino acid composition. Fast atom bombardment mass spectrometry yielded the molecular ion peak and expected fragments for the desired sequences. A typical yield for one of the full-length peptides starting from 2 g of resin at a substitution level of 0.5 mmol of the C-terminal amino acid per gram of resin would be 0.5 g crude peptide, 0.15 g purified.

In general, the signal peptides showed the solubility behavior expected given their highly hydrophobic sequences; somewhat surprisingly, they were soluble in water up to about millimolar concentration. Alcohols, acetonitrile/water mixtures, and aqueous surfactant solutions [such as sodium dodecyl sulfate (SDS)] dissolved all of the peptides, as did dimethyl sulfoxide.

B. Circular Dichroism of Signal Peptides

Circular dichroism (CD) spectra of the four signal peptide fragments (Briggs and Gierasch, 1984) and the full-length signal sequence peptides (Briggs, 1986) indicate that they adopt different conformations in different environments. To compare their behavior in a structure-promoting solution that also resembles the interfacial regions characteristic of a biological membrane, we obtained CD spectra of the signal peptides in aqueous SDS solutions as well as in buffer alone (Fig. 4). All spectra have been analyzed by fitting them to linear combinations of reference curves for α helix, β structure, and "random coil" (Greenfield and Fasman, 1969); the resulting secondary structure distributions are given in Table I.

The CD data indicate a clear distinction between the conformational properties of the wild-type and functional pseudorevertant signal peptides and those of the deletion mutant. While all the peptides display an increased tendency to adopt an α-helical conformation in the SDS solutions, the nonfunctional deletion mutant does so to a lesser extent. All the peptides are predominantly unstructured in buffer by CD. The fraction of helix observed for the wild-type and Pro>Leu pseudorevertant peptides

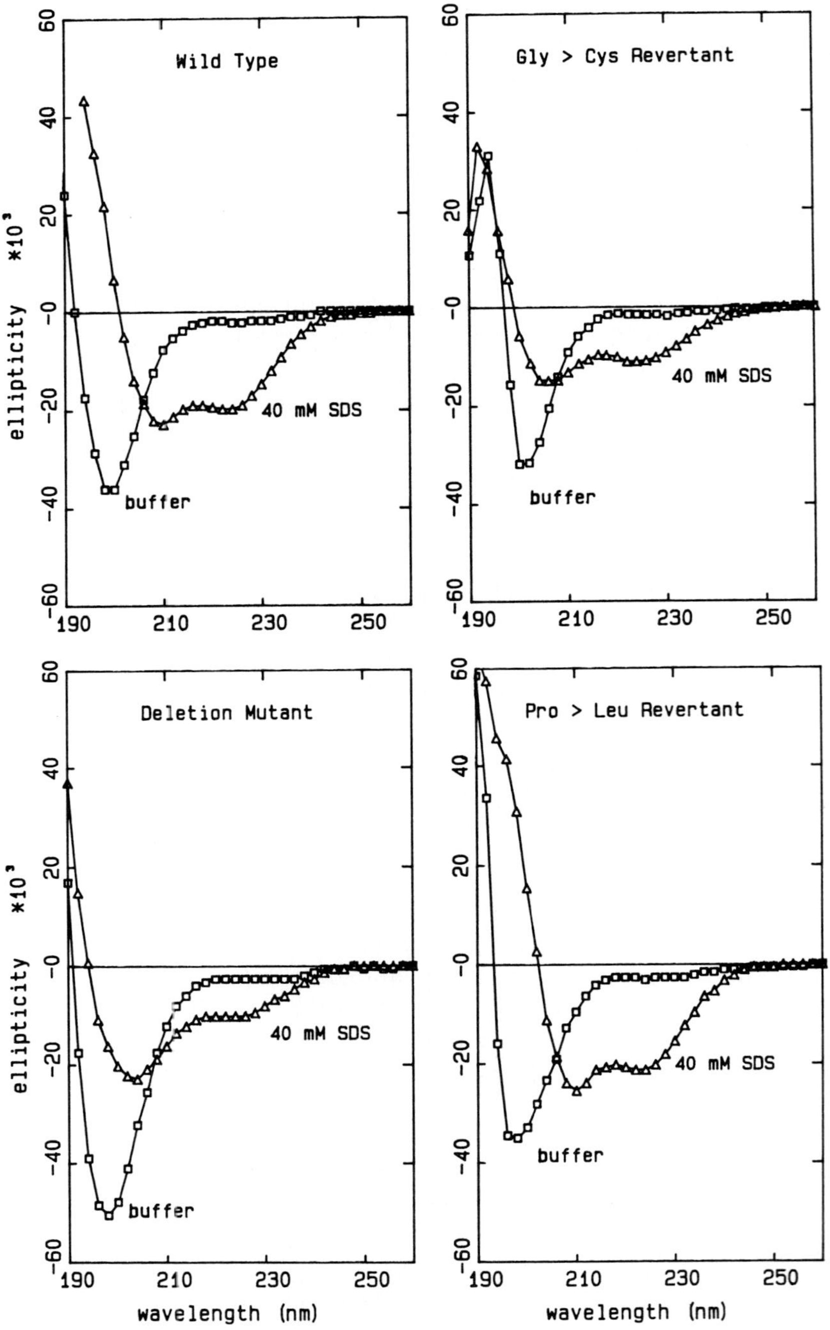
Wild Type
Gly > Cys Revertant
Deletion Mutant
Pro > Leu Revertant
ellipticity *10³
wavelength (nm)
40 mM SDS
buffer

TABLE I

Signal Peptide Conformations from Circular Dichroism[a]

Peptide	Solvent	α Helix (%)	β Structure (%)	"Random" (%)
Wild type	Buffer[b]	7	17	76
	40 m*M* SDS[c]	80	0	20
Deletion	Buffer	8	13	79
mutant	40 m*M* SDS	30	1	69
Gly → Cys	Buffer	9	10	81
revertant	40 m*M* SDS	52	0	48
Pro → Leu	Buffer	20	1	79
revertant	40 m*M* SDS	76	0	24

[a] Observed CD data were fitted to reference spectra (see text).
[b] 5 m*M* Tris, pH 7.3.
[c] In 5 m*M* Tris, pH 7.3.

in SDS solution (~80%) is high for peptides of this length (25 and 21 residues, respectively).

These results support the secondary structural hypothesis put forth by Emr and Silhavy (1983) on the basis of genetic data, namely that adoption of an α-helical conformation by the signal sequence appears to be required at some step in the secretion process. Furthermore, these results argue for a correlation of the structure and function of the signal sequence in isolation from the mature protein and the other components of the secretion apparatus. The CD findings suggest that it is indeed worthwhile to explore further the behavior of the synthetic signal peptides.

III. ARE MEMBRANE INTERACTIONS OF SIGNAL SEQUENCES CORRELATED WITH FUNCTION?

A. Description of Surface Tension Measurements

Interaction of the signal sequence with membrane lipids might play an important role in protein translocation. Although biochemical and bio-

Fig. 4. Circular dichroism spectra of the full-length synthetic LamB signal sequences. Peptide concentrations were approximately 0.2 mg/ml. Ellipticity is expressed as degrees centimeters squared per decimole per residue. Note that all four of the signal peptides show spectral shapes typical of "random coil" in buffer (5 m*M* Tris, pH 7.3), and increasingly helical CD curves in 40 m*M* SDS; the spectra of the revertant and wild-type peptides change to a greater extent than that of the nonfunctional deletion mutant. The results of fitting these spectra to reference CD data (Greenfield and Fasman, 1969) are given in Table I.

physical studies have shown that signal sequences can interact with lipid bilayers *in vitro* (Ohno-Iwashita *et al.*, 1984; Nagaraj, 1984), there is no direct evidence that such associations are essential to function. We addressed this problem by studying the effect of the genetically defined signal peptides on phospholipid monolayers (Briggs *et al.*, 1985).

Phospholipid monolayers have been used to simulate the membrane bilayer (Colacicco, 1970; Verger and Pattus, 1982; Rothfield and Fried, 1975; Fendler, 1982). A surface-active species (a detergent or an amphiphilic peptide or protein, for example) dissolved in the aqueous phase beneath the monolayer can enter the monolayer and interact with the lipids at both the head groups and the hydrocarbon chains. Its insertion into the monolayer is indicated by an increase in the surface pressure of the monolayer (Pethica, 1955).

B. Insertion of Signal Peptides into a Lipid Monolayer

The interaction of the LamB signal sequence fragments (Fig. 3) with a phospholipid monolayer composed of lipids found in the *E. coli* inner membrane was studied (Briggs *et al.*, 1985). To ensure complete mixing of peptides with lipids and to avoid the formation of patches of pure peptide, the peptides were injected beneath a monolayer that was spread to an initial pressure greater than the peptide's equilibrium surface pressure at an air–water interface. The consequent increase in surface pressure is dependent on peptide concentration and indicates that the peptide is interacting with the monolayer lipids (Fig. 5). The maximum increases in surface pressure for the wild-type and Pro>Leu pseudorevertant peptides were 10 and 8 dyn/cm, respectively. At half-maximal surface pressure increase, the peptide concentration is approximately equal to the dissociation constant for monolayer binding. These values are 0.3 and 0.1 μM for the wild type and the Pro>Leu pseudorevertant, respectively. In contrast, the maximal surface pressure increase for the deletion mutant was only 1.5 dyn/cm, and its concentration at half-maximal surface pressure increase was 0.8 μM.

The ability of one surface-active molecule to penetrate a monolayer of another is dependent on the initial surface pressure of the monolayer (Verger and Pattus, 1982). The lipid monolayer pressure above which the penetrating molecule no longer inserts (thus causing a pressure change) is called the critical pressure of insertion and is obtained by measuring the dependence of the surface pressure increase on initial monolayer surface pressure and extrapolating to a pressure increase of zero. Critical pressures of insertion were determined at saturating peptide concentrations (Fig. 6). The critical pressures of insertion of the wild-type and Pro>Leu pseudorevertant peptides are 28 and 26 dyn/cm, respectively, which are

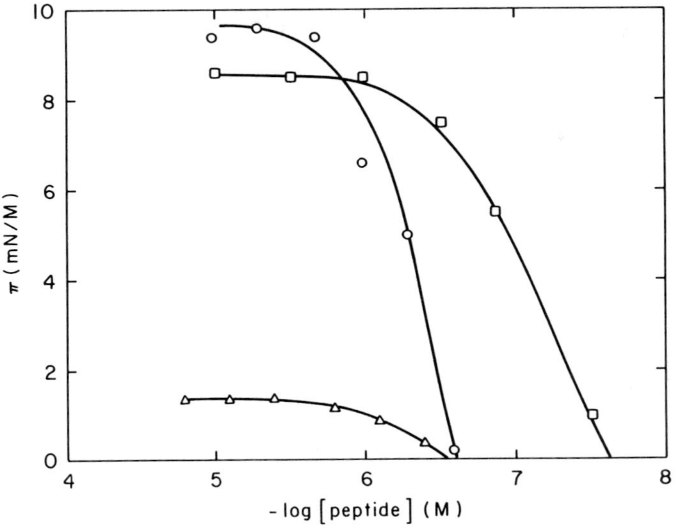

Fig. 5. Increase in surface pressure of phospholipid monolayers as a function of signal peptide concentration (semilogarithmic plot). A monolayer of egg phosphatidylethanolamine and egg phosphatidylglycerol (65 : 35) (Avanti, Birmingham, Alabama) was spread from a benzene solution onto 5 m*M* Tris buffer, pH 7.3, yielding a final surface pressure of 20 dyn/cm after evaporation of the benzene. The peptide was added by injecting a concentrated solution below the lipid–water interface. The surface pressure was measured by the du Nouy ring method with a Fisher Autotensiomat equipped with a platinum–iridium ring. The plateau values are plotted as a function of the peptide concentration for the (○) wild-type, (□) pseudorevertant, and (△) deletion mutant peptides. (From Briggs *et al.*, 1985. Copyright 1985 by the American Association for the Advancement of Science.)

similar to those of membrane-associated proteins such as rat apolipoprotein AI (Phillips and Sparks, 1980) and colicin A (Pattus *et al.*, 1983). In contrast, the critical pressure of insertion for the deletion mutant is 16 dyn/cm, which is similar to that of bovine serum albumin, a soluble protein (Phillips and Sparks, 1980).

The critical insertion pressure values provide a rough measurement of the point at which the forces favoring transfer of the peptide from the subphase to the monolayer are balanced by the compressional forces opposing addition of material to the surface. The insertion of the signal peptides into the phospholipid monolayer is accompanied by an area change. Although a quantitative measurement of this area has not been carried out, it is clear that the insertion energies (critical pressure times molecular area in the monolayer) of the functional signal peptides are significantly greater than that of the nonfunctional peptide. This energy-favoring insertion could contribute to lowering energy barriers to protein translocation *in vivo*.

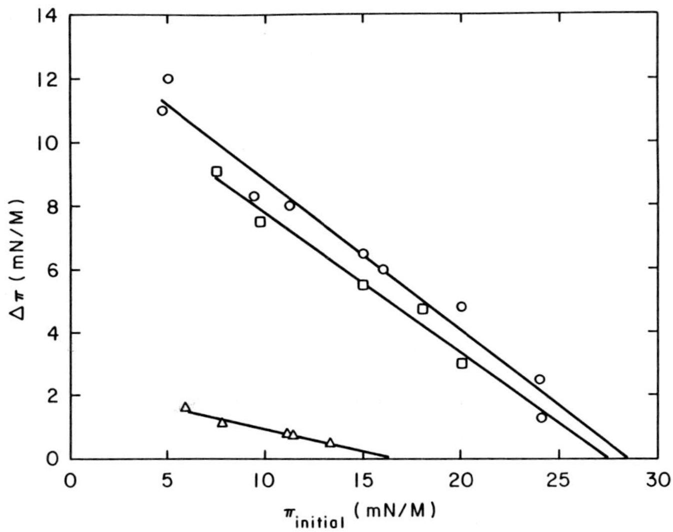

Fig. 6. Determination of critical pressures of insertion. A monolayer of egg phosphatidylethanolamine and egg phosphatidylglycerol was spread (as in Fig. 5) to yield the desired initial surface pressure. Peptide was injected below the lipid surface to a final concentration of 1 μM for the wild-type and Pro>Leu revertant peptides, and 2 μM for the deletion mutant peptide. Surface pressure plateau values are plotted against initial surface pressure for (○) wild-type, (□) pseudcrevertant, and (△) deletion peptides. (From Briggs *et al.*, 1985. Copyright 1985 by the American Association for the Advancement of Science.)

These results show that the ability of these signal peptides to interact with phospholipid monolayers correlates with their *in vivo* activity. A major question in protein translocation is how the signal sequence and the secreted protein interact with the membrane components. Although various proteins are almost certainly necessary for protein export, and may indeed interact with the signal sequence, these data clearly imply that signal sequences also have some contact with the membrane lipids during secretion.

IV. CAN THE CONFORMATIONAL PROPERTIES OF SIGNAL SEQUENCES BE RELATED TO THEIR INTERACTIONS WITH MONOLAYERS?

A. Design of Experiment to Observe Circular Dichroism Spectra of Signal Peptides in Lipid Monolayers

The experiments described above show that two physical properties of the synthetic signal peptides correlate with *in vivo* export function: ten-

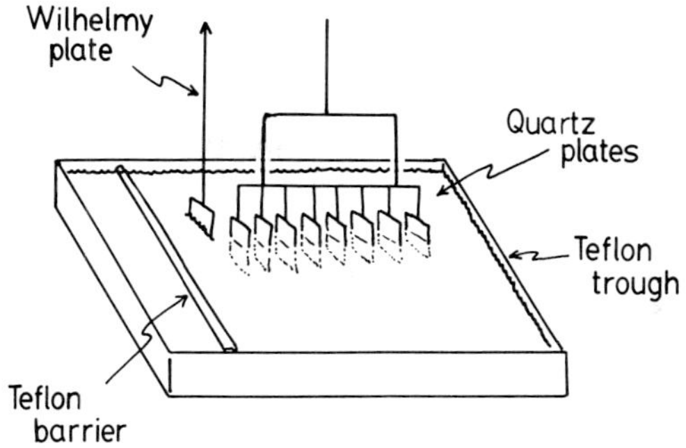

Fig. 7. Diagram of Langmuir trough with setup for coating CD plates. The surface pressure is monitored with the Wilhelmy plate, the quartz plates are immersed in the aqueous solution, the desired lipid monolayer is spread, signal peptide is injected beneath the lipid–water interface, and the quartz plates are withdrawn while maintaining the surface pressure. The quartz plates with a monolayer transferred onto each side are then placed in a specially designed holder to obtain a CD spectrum. (From Briggs *et al.*, 1986. Copyright 1986 by the American Association for the Advancement of Science.)

dency to adopt an α-helical conformation and tendency to insert into lipid monolayers. Both of these properties may arise from the same intrinsic characteristic of the signal sequence. Specifically, interaction with a lipid monolayer reflects amphiphilicity in a molecule. Signal sequences generally possess clustered hydrophilic and hydrophobic regions in their primary structure. Upon folding into an α helix, the hydrophobic primary structure in the core region gains a stronger hydrophobic character as the polar amide groups are sequestered. Therefore, the capacity to fold into a helical conformation would be expected to parallel the surface activity of the peptide. Examination of the conformation of the signal peptides while they are inserted into a lipid monolayer provides a test of this idea. In addition, direct observation of the signal peptides as they interact with the lipid surface should be strongly suggestive of the way the signal sequence would associate initially with the inner surface of the membrane.

Circular dichroism spectroscopy is a very insensitive method. The difference in absorption of right-handed and left-handed circularly polarized light is generally a small fraction of the light absorbed by an optically active chromophore (10^{-2} to 10^{-4}) (Woody, 1985). This intrinsic insensitivity coupled with the low concentration of a solute in a lipid monolayer leads to extreme experimental difficulty in measuring the CD signal of the lipid-resident solute. The strategy for overcoming this difficulty involves coating several quartz plates with a lipid monolayer spread on a film balance and passing the CD beam through an array of monolayers (Fig. 7). In practice, 16 or 32 monolayers were used in the signal peptide measurements. In addition, multiple scans of the CD signal (usually 128) were summed to produce the final spectrum for analysis. Details of the method of obtaining monolayer CD are more fully reported elsewhere (Cornell, 1979).

B. Circular Dichroism Spectra of Signal Peptides Inserted and Adsorbed to Lipid Monolayers

Since the goal of the monolayer CD measurements was to correlate conformation in the lipid environment with the known surface properties of the signal peptides measured by tensiometry, the lipid monolayers were spread to desired initial pressures and the peptides were injected in the aqueous subphase. The peptide concentrations were high enough to ensure maximal insertion [above the plateau in plots of surface pressure change versus concentration (like Fig. 5)]. We monitored surface pressure change after injection of the peptide and maintained the final steady-state pressure while withdrawing the quartz plates from the trough.

The presence of basic residues in prokaryotic signal sequences is thought to be related to an electrostatic interaction with the phospholipid surface. To explore this aspect of the encounter of signal peptides with the lipid monolayer, we injected the full-length wild-type LamB signal peptide beneath a lipid monolayer at a pressure well above the critical insertion pressure of the peptide. Any peptide signal observed in the CD spectrum of the transferred lipid monolayers must arise from noninserted or *adsorbed* signal peptide. Confirming this expectation was the finding of essentially no surface pressure change on injection of the peptide. The resulting CD for signal peptide electrostatically adsorbed on the lipid monolayer shows strong β character (Fig. 8) (Greenfield and Fasman, 1969). It is particularly noteworthy that this spectrum is distinct from the spectrum observed for the wild-type signal peptide in buffer alone. A conformational change has accompanied the interaction of the peptide with the high-pressure lipid monolayer.

When the lipid monolayer was poised at an initial pressure below the critical insertion pressure of the wild-type peptide before injection, the peptide caused the expected surface pressure change, indicating that it inserted. The CD spectrum observed from monolayers transferred to the quartz plates under these conditions (Fig. 8) has the characteristic double-minimum shape reflecting α-helical secondary structure (Greenfield and Fasman, 1969). The spectrum presented in this figure has been corrected for the electrostatically adsorbed (not inserted) peptide. Ultraviolet absorption is used to estimate the amount of peptide adsorbed to the plate, and the CD spectrum obtained under conditions of no insertion is as-

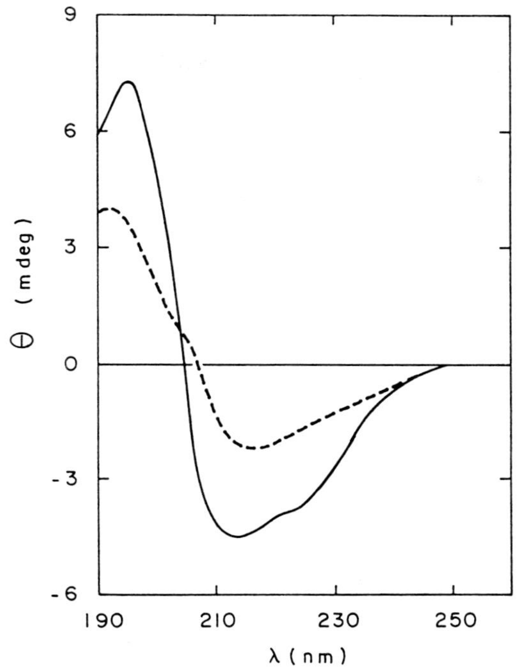

Fig. 8. The CD spectrum of wild-type signal peptide adsorbed to a phospholipid monolayer (dashed line). This sample was formed by injection of the signal peptide below a lipid monolayer poised at a surface pressure above the peptide's critical pressure of insertion. Hence, any peptide observed is interacting with the head group region of the lipids. No surface pressure change was observed subsequent to injection of the peptide. The CD spectrum of wild-type signal peptide inserted into a phospholipid monolayer (solid line). This sample was formed by injection of the signal peptide below a lipid monolayer at an initial surface pressure of 26 dyn/cm. After injection, a surface pressure rise of 14 dyn/cm was observed, indicating insertion of the peptide. Both spectra are results of signal averaging of 128 scans through the CD spectrum of eight quartz plates (16 monolayers). (From Briggs *et al.*, 1986. Copyright 1986 by the American Association for the Advancement of Science.)

sumed to reflect peptide interacting solely with the lipid head groups. We have confirmed the CD observations of signal peptide conformations in lipid monolayers by using Fourier transform-attenuated total reflectance infrared spectroscopy; these results and other details of these experiments will be reported in a full paper (Cornell *et al.*, 1986).

The spectra of wild-type signal peptide in phospholipid monolayers reaffirm the idea that the conformational properties of these peptides and their tendencies to insert into lipid monolayers arise from related origins. The α-helical conformation is adopted by the wild-type signal peptide when it inserts into the monolayer. Perhaps more exciting is the observation of a conformational change from the unstructured state existing in bulk aqueous medium to a β-like structure upon initial interaction with the phospholipid surface.

V. PROPOSED MODEL FOR THE INITIAL INTERACTION OF SIGNAL SEQUENCES WITH THE MEMBRANE

Our results provide a framework for a model for the events occurring *in vivo* when a signal sequence first encounters the membrane. These events can be viewed as involving the ribosome–SRP complex as it associates with the SRP receptor, bringing the exposed signal sequence into proximity with the membrane. Alternatively, in a posttranslational secretion mechanism one can picture the nascent protein, with its signal region accessible to a lipid surface, diffusing toward the membrane. The monolayer insertion studies we have described argue cogently that the signal sequence *will spontaneously insert* into a lipid phase. In fact, we find that the signal peptides do work on insertion. Again, other components are surely involved in protein translocation, but there is no reason not to invoke a direct interaction of signal sequence with lipid. Furthermore, we can postulate from the CD in SDS and especially from the CD in monolayers that an α-helical conformation is adopted by the signal sequence in the lipid phase. However, the lipid surface causes a conformational change to a β-like structure when the signal peptide is not inserted.

A model encompassing all of these ideas is presented in Fig. 9. First, the signal sequence, which has been exposed in some fashion to the aqueous surroundings, interacts electrostatically (via the basic residues as proposed by Inouye and Halegoua, 1980) with the inner membrane surface. Second, the signal sequence adopts a folded β structure with a turn near residues −7 to −10; this structure has hydrophobic character overall and an approximate length of 34 Å (10 residues at a rise per residue of 3.4 Å). We suggest, although we have no evidence for this step, that this is the form of the signal sequence that inserts initially. Next, the environment of the membrane would lead to a conformational change to the more favored α-helical conformation. Concomitant with this conformational change, assuming that the basic residues continue to anchor the N terminus on the inner surface of the membrane, a segment of the mature protein would enter the membrane (10 or more residues). In an α-helical conformation, the signal sequence nearly spans the bilayer (20 residues at 1.5 Å per residue stretches 30 Å). Hence, the cleavage site falls on the

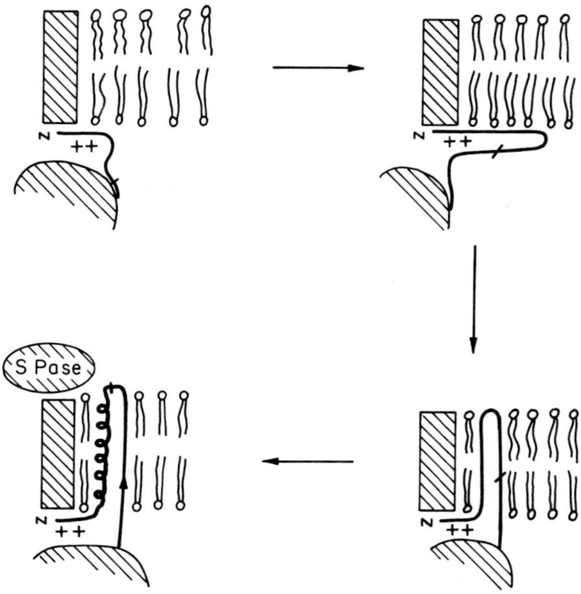

Fig. 9. Model for initial interaction of a signal sequence with a membrane. The steps are described in the text. A slash indicates the site of signal peptide cleavage by the signal peptidase. Note that the signal sequence is viewed as emerging from the ribosome into an aqueous environment, then interacting with the charged surface of the membrane, and subsequently inserting into the hydrophobic region of the membrane. Conformations adopted by the signal sequence in these steps are proposed to be random (aqueous), β-like (membrane surface), and α-helical (inserted). A transient state is suggested wherein the extended, β-like signal peptide inserts and then adopts a helical conformation. *In vivo*, associations with various components of the export apparatus may alter these simplified steps. (From Briggs *et al.*, 1986. Copyright 1986 by the American Association for the Advancement of Science.)

opposite face of the membrane. In the insertion step of this model, the region just C-terminal to the hydrophobic core, which in most signal sequences is more hydrophilic (von Heijne, 1985), resides transiently on the opposite (aqueous) side of the membrane. This arrangement seems physically reasonable. It is also supported by genetic data since charged residues are tolerated in this region but not in the adjacent hydrophobic core (Silhavy *et al.*, 1983).

We conclude by underlining the speculative nature of this model. The biophysical observations are real and establish a correlation of intrinsic properties with *in vivo* function within this family of signal sequences. The extrapolation to a model for the initial steps in signal sequence–

membrane interaction provides a testable working hypothesis which forms the basis for subsequent experiments.

ACKNOWLEDGMENTS

We thank Frank Prendergast of the Mayo Clinic for providing the mass spectrometry data on the signal peptides. This research was supported in part by grants from the National Institutes of Health (GM27616 and GM34962) and the National Science Foundation (DCB-8518695). L.M.G. is a Fellow of the A. P. Sloan Foundation, 1984–86.

REFERENCES

Austen, B. M. (1979). *FEBS Lett.* **103,** 308–313.
Austen, B. M., and Ridd, D. H. (1981). *Biochem. Soc. Symp.* **46,** 235–258.
Benson, S. A., Hall, M. N., and Silhavy, T. J. (1985). *Annu. Rev. Biochem.* **54,** 101–134.
Blobel, G., and Dobberstein, B. (1975a). *J. Cell Biol.* **67,** 835–851.
Blobel, G., and Dobberstein, B. (1975b). *J. Cell Biol.* **67,** 852–862.
Briggs, M. S. (1986). Ph.D. Dissertation, Yale University.
Briggs, M. S., and Gierasch, L. M. (1984). *Biochemistry* **23,** 3111–3114.
Briggs, M. S., Gierasch, L. M., Zlotnick, A., Lear, J., and DeGrado, W. F. (1985). *Science* **228,** 1096–1099.
Briggs, M. S., Cornell, D. G., Dluhy, R., and Gierasch, L. M. (1986). *Science* (in press).
Chen, L., Rhoads, D., and Tai, P. C. (1985). *J. Bacteriol.* **161,** 973–980.
Chou, P. Y., and Fasman, G. D. (1978). *Annu. Rev. Biochem.* **47,** 251–276.
Colacicco, G. (1970). *Lipids* **5,** 636–649.
Cornell, D. G. (1979). *J. Colloid Interface Sci.* **70,** 167–180.
Cornell, D. G., Briggs, M. S., and Gierasch, L. M. (1986). In preparation.
Emr, S. D., and Silhavy, T. J. (1983). *Proc. Natl. Acad. Sci. U.S.A.* **80,** 4599–4603.
Emr, S. D., Hanley-Way, S., and Silhavy, T. J. (1981). *Cell* **23,** 79–88.
Engleman, D. M., and Steitz, T. A. (1981). *Cell* **23,** 411–422.
Fendler, J. H. (1982). "Membrane Mimetic Chemistry." Wiley, New York.
Gilmore, R., and Blobel, G. (1983). *Cell* **35,** 677–685.
Gilmore, R., and Blobel, G. (1985). *Cell* **42,** 497–505.
Gilmore, R., Walter, P., and Blobel, G. (1982). *J. Cell Biol.* **95,** 470–476.
Gilmore, R., Walter, P., Anderson, D., Erickson, A., Müller, M., and Blobel, G. (1984). *In* "Protein Transport and Secretion" (D. Oxender, ed.), pp. 291–296. Alan R. Liss, Inc., New York.
Greenfield, N., and Fasman, G. D. (1969). *Biochemistry* **8,** 4108–4115.
Horiuchi, S., Marty-Mazars, D., Tai, P. C., and Davis, B. D. (1983). *J. Bacteriol.* **154,** 1215–1221.
Inouye, M., and Halegoua, S. (1980). *CRC Crit. Rev. Biochem.* **7,** 339–371.
Inouye, S., Soberon, X., Franchesini, T., Nakamura, K., Itakura, K., and Inouye, M. (1982). *Proc. Natl. Acad. Sci. U.S.A.* **79,** 3438–3441.
Kreil, G. (1981). *Annu. Rev. Biochem.* **50,** 317–348.
Meyer, D. I., Krause, E., and Dobberstein, B. (1982). *Nature (London)* **297,** 647–650.
Mueller, M., and Blobel, G. (1984). *Proc. Natl. Acad. Sci. U.S.A.* **81,** 7421–7425.

Mueller, M., Ibrahimi, I., Chang, C. N., Walter, P., and Blobel, G. (1982). *J. Biol. Chem.* **257,** 11863–11864.
Nagaraj, R. (1984). *FEBS Lett.* **165,** 79–82.
Ohno-Iwashita, Y., Wolfe, P., Ito, K., and Wickner, W. (1984). *Biochemistry* **23,** 6178–6184.
Pattus, F., Martinez, M. C., Dargent, B., Cavard, D., Verger, R., and Lazdunski, C. (1983). *Biochemistry* **22,** 5698–5703.
Perlman, D., and Halvorson, H. O. (1983). *J. Mol. Biol.* **167,** 391–409.
Pethica, B. A. (1955). *Trans. Faraday Soc.* **51,** 1402–1411.
Phillips, M. C., and Sparks, C. E. (1980). *Ann. N. Y. Acad. Sci.* **148,** 122–137.
Prehn, S., Nürnberg, P., and Rapaport, T. (1981). *FEBS Lett.* **123,** 79–84.
Randall, L. L. (1983). *Cell* **33,** 231–240.
Rosenblatt, M., Beaudette, N. V., and Fasman, G. D. (1980). *Proc. Natl. Acad. Sci. U.S.A.* **77,** 3983–3987.
Rothfield, L. I., and Fried, V. A. (1975). *Methods Membr. Biol.* **4,** 277–292.
Shinnar, A. E., and Kaiser, E. T. (1984). *J. Am. Chem. Soc.* **106,** 5106–5107.
Silhavy, T. J., Benson, S. A., and Emr, S. D. (1983). *Microbiol. Rev.* **47,** 313–344.
Takahara, M., Hibler, D. W., Barr, P. J., Gerlt, J. A., and Inouye, M. (1985). *J. Biol. Chem.* **260,** 2670–2674.
Talmadge, K., Kaufman, J., and Gilbert, W. (1980). *Proc. Natl. Acad. Sci. U.S.A.* **77,** 3988–3992.
Verger, R., and Pattus, F. (1982). *Chem. Phys. Lipids* **30,** 189–227.
von Heijne, G. (1983). *Eur. J. Biochem.* **133,** 17–21.
von Heijne, G. (1985). *J. Mol. Biol.* **185,** 99–105.
Walter, P., and Blobel, G. (1981a). *J. Cell Biol.* **91,** 551–556.
Walter, P., and Blobel, G. (1981b). *J. Cell Biol.* **91,** 557–561.
Walter, P., Ibrahimi, I., and Blobel, G. (1981). *J. Cell Biol.* **91,** 545–550.
Walter, P., Gilmore, R., and Blobel, G. (1984). *Cell* **38,** 5–8.
Watson, M. E. E. (1984). *Nucleics Acids Res.* **12,** 5145–5164.
Wickner, W. (1979). *Annu. Rev. Biochem.* **48,** 23–45.
Wickner, W. (1980). *Science* **210,** 861–868.
Woody, R. (1985). *In* "The Peptides: Analysis, Synthesis, Biology" (S. Udenfriend, J. Meienhofer, and V. Hruby, eds.), Vol. 7, pp. 15–114. Academic Press, New York.

12

Studies on the Mechanism of Membrane Fusion

MARY-JANE GETHING,*,[1] ROBERT W. DOMS,† JUDY WHITE,†,[2] AND ARI HELENIUS†

** Cold Spring Harbor Laboratory*
Cold Spring Harbor, New York 11724
and
† Department of Cell Biology
Yale University School of Medicine
New Haven, Connecticut 06510

I. INTRODUCTION

Membrane fusion is a fundamental biological process that mediates constitutive cellular functions such as exocytosis and endocytosis as well as specialized cellular events such as fertilization, muscle development, and synaptic transmission. Enveloped viruses which enter and infect their host cells by fusing with cellular membranes have been extensively employed in studies of membrane fusion (for review, see White *et al.*, 1983). For each virus, fusion is mediated with a characteristic pH dependence by a specific viral membrane glycoprotein. Two genetic approaches, illustrated in this chapter, are currently being employed to analyze the mechanism of the membrane fusion reaction mediated by the hemagglutinin (HA) glycoprotein of influenza virus. The first approach involves studies of variant influenza viruses that induce fusion with raised pH thresholds. Sequence analysis of HAs from the variants locates altered residues that

[1] Present address: Department of Biochemistry and Howard Hughes Medical Institute, University of Texas Health Science Center, Dallas, Texas 75235.

[2] Present address: Department of Pharmacology, University of California, San Francisco, California 94143.

PROTEIN ENGINEERING
Applications in Science,
Medicine, and Industry

play a role in the pH dependence of fusion. The second approach utilizes site-specific mutagenesis of a cloned HA gene to alter the nucleotide sequence coding for selected residues in the "fusion peptide" of the HA molecule. Expression of the mutant genes in simian cells confirms the central role of the fusion peptide and provides insights into the mechanism of HA-induced fusion.

II. STRUCTURE AND FUNCTION OF HEMAGGLUTININ

Hemagglutinin is the major glycoprotein of the influenza virus envelope. It is responsible for the attachment of the virion to sialic acid containing receptors on the host cell surface (Hirst, 1941) and also undergoes a low-pH-induced conformational change that initiates the fusion of the virus envelope with the endosome membrane (Maeda and Onishi, 1980; Huang *et al.,* 1981; Matlin *et al.,* 1981; White *et al.,* 1981; Skehel *et al.,* 1982; Marsh, 1984; Doms *et al.,* 1985). To be active in fusion the HA precursor must be processed by a posttranslational proteolytic cleavage into HA1 and HA2 subunits (Laver, 1971; Klenk *et al.,* 1975; Lazarowitz and Choppin, 1975; White *et al.,* 1981). A new hydrophobic amino terminus, the fusion peptide, is generated on the HA2 subunit. This peptide has been implicated in the fusion activity (Gething *et al.,* 1978; Richardson *et al.,* 1980; Garten *et al.,* 1981; White *et al.,* 1982b) and is highly conserved in HAs from different influenza virus strains (reviewed by Lamb, 1983). The HA molecule in its neutral form is a trimer that projects from the viral envelope as a rod-shaped structure 135 Å in length (Wilson *et al.,* 1981). The hydrophobic fusion peptide in each monomer is tucked into the interface between the subunits of the trimer, approximately 30 Å from where the protein inserts into the lipid bilayer of the virus envelope or the plasma membrane. It has been proposed (Skehel *et al.,* 1982; White *et al.,* 1982b; Daniels *et al.,* 1983, 1985; Doms *et al.,* 1985) that below the threshold pH for fusion, the protonation of one or more amino acid side chains causes partial dissociation of the HA trimer, exposing the fusion peptide, which inserts into the target bilayer. The HA protein would then become an integral component of both the viral and target membranes, presumably bringing them close enough together to fuse.

III. ASSAYS FOR THE LOW-pH-INDUCED CONFORMATIONAL CHANGE AND THE FUSION ACTIVITY OF THE HEMAGGLUTININ MOLECULE

The HA molecule undergoes a conformational change with a pH dependence that parallels that of the fusion activity (Skehel *et al.,* 1982; Doms *et al.*, 1985). This conformational change can be monitored by following the acquisition of protease sensitivity by an ectodomain fragment of HA (BHA) released from the membrane by treatment with bromelain (Brand and Skehel, 1972) and by the ability of this fragment to bind to lipid vesicles or detergents or to aggregate in lipid- or detergent-free solutions (Skehel *et al.*, 1982; Doms *et al.*, 1985). The fusion activity of HA can be manifested experimentally as cell : cell fusion (i.e., polykaryon formation) when monolayers of cells displaying HA on their plasma membranes are transiently exposed to low pH (Maeda and Onishi, 1980; Huang *et al.*, 1981; White *et al.*, 1981, 1982a). Alternatively, fusion activity can be measured quantitatively by red cell-mediated delivery of horseradish peroxidase (HRP) into cells expressing surface HA (Doxsey *et al.*, 1985; Sambrook *et al.*, 1985). Such experiments with cells infected with different influenza viruses have shown that HA-induced cell fusion displays a characteristic pH profile, with the threshold pH varying between 5 and 6 depending on the viral origin of the HA molecule (Huang *et al.*, 1981; White *et al.*, 1983). To confirm the central role of HA in the fusion mechanism, we have used cells expressing HA from a cloned copy of the HA gene inserted into recombinant SV40 or BPV vectors to demonstrate that the HA molecule displays fusion activity in the absence of any other influenza virus-encoded components (White *et al.*, 1982b; Sambrook *et al.*, 1985).

IV. ANALYSIS OF THE HEMAGGLUTININ FROM A VARIANT INFLUENZA VIRUS THAT INDUCES FUSION AT ELEVATED pH

To gain insight into the molecular mechanism of the low-pH-induced fusion reaction mediated by HA, influenza virus variants have been isolated which fuse with pH optima different from those of their parent viruses. In this chapter we summarize results obtained from analysis of the HA from a naturally occurring fusion variant of the X-31 strain (Doms *et al.*, 1986). In another study Rott *et al.* (1984) showed that other variants of this strain of influenza, selected for their ability to undergo activation cleavage and grow in MDCK cells, also display an elevated fusion pH threshold (approximately 0.7 pH unit higher than the wild type). Further variants from other strains of virus have been selected for by growth in the presence of amantadine, a compound that raises endosomal pH. Variant viruses were obtained that mediated fusion at pH values 0.1–0.7 unit higher than the parent strain (Daniels *et al.*, 1985). Sequence analysis of the HA from each variant has identified amino acid residues in HA that play a role in the pH dependence of fusion.

A. Identification and Isolation of Variant Influenza Viruses

In experiments to analyze the acid-induced conformational change in HA that mediates membrane fusion, it was observed that a stock preparation of the X-31 strain of influenza virus appeared to contain a minority population of HA molecules that converted to the low-pH conformation at a pH 0.3 unit higher than that previously reported for the wild-type protein (Doms *et al.*, 1985, 1986). To determine whether the heterogeneity in the response of HA to low pH occurred at the level of individual virus particles, the preparation was examined by negative stain electron microscopy. Egg-grown X-31 virus was incubated at pH 5.6 for 15 min at 37° C (a treatment shown by assays of protease sensitivity to convert 20% of the HA to the low-pH conformation) and stained with sodium silicotungstate (pH 7.4). Inspection of the virus particles in the electron microscope revealed that the HA on individual virions had either completely converted to the acid form, characterized by thin, disordered surface projections (Figs. 1B and 1C), or remained entirely in the neutral form, with well-ordered rectangular HA molecules projecting from the viral surface (Figs. 1A and 1C). No viruses were observed that displayed a mixture of the two morphological forms of spikes on their surfaces. Thus, the stock X-31 virus contained a mixture of different virus types.

Following plaque purification of viruses in MDCK cells, the hemolytic activities of 10 isolates and the original stock virus were assayed to compare the pH thresholds of fusion. The midpoint of hemolysis activity varied between pH 5.3 and 5.6 (Fig. 2). Isolate C22, representing viruses similar to the majority population, and isolate A31, representing a minority population with a significantly elevated threshold pH for hemolysis, were chosen for further characterization. By using a number of assays to monitor the fusion activity and conformational change in the variant and wild-type viruses, a strong correlation was found between the pH dependence of the morphological change in the HA spike, the biochemical manifestation of the conformational change, the ability of the HA to bind to a target membrane, and the induction of membrane fusion (Doms *et al.*, 1986). In every case, the pH values at which these low-pH-induced changes occurred were higher for the variant than for the wild-type HA. To determine whether the differences observed between the wild-type and variant viruses in the *in vitro* fusion assays were manifested during infectious entry of the virions into cells, CV-1 monolayers were incubated with each virus stock in the presence of increasing concentrations of ammonium chloride. This lysosomotropic weak base elevates endosomal

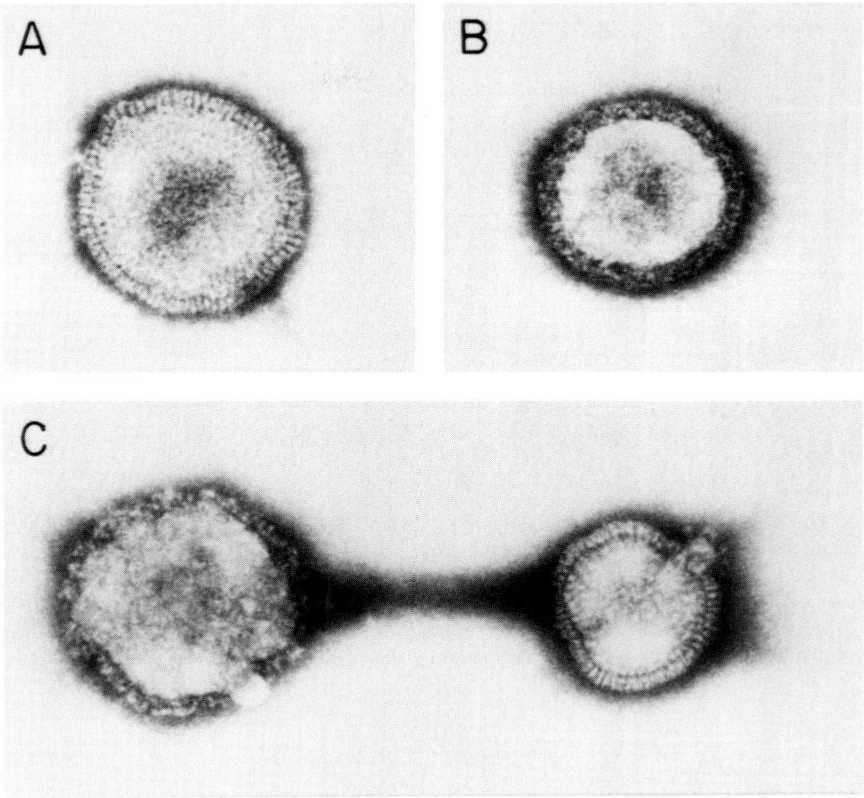

Fig. 1. Negative stain electron microscopy of X-31 virions at pH 5.0, 5.6, and 7.0. Egg-grown X-31 virus was incubated at (A) pH 7.0, (C) pH 5.6, or (B) pH 5.0 for 15 min at 37°C. The samples were placed on carbon films floated from freshly cleaved mica and stained with sodium silicotungstate in 10 m*M* phosphate-buffered saline (PBS), pH 7.0. Magnification ×91,000.

and lysosomal pH (DeDuve *et al.*, 1974) and blocks infection of influenza and other acid-triggered enveloped animal viruses (Jensen and Liu, 1963). Isolate A31 was more resistant than the parent virus to the effects of this agent, as would be expected if its primary mode of antiviral activity is to elevate endosomal pH. (Doms *et al.*, 1986).

B. Cloning and Sequence Analysis of Genes Encoding the Wild-Type and Variant Hemagglutinins

That the variant fusion phenotype of A31 was due to a difference in the HA and not some other viral protein was proved by cloning both the

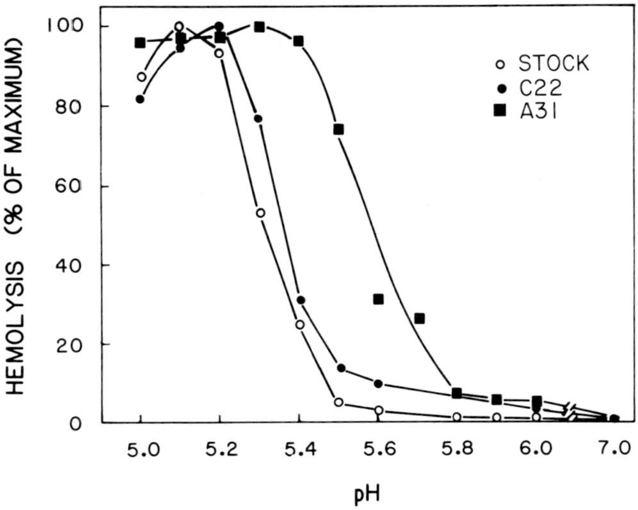

Fig. 2. pH dependence of hemolysis by A31, C22, and parent X-31 viruses. Allantoic fluids from eggs infected with the various viruses were diluted with PBS and incubated with guinea pig red blood cells at 37°C. The samples were acidified to the indicated pH for 15 min, then neutralized by the addition of base. The red cells were removed by low-speed centrifugation and the extent of hemolysis determined by measuring absorbance at 540 nm. Values from four experiments were averaged and expressed relative to the maximum amount of hemolysis observed.

variant and wild-type HAs, expressing them in simian cells, and analyzing their fusion phenotypes (see below and Fig. 6). To analyze the structural basis of the difference in fusion phenotype, double-stranded cDNA copies of the HA viral RNA genes from wild-type or A31 virions were cloned into bacterial plasmids. Restriction fragments of the cDNAs were inserted into M13 phage vectors for sequence analysis by the dideoxy chain termination technique (Sanger *et al.,* 1977, 1980). Deduction of the amino acid sequence of A31 HA from the nucleotide sequence of the gene revealed that three residues were altered from those present in the wild-type protein. The positions of the altered residues in HA are illustrated in Fig. 3, which shows the structure of the HA monomer (Wilson *et al.,* 1981), and in Fig. 4, which relates the various functional regions of the HA molecule to the linear protein and DNA sequence. Two of the substitutions were in the HA1 subunit while the third amino acid change was in HA2. Both substitutions in HA1 had occurred in previously defined antigenic epitopes : glycine 144 to aspartic acid in antigenic site A (Wiley *et al.,* 1981) in the loop region of HA1 and proline 215 to leucine in antigenic region D,

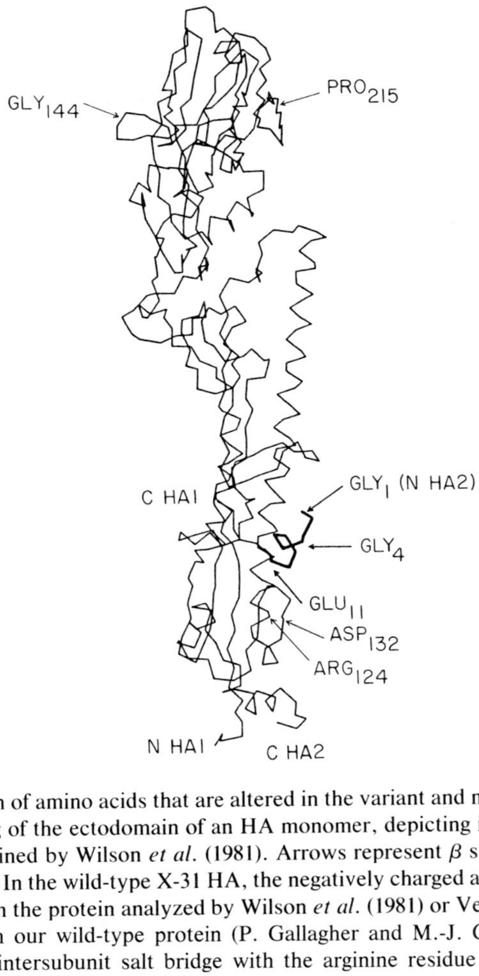

Fig. 3. Location of amino acids that are altered in the variant and mutant hemagglutinin molecules. Drawing of the ectodomain of an HA monomer, depicting its three-dimensional structure as determined by Wilson *et al.* (1981). Arrows represent β structures and barrels represent α helices. In the wild-type X-31 HA, the negatively charged amino acid at position 132 [glutamic acid in the protein analyzed by Wilson *et al.* (1981) or Verhoeyen *et al.* (1980) but aspartic acid in our wild-type protein (P. Gallagher and M.-J. Gething, unpublished results)] forms an intersubunit salt bridge with the arginine residue at position 124 in a second HA2 subunit. In the variant A31, aspartic acid 132 is replaced by an asparagine, disrupting the salt bridge. The other substitutions in A31 HA (at residues 144 and 215 in HA1) are also shown. The fusion peptide at the amino terminus of HA2 is shown by a thicker line. The positions of the residues in the fusion peptide (Gly 1, Gly 4, and Glu 11) that were altered by site-directed mutagenesis of the Japan HA gene are also indicated.

which is located close to the trimer interface region of HA1 (Wilson *et al.*, 1981). The first substitution has been observed in a variety of influenza isolates of the H3 subtype (Both *et al.*, 1983). However, the mutation at residue 215 has not previously been described in H3 HA molecules. The third mutation causes the substitution of an asparagine for aspartic acid at position 132 in the HA2 subunit.

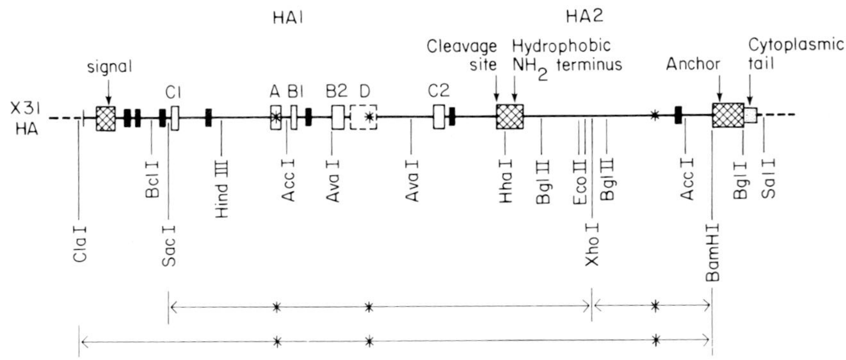

Fig. 4. Schematic diagram of X-31 HA. The linear map relates the various functional regions of the HA molecule to the restriction sites in the gene and shows the positions of the altered residues found in the variant A31 HA. ▩, Hydrophobic sequence; □, antigenic region; ■, glycosylation site; *, altered residue in A31 HA. The restriction fragments that were excised from the A31 HAcDNA and inserted into the SVEXHA vector (Fig. 5) replacing the equivalent wild-type HA sequences are shown (↔).

C. Expression of Wild-Type and Variant Hemagglutinins in Simian Cells Using SV40-Based Vectors

To determine which of the three mutations was responsible for the observed shift in fusion phenotype, it was necessary to express and assay HA proteins that contained various combinations of the altered amino acids. Hybrid HA genes were constructed that contained all three mutations (muHA1,2), only the two substitutions in HA1 (muHA1), or only the single substitution in HA2 (muHA2). Restriction fragments that contained HA sequences encoding three, two, or only one of the amino acid substitutions (Fig. 4) were isolated and inserted into the pSVEXHA vector (Fig. 5), replacing the equivalent wild-type HA sequences. High-titer virus stocks containing each recombinant genome were generated as described previously (Gething and Sambrook, 1981; Doyle *et al.*, 1985) and used to infect CV-1 cells. In every case, at 65 hr postinfection >90% of the CV-1 cells bound erythrocytes, indicating that large amounts of HA molecules were present on the cell surface (Doms *et al.*, 1986). To confirm that each vector expressed the correct wild-type or mutant HA gene, monoclonal antibodies were used to discriminate between HA proteins that contained glycine or aspartic acid at position 144. All of the HA proteins were recognized and immunoprecipitated by monoclonal antibodies directed against either the hinge (antigenic site C, Mab 69/1) or the tip (antigenic site B, Mab 17/2) region (Wiley *et al.*, 1981) of the HA molecule. However, neither muHA1,2 nor muHA1 proteins were recognized

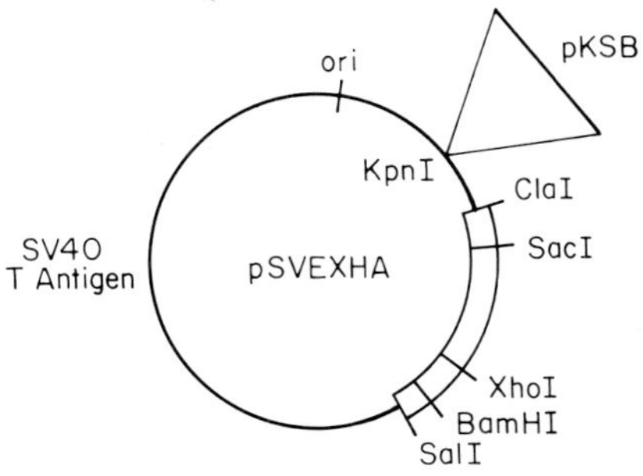

Fig. 5. Diagram of the SVEXHA vector used to express wild-type and variant HAs in CV-1 cells. The boxed section of the circle represents the X-31 HA gene; the single-line section of the circle represents SV40 sequences including the origin of replication (*ori*); the triangle represents plasmid sequences (described in Doyle *et al.*, 1985).

by Mab 11/4, which is directed against the loop region (antigenic site A), confirming the presence of the substitution at residue 144 in these variant HA molecules. As expected, this monoclonal antibody did precipitate the wild-type and muHA2 proteins (Doms *et al.*, 1986).

D. Analysis of the Fusion Activity of the Wild-Type and Variant Hemagglutinins

When the pH dependence of fusion mediated by the wild-type and chimeric HA molecules was analyzed (Fig. 6), it was found that the HA molecules that contained the mutation in HA2 (muHA2 and muHA1,2) induced cell : cell fusion at a pH 0.2 unit higher than either the wild-type or muHA1. The pH dependence of fusion was also determined by binding and fusing red blood cells to the surface of infected CV-1 cells (Doxsey *et al.*, 1985). The extent of rbc : CV-1 fusion was determined by assaying by benzidine staining for the delivery of hemoglobin into the CV-1 cytoplasm. The pH dependence of delivery was the same as that observed for cell : cell fusion (Fig. 6). Thus the mutation at residue 132 in HA2 was solely responsible for the shift in fusion threshold to a higher pH.

E. Functional Implications of the Structural Locations of Residues Altered in Variant Hemagglutinins

Examination of the three-dimensional structure of HA (Wilson *et al.*, 1981) reveals that residue 132 is located near the base of the molecule, approximately 20 Å from where the polypeptide enters the lipid bilayer of the viral envelope or the plasma membrane (Fig. 3). In the wild-type

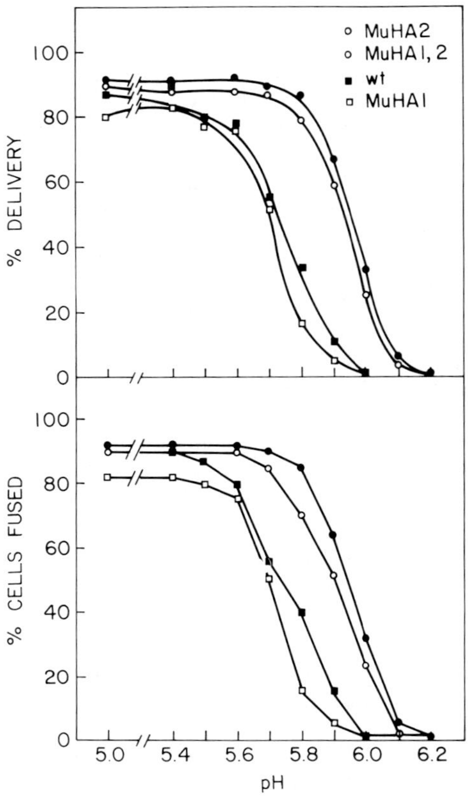

Fig. 6. pH dependence of HA-induced cell : cell and rbc : cell fusion in CV-1 cells infected with SVEXHA expression vectors. CV-1 cells were infected with SVXHA recombinant viruses containing the wild-type, variant, or chimeric HA genes. At 60 hr postinfection, cell surface HA was activated by treatment with trypsin. The rbc : cell fusion was assayed by staining for the delivery of benzidine into the CV-1 cytoplasm (upper panel); cell : cell fusion was assayed by polykaryon formation (lower panel). Details of the assays are given in Doms *et al.* (1986).

HA, aspartic acid 132 forms a salt bridge with arginine 124 in an adjacent HA2 subunit in the trimer (Daniels *et al.*, 1985). The residues involved in this salt link are highly conserved in the HAs from many different strains of influenza, suggesting that this is an important structural interaction. A different fusion variant described by Daniels *et al.* (1985), in which phenylalanine 9 of HA2 was replaced with a leucine, may exert its effects at least in part by destablizing the same interaction. Both the substitutions at residues 9 and 132 and several others described by Daniels *et al.* affect residues that are involved in stabilizing the HA trimer at neutral pH. As suggested by those authors, substitutions which disrupt these interactions may destabilize the trimer, allowing it to undergo the fusion-inducing conformational change at a higher pH. These findings are consistent with the hypothesis that the HA trimer undergoes at least partial dissociation at acid pH (Skehel *et al.*, 1982; Daniels *et al.*, 1985; Doms *et al.*, 1985), though direct evidence for this has yet to be obtained.

V. SITE-DIRECTED MUTAGENESIS OF THE FUSION PEPTIDE OF HEMAGGLUTININ

As discussed above, most of the residues that were altered in the various fusion variants were located along the interface between the individual subunits of the HA trimer (Daniels *et al.*, 1985; Doms *et al.*, 1986). However, others, including three amino acids located within the fusion peptide, stabilized the unexposed location of the amino terminus of HA2 (Rott *et al.*, 1984; Daniels *et al.*, 1985). It was notable that the three amino acids located in the fusion peptide had undergone conservative substitutions; none had been altered to charged amino acids. This probably reflects the fact that mutant viruses that could not carry out the fusion reaction at any pH could not enter and infect cells and therefore could not be propagated. In an attempt to obtain mutant HAs that might be inactive or disabled for fusion and to probe the consequences of altering the hydrophobicity and length of the fusion peptide, we employed oligonucleotide-directed mutagenesis to engineer HA molecules that contain specific nonconservative amino acid changes in the fusion peptide (Gething *et al.*, 1986).

A. Construction of Hemagglutinin Fusion Mutants

Figure 7 shows the amino acid alterations that were achieved by introducing single nucleotide alterations into the cloned DNA sequence encoding the fusion peptide of the HA from the A/Japan/305/57 strain of influenza virus. The rationale for our choices was as follows. In the neutral conformation of the HA molecule, the fusion peptide is tucked into the interface between the subunits of the trimer with the amino nitrogen of residue 1 (Gly) and the amide nitrogens of residues 4, 5, and 6 (Gly, Ala, Ile) of HA2 forming hydrogen bonds to oxygen atoms of residue 112 (Asp) in the long helix of HA2 (Wilson *et al.*, 1981; Daniels *et al.*, 1985). Substitution of glutamic acid for glycine residues at positions 1 and 4 (mutants M1 and M4) introduces negative charges into the fusion peptide, which might disrupt the normal hydrogen-bonding interactions. The M4 substi-

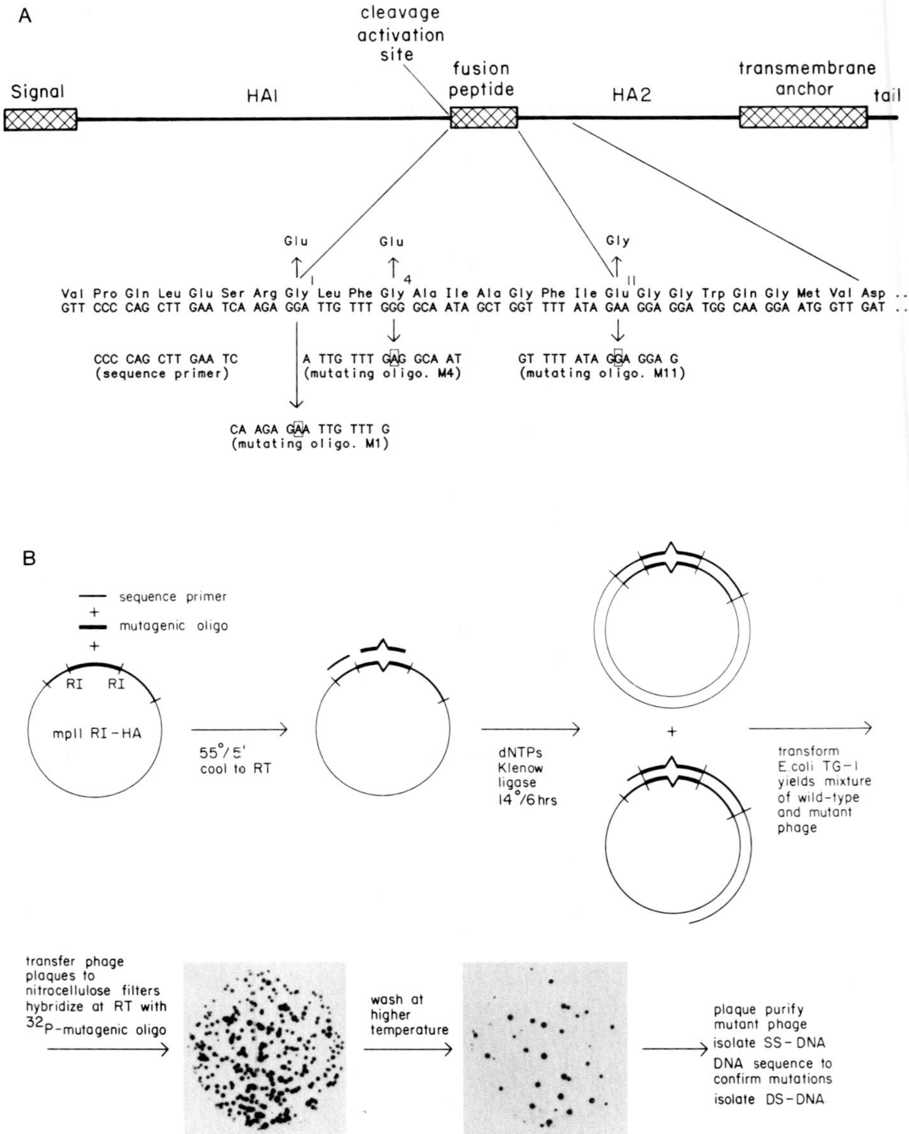

Fig. 7. Oligonucleotide-directed mutagenesis of the fusion peptide of HA. (A) Linear diagram of the HA molecule showing the three hydrophobic domains of the protein as cross-hatched regions. The amino acid sequence and the encoding nucleic acid sequence are displayed for the region of HA that includes the fusion peptide. The mismatched oligonucleotides that were chosen to introduce specific amino acid changes in this region are shown, together with the matched oligonucleotide that was used as primer for DNA sequence analysis. (From Gething *et al.*, 1986.) (B) The *Eco*RI restriction fragment of the Japan HA

tution would also significantly shorten the stretch of apolar amino acids. Substitution of the glutamic acid at residue 11 by a glycine (mutant M11) extends the apolar sequence to 18 residues, permitting additional testing of the role played by the length of the hydrophobic sequence. The positions of these residues in the HA monomer are illustrated in Fig. 3.

Three oligonucleotides were designed, synthesized by phosphate triester chemistry, and utilized to introduce single mutations in HA sequences that had been cloned into an M13 vector (See Fig. 7 and Gething *et al.*, 1986 for a description of the mutagenesis procedures). Once the presence of the desired mutations had been confirmed by DNA sequence analysis, restriction fragments containing each altered sequence were used to replace wild-type sequences in an SV40–HA recombinant virus vector (Fig. 8). High-titer virus stocks containing the recombinant genomes were developed and used to infect simian CV-1 cells for expressing of the wild-type and mutant HA proteins (Gething and Sambrook, 1981; Doyle *et al.*, 1985).

B. Synthesis and Cell Surface Expression of Wild-Type and Mutant Hemagglutinins

Before analyzing the fusion phenotype of the mutant HA proteins, experiments were performed to characterize the HA molecules and to ensure that the mutants were not grossly altered in their structure or level of expression (Gething *et al.*, 1986). Solid-phase radioimmunoassays demonstrated that the wild-type and mutant proteins were expressed at very similar levels in CV-1 cells infected with the SV40–HA vectors, i.e., approximately 10^8 molecules per infected cell at 60 hr postinfection. To further characterize the mutant HAs, infected cells were labeled with [^{35}S]methionine. Following incubation for 2 hr to allow expression of the proteins on the cell surface, the intact monolayers were treated with trypsin to cleave HA0 into HA1 and HA2 subunits before cell extracts were prepared, immunoprecipitated, and analyzed by electrophoresis through sodium dodecyl sulfate (SDS)–polyacrylamide gels. The results shown in Fig. 9 indicate that the mutant proteins were transported to the

---

gene (Gething *et al.*, 1980) was cloned into M13mp11 phage DNA and the single-stranded virion RNA was used as template for mutagenesis by the double-primer method (Zoller and Smith, 1984), using each of the mismatched oligonucleotides shown in panel A. Phage plaques containing mutant HA sequences were identified by hybridization with ^{32}P-labeled oligonucleotide. The *Eco*RI restriction fragments containing the mutant sequences were excised from the double-stranded replicative form of the phage DNAs and used to replace the equivalent wild-type sequences in the SVEHA3 vector. Details are given in Gething *et al.* (1986).

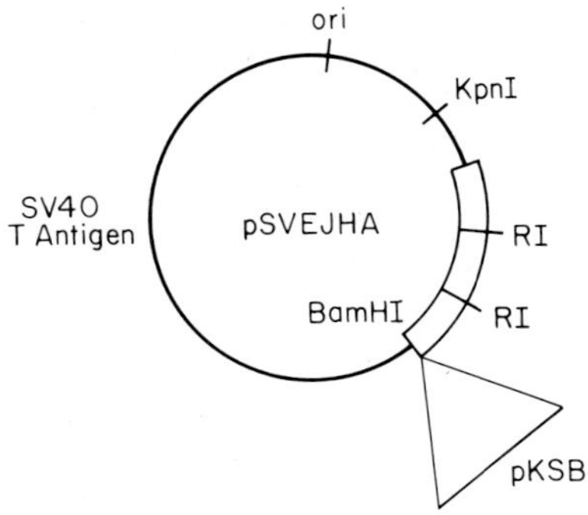

Fig. 8. Diagram of the SVEJHA vector used to express wild-type and variant HAs in CV-1 cells. The boxed section of the circle represents the Japan HA gene; the single-line section of the circle represents SV40 sequences including the origin of replication (*ori*); the triangle represents plasmid sequences (described in Gething *et al.*, 1986).

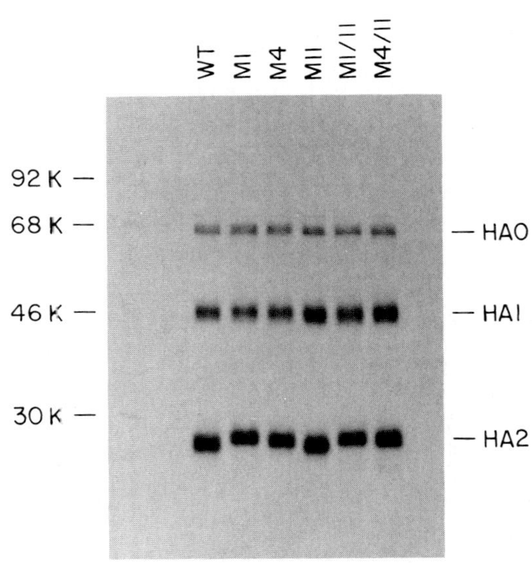

Fig. 9. Comparison of wild-type and mutant HA proteins by SDS–PAGE. CV-1 cells infected for 36 hr with recombinant SV40-HA virus stocks were labeled with [^{35}S]methionine for 15 min. The cells were then washed and incubated in medium containing an excess of nonradioactive methionine for 2 hr to allow expression of the labeled wild-type or mutant proteins on the cell surface. The intact monolayers were then treated for 15 min at 37°C with DME containing 5 mg/ml trypsin before cell extracts were prepared and analyzed by immunoprecipitation and SDS–PAGE as described in Gething *et al.* (1986).

cell surface (where they became accessible to trypsin in the medium) with the same efficiency as the wild-type HA. Although the HA0 and HA1 bands of all the proteins had very similar or identical mobilities, minor differences were detected in the mobilities of the HA2 bands. These differences may result from the alteration in charge of the HA2 subunit consequent on the nonconservative amino acid substitutions (Seeburg *et al.*, 1984), or they may reflect minor alterations in the addition of terminal sugars to the single oligosaccharide side chain on HA2. Hemagglutination of red blood cells to the surfaces of infected cells provided further evidence of efficient cell surface expression (Fig. 10). All the mutant proteins were recognized by a panel of monoclonal antibodies directed against the A/Japan/305/57 HA and a closely related HA from the H2 virus subtype A/Guiyang/1/57 (Yamada *et al.*, 1984). In particular, all of the mutant HAs were precipitated by two of the monoclonal antibodies which recognize native HA but do not recognize the denatured molecule. These results indicate that the site-specific mutagenesis of HA had caused no gross alterations in the protein structure. Furthermore, soluble BHA fragments prepared from cells expressing the mutant proteins migrated on sucrose gradients with the mobility (9 S) characteristic of the native trimeric ectodomain. Finally, the mutant BHAs all displayed similar resistance to proteolytic degradation at neutral pH—a criterion for the intact structure.

C. Fusion Activities of Mutant Hemagglutinins

CV-1 cells that express HA on their plasma membranes can be induced to fuse with red blood cells (Doxsey *et al.*, 1985) or with each other (White *et al.*, 1982b) following a brief exposure to low pH. Both types of fusion assays were used to monitor the effects of substitutions of amino acids in the fusion peptide (Gething *et al.*, 1986).

The rbc : cell fusion was measured by loading erythrocyte ghosts with horseradish peroxidase and monitoring delivery of the HRP into CV-1 cells by staining fixed cells with diaminobenzidine (DAB). The results of such experiments with CV-1 cells expressing the wild-type or mutant HAs are illustrated as light micrographs of fixed and stained cells in Fig. 10 and shown graphically in Fig. 11. Cells expressing the M11 mutant HA showed a profile almost identical to that of the wild-type protein, indicating that replacing the glutamic acid at position 11 with glycine had no effect on the pH dependence or efficiency of rbc : cell fusion. By contrast, both the pH profile and efficiency of rbc : cell fusion were altered for cells expressing the M4 HA (glycine$_4$ to glutamic acid). The M1 mutation (glycine$_1$ to glutamic acid) had the striking effect of abolishing the fusion activity of HA; cells expressing the M1 mutant protein showed little or no

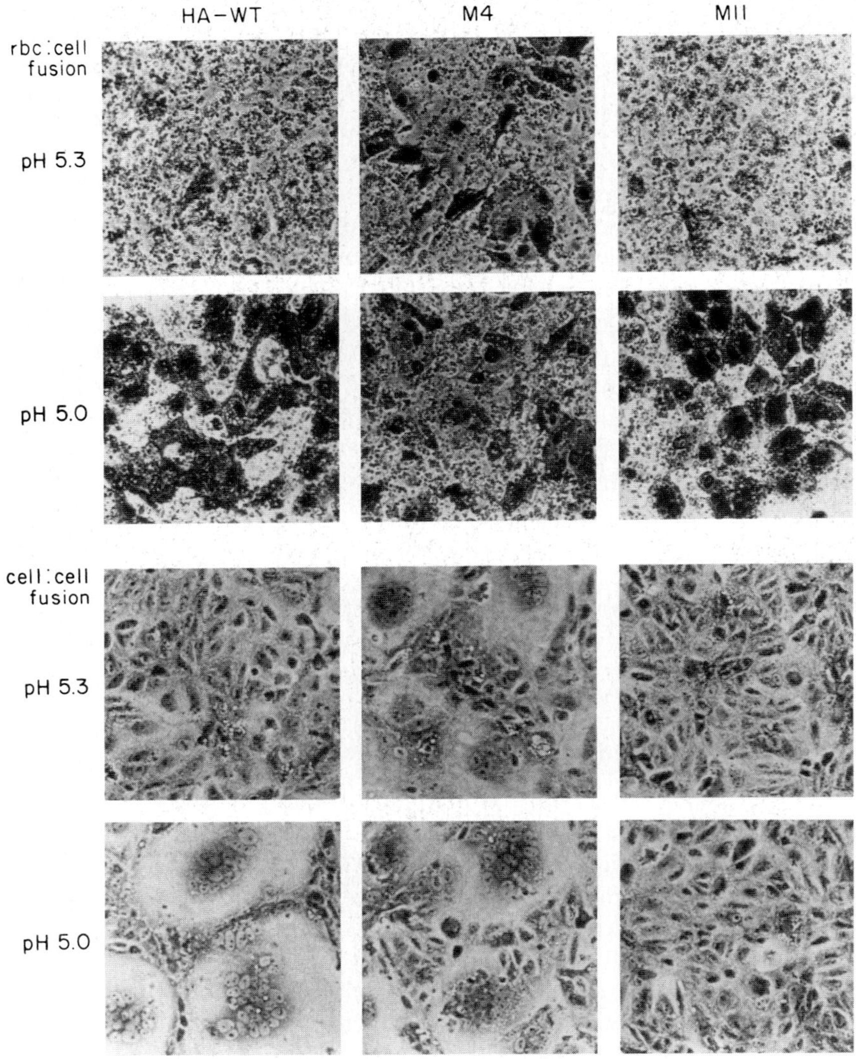

Fig. 10. Low-pH-induced rbc : cell and cell : cell fusion of CV-1 cells expressing wild-type or mutant HA glycoproteins. CV-1 cells that had been infected for 54–60 hr with SV40–HA recombinant vectors containing wild-type or mutant HA genes were treated with trypsin to cleave cell surface HA into HA1 and HA2 subunits. (Upper panels) Human erythrocytes loaded with HRP were bound to the infected cells and then the pH of the medium was transiently decreased to pH 5.3 or 5.0. After incubation in culture medium for 45 min, the cells were fixed with glutaraldehyde and stained with diaminobenzidine to identify CV-1 cells that had fused with HRP-loaded erythrocytes. (Lower panels) The infected cells were treated transiently with medium at pH 5.0 or 5.3. After incubation in culture medium for 3 hr to allow clumping of nuclei in polykaryons, the cells were fixed and photographed.

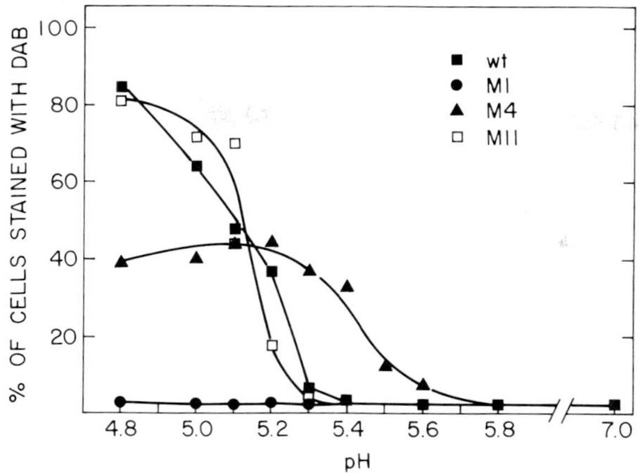

Fig. 11. pH profiles of rbc : cell fusion induced by wild-type and mutant HA proteins. CV-1 cells were infected with SV40–HA vectors containing the wild-type or mutant HA genes. At 54 hr postinfection, the cells were treated with trypsin and tested at different pH values for their ability to fuse with HRP-loaded erythrocytes. The cells were then fixed with glutaraldehyde, stained with diaminobenzidine, and scored for the extent of fusion as described in Gething *et al.* (1986).

DAB staining at any pH, even though they bound red blood cells with the same efficiency as the cells expressing wild-type HA.

The results of a parallel study using polykaryon formation as an assay for HA-mediated cell : cell fusion are also shown in Fig. 10. The results obtained with cells expression the M1 and M4 mutant HAs paralleled those seen in rbc : cell fusion assays. The M1 HA induced no polykaryon formation at any pH (data not shown), consistent with the inability of M1 to mediate HRP delivery. Cells expressing the M4 mutant fused with a higher pH threshold (pH 5.6). Approximately 50% fusion was achieved at pH 5.3, but the level of fusion was not further increased at lower pH values. Most surprisingly, the M11 HA was unable to induce polykaryon formation at or above pH 5.0, despite being able to mediate rbc : cell fusion with the same pH profile and efficiency as wild-type HA.

D. Effect of the Mutations on the Conformational Change and Lipid-Binding Properties of Hemagglutinin

The HA trimer has been shown to undergo a conformational change under the conditions that trigger membrane fusion (Skehel *et al.*, 1982;

Doms *et al.*, 1985). A water-soluble fragment of HA, termed BHA, undergoes a conformational change indistinguishable from that observed in intact HA but more convenient to study in detail (Doms *et al.*, 1985). A consequence of the low-pH-induced conformational alteration is that the BHA molecule, which is resistant to protease at neutral pH, becomes sensitive to digestion by proteinase K. The pH dependence of the protease sensitivity parallels that of the fusion activity (Doms *et al.*, 1985). A second consequence of the low-pH-induced conformational change in BHA is that the molecule acquires amphiphilic properties (Skehel *et al.*, 1982; Doms *et al.*, 1985). As a result, the fragment will bind to other amphiphiles such as nonionic detergents and artificial lipid bilayers.

All of the mutant proteins were resistant to protease at neutral pH and were capable of undergoing the conformational change, with maximal protease sensitivity achieved after incubation at pH 4.8. The conformational change in M1 BHA occurred at a lower pH than in the wild-type, since the protease sensitivity profile was shifted 0.2 pH unit in the acid direction. The conformational change in the M4 BHA occurred more readily than that in the wild-type protein; the pH profile of protease sensitivity was shifted 0.3 unit in the alkaline direction. Finally, the protease sensitivity of the M11 BHA followed the same pH profile as the wild-type protein.

The extent of binding of the various BHAs to liposomes varied with pH in a manner identical to that observed for the protease sensitivity of each molecule. The association of BHA with liposomes is hydrophobic in nature, and, once bound, the protein cannot be removed by high or low salt, chaotropic agents, or urea (Doms *et al.*, 1985). It can, however, be eluted by elevated pH (>10.5). The wild-type and mutant BHAs showed identical characteristics in their interaction with lipid bilayers. Thus we were unable to detect any differences in the nature of the interaction with liposomes of the wild-type or the mutant proteins.

The kinetics of conversion to protease sensitivity of the wild-type, M4, and M11 HA molecules closely paralleled those of acquisition of amphiphilic character. The M11 BHA showed the same kinetics as the wild type, with 50% of the maximal values reached 3 min after acidification. The M4 BHA converted to the low-pH form more rapidly. The M1 BHA was the only protein that displayed different kinetics for protease sensitivity and lipid binding. Compared to the wild-type BHA, M1 displayed a lower rate of conversion to protease sensitivity and an even lower rate of interaction with lipid, suggesting that its mode of interaction with the target bilayer may be altered from that of the wild-type HA or the other mutant proteins.

TABLE I

Summary of the Fusion Phenotypes of the Wild-Type and Mutant Hemagglutinin Proteins[a]

	WT	M1	M4	M11
Threshold pH of rbc : cell fusion	5.3	*[b]	5.7	5.3
Threshold pH of cell : cell fusion	5.3	*	5.6	4.6
Efficiency at pH 5 of rbc : cell fusion (%)	90	0	50	90
Efficiency at pH 5 of cell : cell fusion (%)	90	0	50	0
pH at which 50% of BHA is converted to protease sensitivity and lipid binding	5.3	5.0	5.5	5.3
Time at pH 4.8 for 50% conversion to protease sensitivity or lipid binding (min)	3 3	6 16	1 1	3 3

[a] These values are summarized from the detailed data reported in Gething *et al.* (1986).
[b] *, No fusion detected at any pH tested.

E. Summary of the Fusion Phenotypes of the Wild-Type and Mutant Hemagglutinin Proteins

The fusion phenotypes of the wild-type and mutant HAs are summarized in Table I. The wild-type HA protein undergoes rapid conversion below pH 5.3 to a conformation that is sensitive to degradation with protease and can interact with lipid. Concomitant with this conformational change, wild-type HA expressed on the surface of a CV-1 cell can mediate fusion with high efficiency either between the CV-1 cell and erythrocytes, or between adjacent CV-1 cells to form polykaryons.

Substitution of a glutamic acid for the glycine residue at the amino terminus of the fusion peptide abolished all the fusion activity of the HA protein. However, the M1 HA could still undergo a conformational change at low pH that resulted in protease sensitivity and lipid binding capability. The lower pH at which this conformational change occurred and the greater time taken for the conversion at pH 4.8 suggested that the mutation had in some way stabilized a neutral conformation of the protein. The M1 mutant has provided the first indictation that the conformational change can be temporarily separated from lipid binding; in all previous studies these events have appeared to occur simultaneously (Skehel *et al.*, 1982; Doms *et al.*, 1985). The M1 mutant also allows separation of the lipid binding and bilayer fusion stages of the process, indicating that HA does more than simply bring the two membranes close together. It is possible that the fusion peptide must assume a precise structure in the target membrane in order to cause sufficient destabilization to promote fusion.

Substitution of a glutamic acid for the glycine residue at position 4 of the fusion peptide (which decreases the length of the apolar stretch to six amino acids) increased the rate and raised the threshold pH at which the low-pH-induced conformational change occurred in the M4 mutant protein. Thus the mutation had destabilized the neutral conformation of HA in a similar fashion to the amino acid alterations identified in the variant viruses described above. The raised pH threshold for the conformational change was reflected in a raised pH threshold for rbc : cell or cell : cell fusion. However, the fusion mediated by this mutant HA never became as efficient as that caused by the wild-type protein; even at pH < 5.0 only 50% fusion was observed. This result suggests that the amino acid at position 4 in HA2 may play two roles—a structural role in maintaining the fusion peptide in its neutral conformation and another role in the lipid destabilization phase of fusion.

Substitution of a glycine residue for the glutamic acid residue at position 11 of HA2 (which increases the length of the hydrophobic sequence to 18 amino acids) had no effect on the rate or pH dependence of the conformational change in the M11 mutant protein. Furthermore, the mutation had no effect on the initial binding of the protein to the target lipid bilayer, suggesting that the length of the apolar sequence is not critical for this interaction. The M11 HA induced rbc : cell fusion with the same pH dependence and efficiency as wild-type HA. However, the mutant was greatly impaired in mediating cell : cell fusion. This phenotype provides the first instance of a distinction between rbc : cell fusion and cell : cell fusion. The M11 HA is competent to mediate fusion of the CV-1 and erythrocyte cell membranes over the small areas necessary for injection of HRP into the CV-1 cytoplasm, but except at very low pH it appears to be unable to cause lipid bilayer destabilization over areas sufficient to allow polykaryon formation.

VI. CONCLUSIONS

Analyses of variant HAs that we (Doms *et al.*, 1986) and others (Rott *et al.*, 1984; Daniels *et al.*, 1985) have described and of mutant HAs with

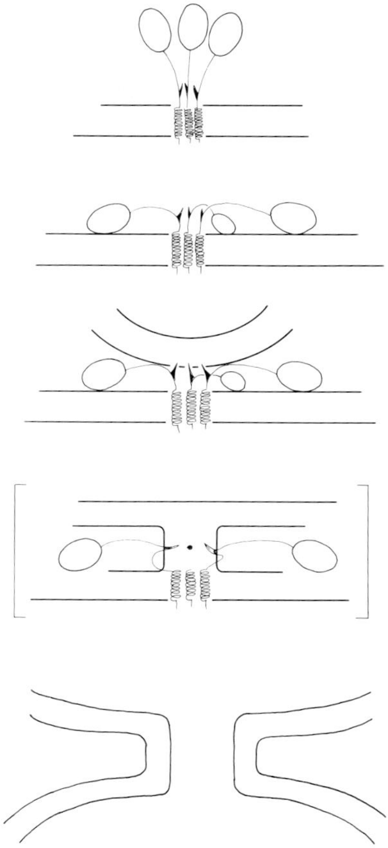

Fig. 12. Possible stages of the mechanism of HA-mediated membrane fusion.

site-directed alterations in the fusion peptide (Gething *et al.*, 1986) have allowed us to delineate several stages of the mechanism of HA-mediated membrane fusion. These stages are illustrated in Fig. 12 and include (1) the low-pH-induced conformational change in HA that exposes the fusion peptide, (2) the interaction of the fusion peptide and/or other regions of the HA molecule with the target lipid bilayer, (3) destabilization of the lipid bilayer and membrane coalescence over small areas, and (4) membrane coalescence and separation over large areas resulting in polykaryon formation.

Although this study has provided insights into the sequence of events that lead to fusion, precise molecular details of the role of HA in the fusion mechanism remain to be elucidated. We hope that future experiments that combine the power of recombinant DNA techniques with the elegance of the model system provided by influenza HA will reveal further details of the molecular mechanism of HA-induced fusion.

ACKNOWLEDGMENTS

We thank Don Wiley, Ian Wilson, and Joe Sambrook for helpful discussions. The work at Cold Spring Harbor was supported by grant NIH AI19630 to M.J.G. and at Yale by grant NIH AI18582 to A.H. J.W. was supported by a Swebelius Cancer Research Award.

REFERENCES

Both, G. W., Sleigh, M. J., Cox, N. J., and Kendal, A. P. (1983). *J. Virol.* **48,** 52–60.

Brand, C. M., and Skehel, J. J. (1972). *Nature (London) New Biol.* **238,** 145–147.

Daniels, R. S., Douglas, A. R., Skehel, J. J., Waterfield, M. D., Wilson, I. A., and Wiley, D. C.(1983). *In* "The Origin of Pandemic Influenza Viruses" (W. G. Laver, ed.), pp. 1–7. Am. Elsevier, New York.

Daniels, R. S., Downie, J. C., Hay, A. J., Knossow, M., Skehel, J. J., Wang, M. L., and Wiley, D. C. (1985). *Cell* **40,** 431–439.

DeDuve, C., DeBarsy, T, Poole, B, Trouet, A, Tulkens, P., and Van Hoof, F. (1974). *Biochem. Pharmacol.* **23,** 2495–2531.

Doms, R. W., Helenius, A. H., and White, J. (1985). *J. Biol. Chem.* **260,** 2973–2981.

Doms, R. W., Gething, M. J., Henneberry, J., White, J., and Helenius, A. (1986). *J. Virol.* **57,** 603–613.

Doxsey, S. J., Sambrook, J., Helenius, A., and White, J. (1985). *J. Cell Biol.* **101,** 19–27.

Doyle, C., Roth, M. J., Sambrook, J., and Gething, M.-J. (1985). *J. Cell Biol.* **100,** 704–714.

Garten, W., Bosch, F.-X., Linder, D., Rott, R., and Klenk, H.-D. (1981). *Virology,* **115,** 361–374.

Gething, M.-J., and Sambrook, J. (1981). *Nature (London)* **293,** 620–625.

Gething, M. J., White, J. M., and Waterfield, M. D. (1978). *Proc. Natl. Acad. Sci. U.S.A.* **75,** 2737–2740.

Gething, M. J., Bye, J., Skehel, J. J., and Waterfield, M. D. (1980). *Nature (London)* **287,** 301–306.

Gething, M.-J., Doms, R. W., York, D., and White, J. (1986). *J. Cell Biol.* **102,** 11–23.

Helenius, A., Mellman, I., Wall, D., and Hubbard, A. (1983). *Trends Biochem. Sci.* **8,** 245–250.

Hirst, G. K. (1941). *Science* **94,** 22–23.

Huang, R. T. C., Rott, R., and Klenk, H.-D. (1981). *Virology* **110,** 243–247.

Jensen, E. M., and Liu, O. C. (1963). *Proc. Soc. Exp. Biol. Med.* **112,** 456–459.

Klenk, H.-D., Rott, R. Orlich, M., and Blodorn, J. (1975). *Virology* **68,** 426–439.

Lamb, R. A. (1983). *In* "Genetics of Influenza Viruses" (P. Palese and D. W. Kingsbury, eds.), pp. 21–69. Springer-Verlag, Berlin and New York.

Laver, W. G. (1971). *Virology* **45,** 275–288.

Lazarowitz, S., and Choppin, P. W. (1975). *Virology* **68,** 440–454.

Maeda, T., and Onishi, S. (1980). *FEBS Lett.* **122,** 283–287.

Marsh, M. (1984). *Biochem. J.* **218,** 1–10.

Matlin, K., Reggio, H., Helenius, A., and Simons, K. (1981). *J. Cell Biol.* **91,** 601–613.

Richardson, C., Scheid, A., and Choppin, P. (1980). *Virology* **105,** 205–222.

Rott, R., Orlich, M., Klenk, H.-D., Want, M. L., Skehel, J. J., and Wiley, D. C. (1984). *EMBO J.* **3,** 3329–3332.

Sambrook, J., Rodgers, L., White, J., and Gething, J. J. (1985). *EMBO J.* **4,** 91–103.

Sanger, F., Nicklen, S., and Coulson, A. R. (1977). *Proc. Natl. Acad. Sci. U.S.A.* **74,** 5436–5467.

Sanger, F., Coulson, A. R., Barrell, B. G., Smith, A. J. H., and Roe, B. A. (1980). *J. Mol. Biol.* **143,** 161–178.

Seeburg, P. H., Colby, W. W., Capon, D. J., Goeddel, D. V., and Levinson, A. D. (1984). *Nature (London)* **312,** 71–75.

Skehel, J. J., Bayley, P. M., Brown, E. B., Martin, S. R., Waterfield, M. D., White, J. M., Wilson, I. A., and Wiley, D. C. (1982). *Proc. Natl. Acad. Sci. U.S.A.* **79,** 968–972.

VanRompuy, L., Jou, W. M., Verhoeyen, M., Huylebroeck, D., and Fiers, W. (1983). *Trends Biochem. Sci.* **8,** 414–417.

Verhoeyen, M., Fang, R., Jou, W. M., Devos, R., Huylebroeck, D., Saman, E., and Fiers, W. (1980). *Nature (London)* **286,** 771–776.

White, J., and Helenius, A. (1980). *Proc. Natl. Acad. Sci. U.S.A.* **77,** 3273–3277.

White, J., Matlin, K., and Helenius, A. (1981). *J. Cell Biol.* **89,** 674–679.

White, J., Kartenbeck, J., and Helenius, A. (1982a). *EMBO J.* **1,** 217–222.

White, J., Helenius, A., and Gething, M. J. (1982b). *Nature (London)* **300,** 658–659.

White, J., Kielian, M., and Helenius, A. (1983). *Q. Rev. Biophys.* **16,** 151–195.

Wiley, D. C., Wilson, I. A., and Skehel, J. J. (1981). *Nature (London)* **289,** 373–378.

Wilson, I. A., Skehel, J. J., and Wiley, D. C. (1981). *Nature (London)* **289,** 366-373.

Yamada, A., Brown, L. E., and Webster, R. G. (1984). *Virology* **138,** 276–286.

Zoller, M. J., and Smith M. (1984). *DNA* **3,** 479–488.

13

Expression and Site-Specific Mutagenesis of an Integral Membrane Protein, Bacterio-Opsin

SADASHIVA S. KARNIK,* MARK S. BRAIMAN,* BETTY H. CHAO,* ROBERT DUNN,* MARIE A. GILLES-GONZALEZ,* NEIL R. HACKETT,* ERNEST JAY,† K. ANNE KRONIS,* MICHAEL NASSAL,* AND H. GOBIND KHORANA*

** Departments of Biology and Chemistry*
Massachusetts Institute of Technology
Cambridge, Massachusetts 02139
and
† Department of Chemistry
University of New Brunswick
Frederickton, New Brunswick, Canada E3B 6E2

I. INTRODUCTION

Halobacterium halobium, an archaebacterium, synthesizes three distinct light-transducing membrane proteins. In all of these the chromophore consists of retinaldehyde linked to a lysine as a Schiff base (Stoeckenius and Bogomolni, 1982). Halorhodopsin, a chloride ion pump, and the phototactic sensory rhodopsin are relatively minor components of this extreme halophile, while bacteriorhodopsin (BR) is a major integral membrane protein. The latter appears in a specialized membrane, the purple membrane, which is found as two-dimensional crystalline patches in the cytoplasmic membrane. The BR carries out light-dependent translocation of protons, generating an electrochemical gradient which is used by the

PROTEIN ENGINEERING
Applications in Science,
Medicine, and Industry

organism to synthesize ATP. This process, which provides a source of energy alternative to respiration, is especially active under anaerobic conditions and in light.

Bacteriorhodopsin consists of a single polypeptide chain of 248 amino acids (M_r 26,000) whose sequence has been determined (Khorana *et al.*, 1979; Ovchinnikov *et al.*, 1979). Based on amino acid sequence and electron diffraction and protein cross-linking data, a secondary structure model has been proposed (Engelman *et al.*, 1980; Huang *et al.*, 1982) which is shown in Fig. 1. All-*trans*-retinal is linked to the ε-amino group of Lys 216 as a Schiff base (Bayley *et al.*, 1981). On illumination, the protein displays a multistep photochemical cycle during which a proton is translocated across the membrane. The first and only light-dependent step involves isomerization of all-*trans*-retinal in light-adapted BR (BR570) to the 13-*cis* form (M412) (Stoeckenius and Bogomolni, 1982).

Because bacteriorhodopsin is a relatively small integral membrane protein that carries out an important biological function, it is very attractive for structure–function studies. We are interested in studying (a) interactions between the retinal chromophore and the amino acid side chains in order to understand the photochemical cycle; (b) the mechanism of proton translocation, and (c) the interactions that stabilize the helices within the membrane.

Studies on bacterio-opsin protein (BO) have revealed additional attractive features. The protein can be completely freed from the endogenous lipids and can be reconstituted into phospholipid vesicles with full proton pumping activity. Indeed, a variety of phospholipids, synthetic and endogenous, may be used for reconstitution (Huang *et al.*, 1980). Further studies have shown that BR can be completely denatured and subsequently can refold quantitatively to the correct three-dimensional structure. Refolded BO regenerates the chromophore and shows full proton pumping activity (Huang *et al.*, 1981; London and Khorana, 1982; Liao *et al.*, 1983).

For structure–function studies, a specific and versatile approach is to replace a selected amino acid in the protein with another. This is best carried out by making predetermined mutations in the gene, expressing the mutated genes under nonselective conditions, and analyzing the function of purified protein *in vitro*. With this aim, we have cloned the gene for BO and developed suitable vectors to express the gene in *Escherichia coli*. Procedures have been developed for the purification of the small amounts of protein produced in *E. coli*. We describe the production of BO mutant proteins and their phenotypes assayed by the approach described above, and we discuss the usefulness of this approach for studying integral membrane proteins in general.

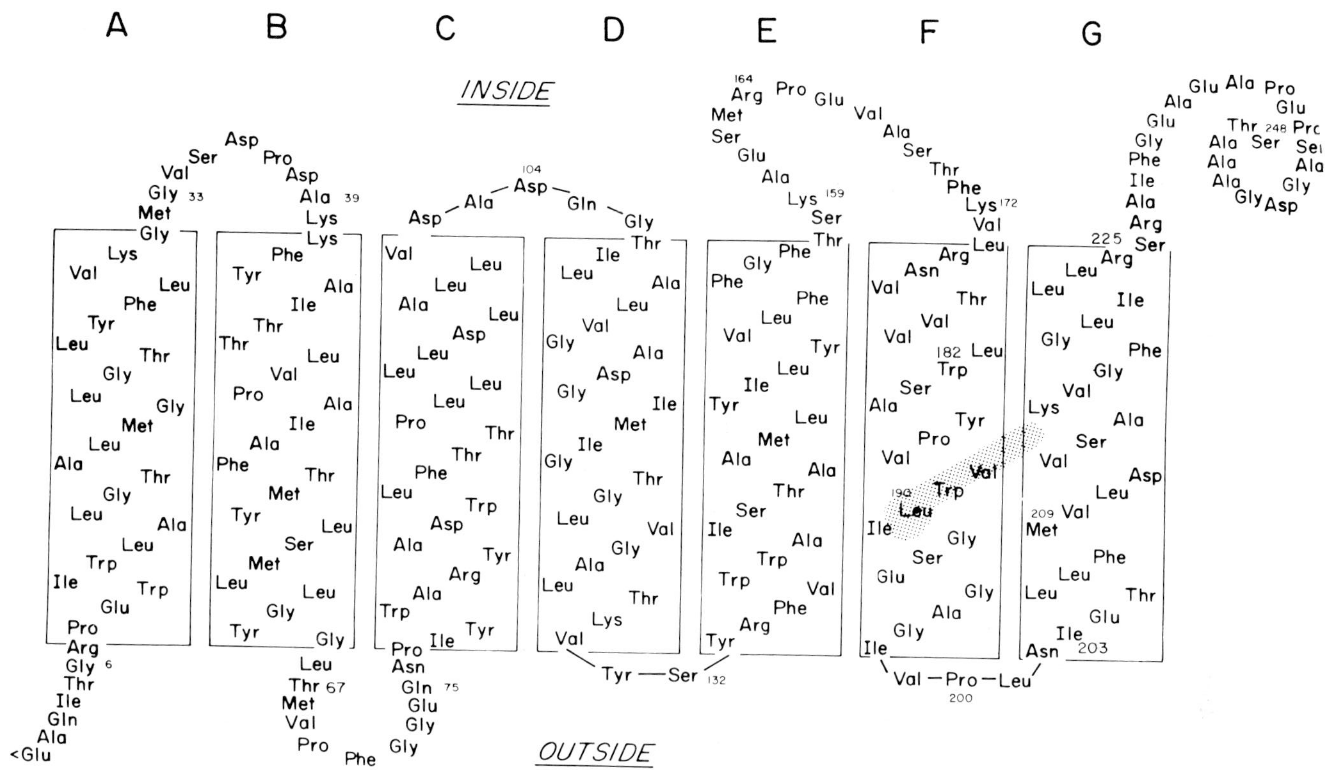

Fig. 1. Proposed arrangement of the polypeptide chain of bacteriorhodopsin across the bilayer. The shaded area respresents the location and orientation of retinal in the protein. The sizes of helices (A–G) and that of loops are uncertain.

```
GGGTGCAACCGTGAAGTCCGCCACGACCGCGTCACGACAGGAGCCGACCAGCGACACCCAGAAGGTGCGAACGGTTGAGTGCCGCAACGA -271
-360

TCACGAGTTTTTCGTGCGCTTCGAGTGGTAACACGCGTGCACGCATCGACTTCACCGCGGGTGTTTCGACGCCAGCCGGCCGTTGAACCA -181
-270

GCAGGCAGCGGGCATTTACAGCCGCTGTGGCCCAAATGGTGGGGTGCGCTATTTTGGTATGGTTTGGAATCCGCGTGTCGGCTCCGTGTC -91
-180

TGACGGTTCATCGGTTCTAAATTCCGTCACGAGCGTACCATACTGATTGGGTCGTAGAGTTACACACATATCCTCGTTAGGTACTGTTGC -1
-90

Met Leu Glu Leu Leu Pro Thr Ala Val Glu Gly Val Ser ↓Gln Ala Gln Ile Thr Gly Arg Pro Glu Trp Ile Trp Leu Ala Leu Gly Thr
ATGTTGGAGTTATTGCCAACAGCAGTGGAGGGGGTATCGCAGGCCCAGATCACCGGACGTCCGGAGTGGATCTGGCTAGCGCTCGGTACG 90
1

Ala Leu Met Gly Leu Gly Thr Leu Tyr Phe Leu Val Lys Gly Met Gly Val Ser Asp Pro Asp Ala Lys Lys Phe Tyr Ala Ile Thr Thr
GCGCTAATGGGACTCGGGACGCTCTATTTCCTCGTGAAAGGGATGGGCGTCTCGGACCCAGATGCAAAGAAATTCTACGCCATCACGACG 180
91

Leu Val Pro Ala Ile Ala Phe Thr Met Tyr Leu Ser Met Leu Leu Gly Tyr Gly Leu Thr Met Val Pro Phe Gly Gly Glu Gln Asn Pro
CTCGTCCCAGCCATCGCGTTCACGATGTACCTCTCGATGCTGCTGGGGTATGGCCTCACAATGGTACCGTTCGGTGGGGAGCAGAACCCC 270
181

Ile Tyr Trp Ala Arg Tyr Ala Asp Trp Leu Phe Thr Thr Pro Leu Leu Leu Leu Asp Leu Ala Leu Leu Val Asp Ala Asp Gln Gly Thr
ATCTACTGGGCGCGGTACGCTGACTGGCTGTTCACCACGCCGCTGTTGTTGTTAGACCTCGCGTTGCTCGTTGACGCGGATCAGGGAACG 360
271

Ile Leu Ala Leu Val Gly Ala Asp Gly Ile Met Ile Gly Thr Gly Leu Val Gly Ala Leu Thr Lys Val Tyr Ser Tyr Arg Phe Val Trp
ATCCTTGCGCTCGTCGGTGCCGACGGCATCATGATCGGGACCGGCCTGGTCGGCGCACTGACGAAGGTCTACTCGTACCGCTTCGTGTGG 450
361

Trp Ala Ile Ser Thr Ala Ala Met Leu Tyr Ile Leu Tyr Val Leu Phe Phe Gly Phe Thr Ser Lys Ala Glu Ser Met Arg Pro Glu Val
TGGGCGATCAGCACCGCAGCGATGCTGTACATCCTGTACGTGCTGTTCTTCGGGTTCACCTCGAAGGCCGAAAGCATGCGCCCCGAGGTC 540
451

Ala Ser Thr Phe Lys Val Leu Arg Asn Val Thr Val Val Leu Trp Ser Ala Tyr Pro Val Val Trp Leu Ile Gly Ser Glu Gly Ala Gly
GCATCCACGTTCAAAGTACTGCGTAACGTTACCGTTGTGTTGTGGTCCGCGTATCCCGTCGTGTGGCTGATCGGCAGCGAAGGTGCGGGA 630
541

Ile Val Pro Leu Asn Ile Glu Thr Leu Leu Phe Met Val Leu Asp Val Ser Ala Lys Val Gly Phe Gly Leu Ile Leu Leu Arg Ser Arg
ATCGTGCCGCTGAACATCGAGACGCTGCTGTTCATGGTGCTTGACGTGAGCGCGAAGGTCGGCTTCGGGCTCATCCTCCTGCGCAGTCGT 720
631

Ala Ile Phe Gly Glu Ala Glu Ala Pro Glu Pro Ser Ala Gly Asp Gly Ala Ala Ala Thr Ser ↓Asp ***
GCGATCTTCGGCGAAGCCGAAGCGCCGGAGCCGTCCGCCGGCGACGGCGCGGCCGCGACCAGCGACTGATCGCACACGCAGGACAGCCCC 810
721

ACAACCGGCGCGGCTGTGTTCAACGACACACGATGAGTCCCCCACTCGGTCTTGTACTC
811                                                       869
```

Fig. 2. Nucleotide sequence of *H. halobium* DNA containing bacterio-opsin gene. The amino acid sequence of BO precursor is shown. The arrows indicate the sites of processing to generate the mature protein.

II. CLONING AND EXPRESSION OF BACTERIO-OPSIN GENE

A. Bacterio-Opsin *(bop)* Gene

As a first step in developing the recombinant DNA approach we have characterized the gene for bacterio-opsin (*bop*) from *H. halobium* (Dunn *et al.*, 1981). The gene has been identified on a 5.3-kilobase (kb) *Pst*I fragment of *H. halobium* DNA. The sequence of the structural gene and neighboring region has been determined (Fig. 2). The amino acid sequence of the protein deduced from the gene sequence is in complete agreement with that obtained by protein sequencing (Khorana *et al.*,

1979). In addition, the DNA sequence shows the presence of an unusual precursor sequence of 13 amino acids at the N-terminus and an additional aspartate at the carboxyl end. The precursor sequence must be cleaved to give the amino terminal pyroglutamic acid in a cotranslational fashion (Seehra and Khorana, 1984). No information is available on the processing of the carboxy-terminal Asp residue.

B. Bacterio-Opsin Expression Plasmids

1. Vectors Encoding BO with N-Terminal Peptide Extensions

The plasmid pEN-22 (Fig. 3A) contains the bacterio-opsin coding sequence starting at an *Eco*RI site introduced at the amino terminus of the mature protein and about 300 nucleotides of *H. halobium* DNA downstream from the termination codon. The expression vectors constructed by Nakamura *et al.* (1982) were specifically designed (Fig. 3B) for expression of secreted proteins in *E. coli*. In the pIN-II series of vectors, the *lac* UV5 promoter–operator fragment has been introduced immediately after the constitutive lipoprotein (*lpp*) promoter. Thus, the system is inducible and is under the control of both *lpp* and *lac* promoters (Fig. 3B). The Shine–Delgarno sequence and initiator codon are derived from the *lpp* gene. Bacterio-opsin expression plasmids were prepared by insertion of an *Eco*RI–*Bam*HI fragment from pEN-22 containing the *bop* gene into pIN-IIA, pIN-IIB, and pIN-IIC. The resulting plasmids, pLBA, pLBB, and pLBC, encode the BO derivatives with extensions at the amino terminus as shown in Fig. 3C. The gene from plasmid pLBA encodes six additional amino acids. Plasmid pLBB-encoded protein has a total of 24 additional amino acids at the amino terminus of bacterio-opsin. The pLBC plasmid specifies the complete lipoprotein signal sequence and eight amino acids of mature lipoprotein.

Figure 4A shows the relative amounts of bacterio-opsin fusion proteins produced from the three expression vectors. The amounts of proteins expressed in *E. coli* have been very small (0.02–0.05% of total membrane protein). Plasmid pLBC appeared to produce more protein per cell, but induction of this protein was toxic to cells. Expression vector pLBB was used in all further work.

2. Vectors Encoding BO without Amino Acid Extensions

For structure–function studies the protein produced in *E. coli* should have minimal differences from the native BO. The protein with the 24-amino-acid amino-terminal extension specified by pLBB is not desirable

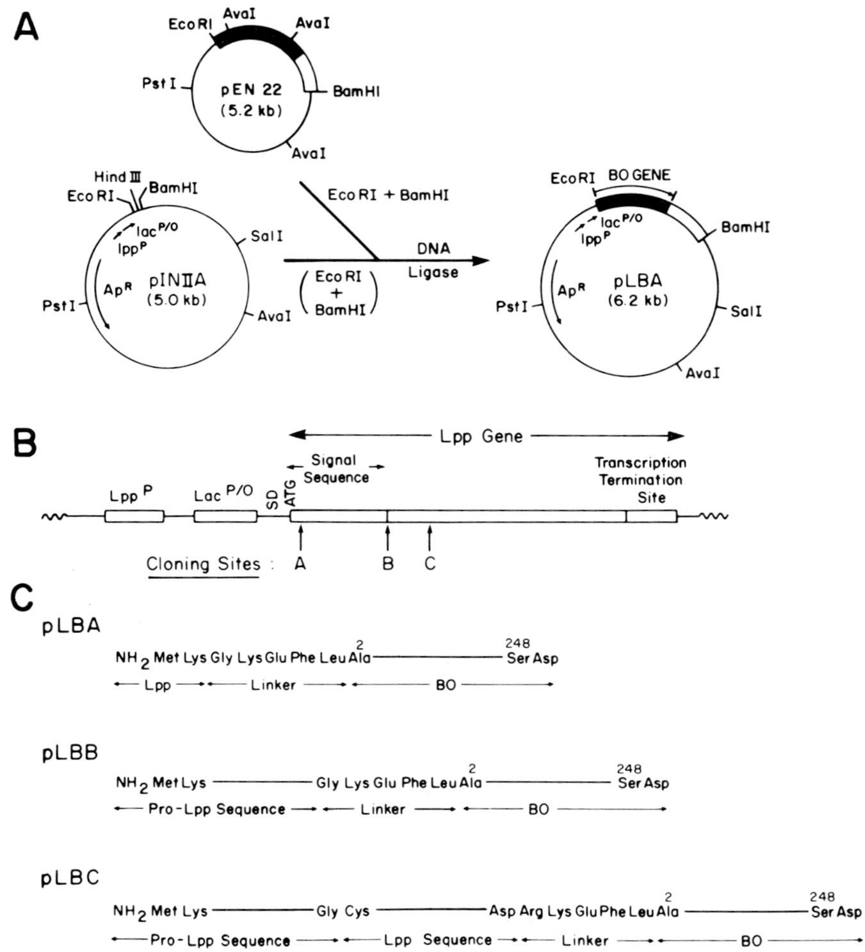

Fig. 3. Expression vectors encoding bacterio-opsin fusion proteins. The plasmid pEN-22 containing the B . gene without precursor sequence and an *Eco*RI restriction site at the amino terminus was constructed (Dunn *et al.*, 1983). The vectors pIN-IIA, pIN-IIB, and pIN-IIC were obtained from Dr. M. Inouye. The BO gene from pEN-22 was inserted into pIN-IIA to form the plasmid pLBA (A). Plasmids pLBB and pLBC were produced by using pIN-IIB and pIN-IIC, respectively (B). The N-terminal extensions expected following the expression of BO gene in the three plasmids are shown (C).

(see below). Therefore, in developing new vectors the aim was to express BO without additional amino acids.

The *bop* gene was modified to code for the mature protein. At the 3′ end a DNA duplex coding for Ala-Ala-Thr-Ser followed by TAA and TGA

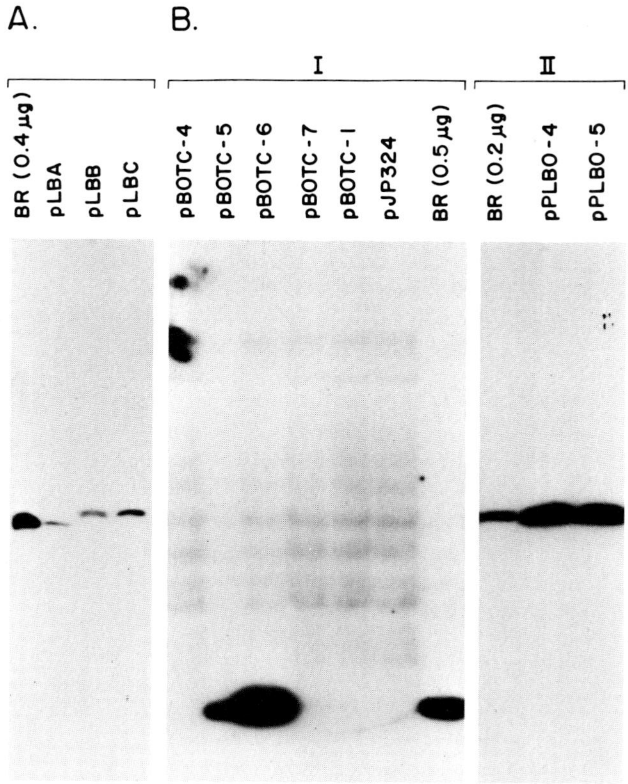

Fig. 4. Immunological characterization of bacterio-opsin formed *in vivo* from different expression vectors. Cells harboring the plasmids designated were grown and lysed. The lysates were fractionated by electrophoresis. The separated proteins were transferred to nitrocellulose sheets and reacted with monoclonal anti-BR-antibodies (Kimura *et al.*, 1982). The antigen–antibody complex was visualized by reaction with ^{125}I-labeled protein A and autoradiography. The BO-related fusion proteins (A) and BO with N-terminal methionine produced from constitutive (BI) and inducible (BII) expression vectors are shown.

codons was used to replace a restriction fragment to eliminate the C-terminal Asp codon. The duplex SK-1 (Fig. 5A) was introduced at the 5′ end of the gene. The modified BO gene now included the entire coding sequence between *Hin*dIII and *Bam*HI sites as in pBOTC (Fig. 5A).

The expression vector we used in this study contains a synthetic P25 promoter of phage T5 (Jay *et al.*, 1984) and a synthetic ribosome binding site which is optimized for high-level expression of heterologous genes in *E. coli.* The bacterio-opsin gene was introduced between the synthetic ribosome binding site and the *tet* gene derived from pBR322 (Fig. 5A).

A

Plasmid clone	Oligo	Sequence	Restriction sites	Intent	Local GC content	Secondary structure potential	Extent
BOTC-1	SK-1	met gln ala gln ile thr gly arg AGCTT ATG CAG GCC CAG ATC ACC GGA CGT A TAC GTC CGG GTG TAG TGG CC	HindIII-AatII	Wild Type	28/60	—	–
BOTC-4	SK-4	agctt atg caA gcT caA atT acT gga cgt a tac gtT cgA gtT taA tgA cc	HindIII-AatII	Third position of the codons changed	23/60	Moderate	–
BOTC-5	SK-5	agctt utg caA gcA caA atT acA gga cgt a tac gtT cgT ctT taA tgT cc	HindIII-AatII	"	23/60	Moderate	+
BOTC-6	SK-6	agctt atg GCT caA atT acT gga cgt a tac CGA gtT taA tgA cc	HindIII-AatII	Gln at position 2 deleted	21/60	Good	+++
BOTC-7	SK-7	agctt atg AAA gcT caA atT acT gga cgt a tgc TTT cgA gtT taA tgA cc	HindIII-AatII	Lys at position 2 for Gln	21/60	Better	–

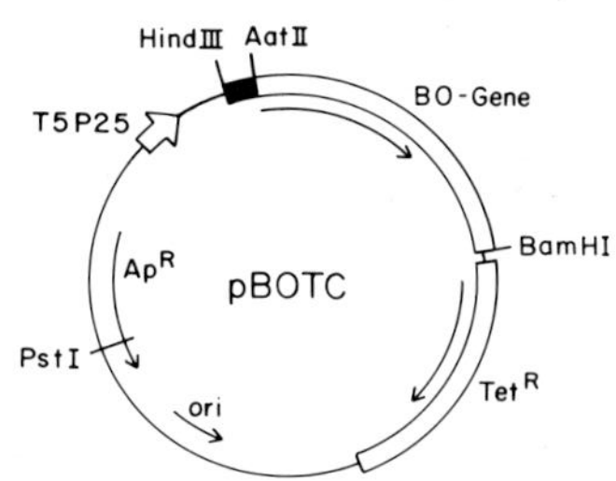

Fig. 5. Vectors encoding mature BO protein. The expression vector pJP324 was obtained from Dr. E. Jay (Jay *et al.*, 1984) and the BO gene was introduced into it as a *Hin*dIII–*Bam*HI fragment. Nucleotide changes introduced to improve the translation initiation were accomplished by replacing the *Hin*dIII–*Aat*II fragment on the plasmid with the oligonucleotides shown (A). The inducible vector was obtained from Remaut *et al.* (1981). The BO gene with synthetic ribosome binding site was introduced as an *Eco*RI–*Bam*HI fragment about 150 nucleotides downstream of the promoter (B).

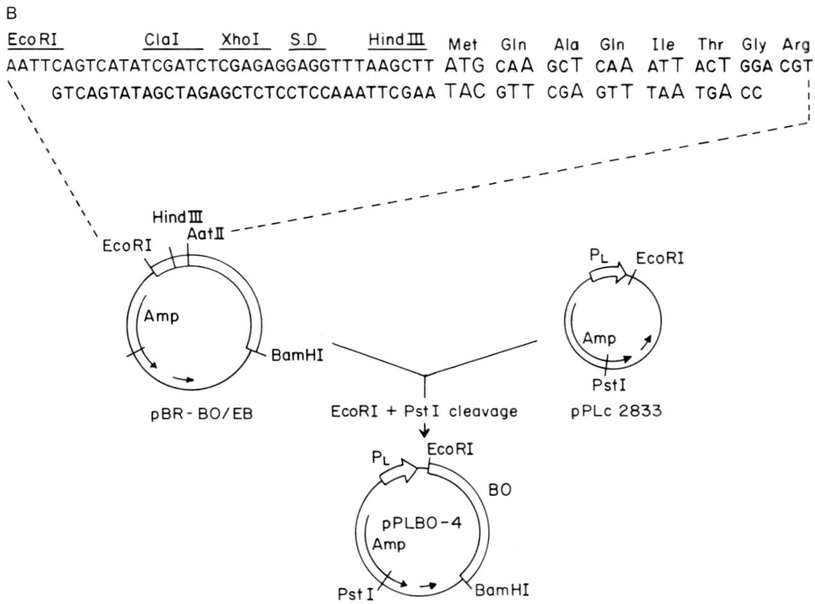

Fig. 5. Continued

The *tet* gene expression requires transcription initiation at T5P25 promoter in this vector, and hence selection for *tet* gene function could be used to maintain the promoter activity.

The *bop* gene without any sequence changes (pBOTC-1) did not produce detectable amounts of protein. The same gene cloned in front of two combinations of promoter–ribosome binding site showed that the gene is transcribed and the mRNA is stable (data not shown). However, no protein was detected in the cell extracts. The block in expression appeared to be poor initiation of translation. Therefore, attempts were made to improve the nucleotide sequence context of the translation initiation site.

Synthetic DNA duplexes carrying codon alterations (Fig. 5A) were introduced between the *Hin*d III site and the *Aat*II site. Plasmid clones pBOTC-5 and pBOTC-6 express BO constitutively (Fig. 4B,I). Analysis of the secondary structure of mRNA produced in different constructions shows that a stem–loop structure could form with different stabilities in all the clones. The transcript of *bop* gene in pBOTC-1 does not show a potential to fold similar to that of transcripts of other clones, hence the folded structure of mRNA appears to be important for translational efficiency.

Vectors which constitutively expressed bacterio-opsin could be main-

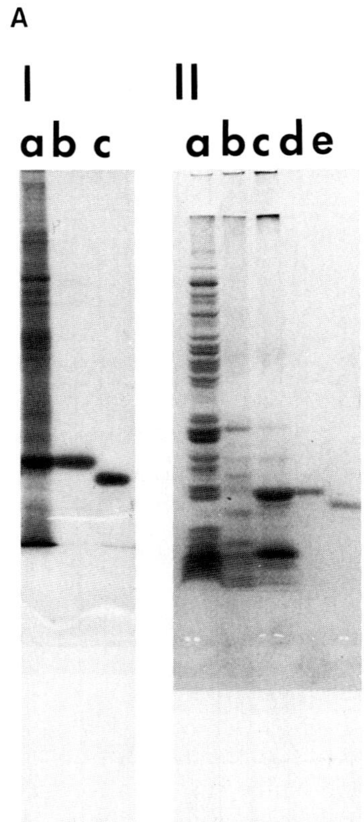

Fig. 6. Purification profile and chromophore regeneration of *E. coli*-produced BO fusion proteins. The *E. coli* cells producing BO were lysed by French press and the crude membrane fraction (A,II,a) was separated by centrifugation through sucrose. The membrane fraction was washed several times with urea and reacted with antiBR-antibody linked to Sepharose in the presence of Triton X-100. The bound protein was released by acid wash (A,I,a). Alternatively, the crude membrane pellet was extracted with a chloroform–methanol–triethylamine mixture (A,II,e). The crude BO obtained by both procedures was further purified by preparative gel electrophoresis (A,I,b and A,II,d) and dissolved in SDS (1%). Proteins remaining in the organic phase after displacement of BO (A,II,b) and BR standards are included for comparison (A,I,c and A,II,e). The protein in SDS was mixed with DMPC, CHAPS, NaPi, buffer, and retinal to regenerate the chromophore (B). The regeneration yield was calculated from protein concentration (A_{280}) and molar extinction coefficient of BR548 (London and Khorana, 1982).

tained for about 50 generations, after which nonexpressing variants outgrew them. This problem was solved by using an inducible promoter system. The ribosome binding sequence and *bop* gene from pBOTC-4 and pBOTC-5 were transferred to pPLC28 (Remaut *et al.*, 1981), which con-

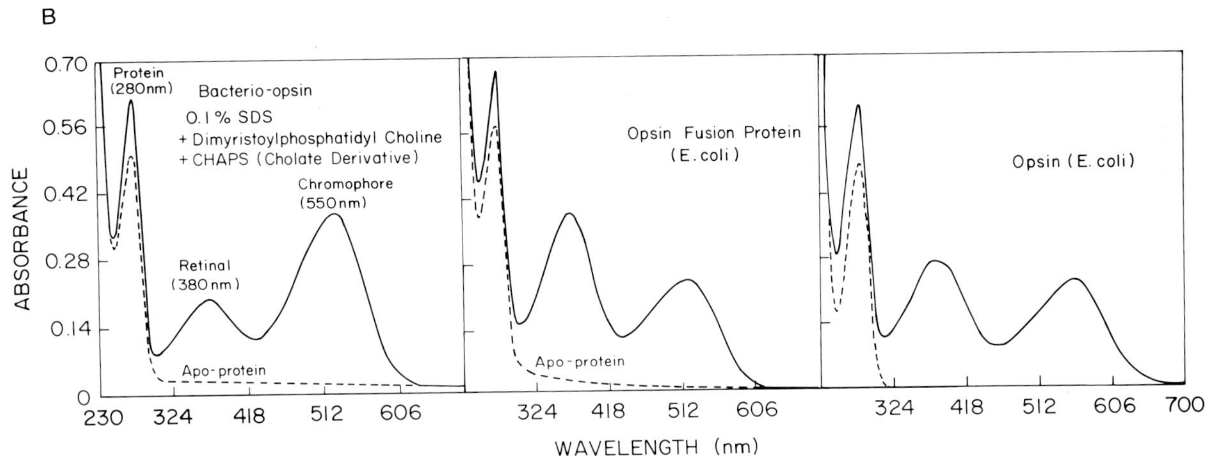

Fig. 6. Continued

tains the leftward promoter of phage λ (Fig. 5B). Both sequences are translated efficiently to yield BO, which constitutes approximately 0.8–1.0% of total cellular protein (Fig. 4B,II).

III. PURIFICATION AND RECONSTITUTION OF *E. Coli*-PRODUCED BACTERIO-OPSIN

Bacterio-opsin produced in *E. coli* is predominantly "associated" with membranes. Initial methods of purification involved a high-salt–urea wash of membrane fraction and affinity purification of Triton-solubilized membrane protein on a monoclonal antibody column (Kimura *et al.*, 1982). More than 80% pure protein (Fig. 6A,I) was obtained. However, the yield of the protein was relatively low with this method. An alternative purification procedure which is rapid and efficient has been developed. This involves enrichment of *E. coli*-produced BO by extraction of the membranes with organic solvents followed by preparative gel electrophoresis or antibody column purification (Fig. 6A,II).

Bacterio-opsin purified from *E. coli* membrane has been reconstituted from sodium dodecyl sulfate (SDS) into mixed micelles (Fig. 6B) by procedures described earlier (Liao *et al.*, 1983) to generate a native chromophore (550 nm) capable of light/dark adaptation. The presence of the 24-amino-acid amino-terminal extension does not affect the spectrum, but the extent of regeneration is only about 50%. The rate of folding is lower than that of BO from purple membrane. Bacterio-opsin without the N-terminal extension regenerates the chromophore to about 65% of theoretical yield, and the rate of regeneration is comparable to that of the protein purified from purple membrane. Bacterio-opsin mutants will be recloned in this system for further work.

IV. MUTAGENESIS OF BACTERIO-OPSIN GENE

Selection of bacteriorhodopsin point mutants *in vivo* is hindered by a high frequency of spontaneous mutations caused by insertion elements (DasSarma *et al.*, 1983). In addition, there is no genetic transformation system available as yet in *H. halobium*. This rules out random mutagenesis as a possibility for defining functional domains. However, the available information on the structure of the protein and current postulates on the mechanism of proton translocation allow reasonable predictions of amino acid residues that could be evaluated in the function of the protein. Hence, we have chosen site-directed mutagenesis and expression of *bop* gene in *E. coli,* where the protein does not confer a selectable phenotype on the host. Four different methods of mutagenesis have been used. The details of the procedures have been described elsewhere (Gilles-Gonzales *et al.*, 1986) and only a summary of the results is presented here (Fig. 7).

1. Amber mutations in the bacterio-opsin gene were produced (McCoy and Khorana, 1983) by passage through an *E. coli* mutator strain. The mutants (Fig. 6) were selected by the phenotypic reversion of mutation in an amber suppressor host. Precise localization of mutations was done first by sizing amber-terminated peptides in an *in vitro* transcription–translation system, and second by DNA sequencing.
2. Short internal deletions were introduced at predetermined sites by restriction endonuclease, exonuclease III, and S1 digestions. This approach was intended to test the structural models and to investigate the sizes of loops in the aqueous environment and the domains in the lipid bilayer.
3. Oligonucleotide-directed point mutagenesis by mismatched oligonucleotides, as described by Dalbadie-McFarland *et al.* (1982), was used for the production of a single nucleotide change (A to C) in the *bop* gene, which changes Ile 203 to Leu (Lo *et al.*, 1984).
4. Replacement of restriction fragments by mutagenic synthetic DNA duplex (Lo *et al.*, 1984) was used extensively. Restriction fragments in the region of interest are excised from the gene and replaced by totally synthetic counterparts. The majority of the point mutants were created by this procedure. A number of mutants were generated in the portion of the *bop* gene coding for helices F and G of the protein. These mutations are desirable for studying the retinal binding pocket and interaction of the chromophore with the protein.

V. PHENOTYPES OF BACTERIORHODOPSIN MUTANTS

Phenotypes of the mutant proteins have been determined by (1) the ability of the BO mutant to regenerate BR-like chromophore, and (2) absorption spectral characteristics of the chromophore generated. Three classes of mutants have been observed (Fig. 7). Class I mutants regenerate the chromophore and, in addition, show no effect on the absorption spectrum, e.g., Tyr 185 → Phe, Ser 193 → Ala, and Phe 219 → Leu. Class II mutants also regenerate the chromophore but show an absorption spectral shift, e.g., Trp 189 → Phe. These residues appear to interact with the retinal. It will be interesting to see the effects of these two groups of mutants on proton translocation. Class III mutants fail to regenerate the chromophore, e.g., Lys 216 → Val and Pro 186 → Leu. Structural analy-

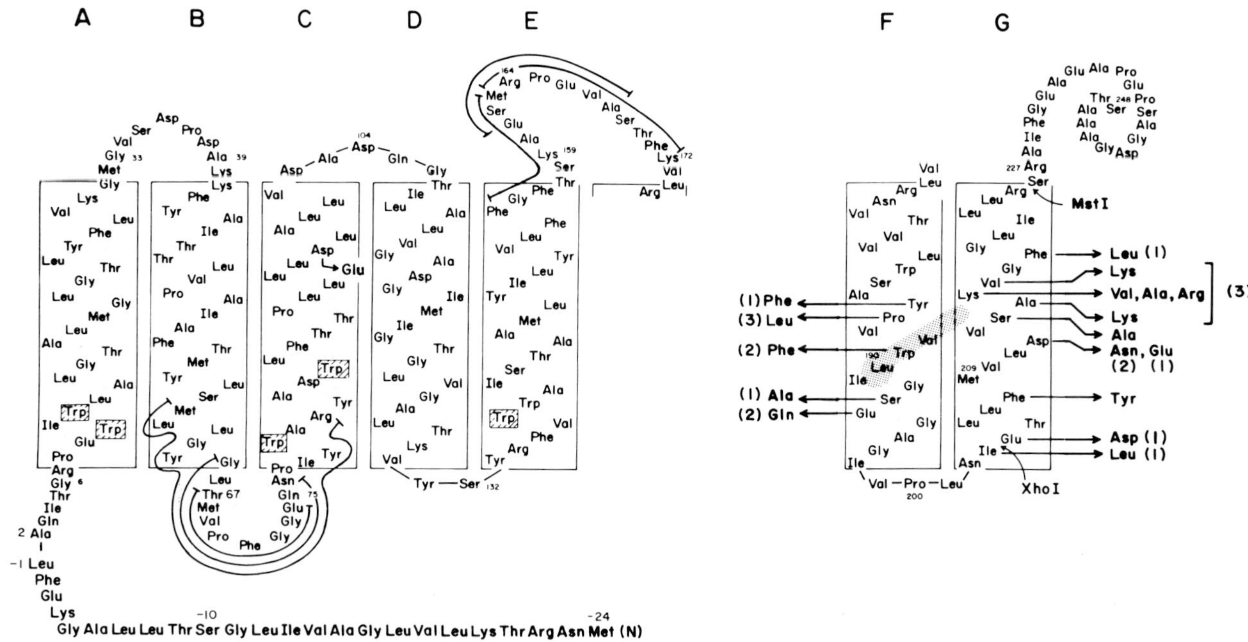

Fig. 7. Catalog of BO mutants available at present and spectral characteristics of some of the mutants. The analyses were done in the fusion protein system. The amber mutants (▨) and deletion mutants (⊢⊣) have not yet been analyzed. Chromophore regeneration characteristics of some of the amino acid substitution (→) mutants have been determined. Class I (1) mutants regenerate chromophore with the same absorption maximum as the wild type. Class II (2) mutants regenerate a chromophore and give a spectral shift in the absorption maximum. Class III (3) mutants do not regenerate the chromophore.

sis is required to evaluate the effect of these mutations on the renaturation of the mutant proteins. It is clear that in-depth studies of these and additional mutants should provide insights into the requirements for folding and stability of tertiary structure of the protein in the membrane. By examination of mutants defective in proton translocation, conclusions regarding the path of proton movement should be possible.

VI. CONCLUDING REMARKS

Bacteriorhodopsin is an attractive integral membrane protein for mechanistic studies. Its amino acid sequence and models for tertiary structure are available. The protein is robust and amenable to structural and biochemical studies. To pursue a psecific and versatile approach to amino acid replacement in structure–function studies, we have cloned the gene and expressed it in *E. coli*. The protein produced has been purified and reconstituted into phospholipid mixed micelles and vesicles. We have begun characterization of mutant proteins.

Folding pathways of water-soluble proteins have been intensively studied, while very little is known about the folding of membrane proteins. Although the properties of several membrane proteins have been studied under different denaturing conditions, no polytopic integral membrane protein except BR has been completely denatured and subsequently renatured into functional molecules. Our studies show that bacterio-opsin can fold correctly under appropriate conditions and regain functionality quantitatively. This presents a variety of new opportunities for the study of structure and function of membrane proteins in general. In this chapter we have shown that it is possible to express heterologous membrane protein in *E. coli*. Our study establishes that, by proper manipulation, an integral membrane protein can be produced without the presence of an added signal sequence or a presequence. The membrane-associated form of the protein can be purified and renatured to yield a functional molecule. The use of recombinant DNA techniques to construct and express mutants of BO makes it possible now to assess unequivocally the role of individual amino acids in the path of protons across the membrane and their role in the photocycle.

ACKNOWLEDGMENTS

This work was supported by grants GM28289 and AI11479 from the National Institutes of Health, grant PCM78-13713 from the National Science Foundation, and contract N00014-82-K-0668 from the Office of Naval Research, Arlington, Virginia.

REFERENCES

Bayley, H., Huang, K. S., Radhakrishnan, R., Ross, A. H., Takagaki, Y., and Khorana, H. G. (1981). *Proc. Natl. Acad. Sci. U.S.A.* **78,** 2225–2229.

Dalbadie-McFarland, G., Cohen, L. W., Riggs, A. D., Morin, C., Itakura, K., and Richards, J. H. (1982). *Proc. Natl. Acad. Sci. U.S.A.* **79,** 6409–8413.

DasSarma, S., RajBhandary, U. L., and Khorana, H. G. (1983). *Proc. Natl. Acad. Sci. U.S.A.* **80,** 2201–2205.

Dunn, R. J., McCoy, J. M., Simsek, M., Majumdar, A., Chang, S. H., RajBhandary, U. L., and Khorana, H. G. (1981). *Proc. Natl. Acad. Sci. U.S.A.* **78,** 6744–6748.

Dunn, R. J., Hackett, N. R., Huang, K.-S., Jones, S., Khorana, H. G., Lee, D.-S., Liao, M.-J., Lo, K.-M., McCoy, J., Noguchi, S., Radhakrishnan, R., and RajBhandary, U. L. (1983). Cold Spring Harbr. Symp. Quant. Biol. XLVIII, pp. 853–862.

Engelman, D. M., Henderson, R., McLachlan, A. D., and Wallace, B. A. (1980). *Proc. Natl. Acad. Sci. U.S.A.* **77,** 2023–2027.

Gilles-Gonzalez, M. A., Hackett, N. R., Jones, S. S., Khorana, H. G., Lee, D.-S., Lo, K.-S., and McCoy, J. M. (1986). *In* "Methods in Enzymology" (S. Fleischer and B. Fleischer, eds.), Vol. 125. Academic Press, New York (in press).

Huang, K.-S., Bayley, H., and Khorana, H. G. (1980). *Proc. Natl. Acad. Sci. U.S.A.* **77,** 323–327.

Huang, K.-S., Bayley, H., Liao, M.-J., London, E., and Khorana, H. G. (1981). *J. Biol. Chem.* **256,** 3802–3809.

Huang, K.-S., Radhakrishnan, R., Bayley, H., and Khorana, H. G. (1982). *J. Biol. Chem.* **257,** 13616–13623.

Jay, E., Rommens, J., Pomeroy-Cloney, L., Macknight, D., Wallace, C. L., Wisharl, P., Harrison, D., Lui, W.-Y., Asundi, V., Dawood, M., and Jay, F. (1984). *Proc. Natl. Acad. Sci. U.S.A.* **81,** 2290–2294.

Khorana, H. G., Gerber, G. E., Herlihy, W. C., Gray, C. P., Arderegg, R. J., Nihei, K., and Beimann, K. (1979). *Proc. Natl. Acad. Sci. U.S.A.* **76,** 5046–5050.

Kimura, K., Mason, J. L., and Khorana, H. G. (1982). *J. Biol. Chem.* **257,** 2859–2867.

Liao, M.-J., London, E., and Khorana, H. G. (1983). *J. Biol. Chem.* **258,** 9949–9955.

Lo, K.-M., Jones, S. J., Hackett, N. R., and Khorana, H. G. (1984). *Proc. Natl. Acad. Sci. U.S.A.* **81,** 2285–2289.

London, E., and Khorana, H. G. (1982). *J. Biol. Chem.* **257,** 7003–7011.

McCoy, J. M., and Khorana, H. G. (1983). *J. Biol. Chem.* **258,** 8456–8461.

Nakamura, K., Masui, Y., and Inouye, M. (1982). *J. Mol. Appl. Genet.* **1,** 289–299.

Ovchinnikov, Y. A., Abdulaev, N. G., Feigina, M. Y., Kiselev, A. V., and Lobanov, N. A. (1979). *FEBS Lett.* **100,** 219–224.

Remaut, E., Stanssens, P., and Fiers, W. (1981). *Gene* **15,** 81–93.

Seehra, J. S., and Khorana, H. G. (1984). *J. Biol. Chem.* **259,** 4187–4193.

Stoeckenius, W., and Bogomolni, R. A. (1982). *Annu. Rev. Biochem.* **52,** 487–615.

14

Stability Mutants of Staphylococcal Nuclease: A Correlation between Nuclease Activity in an Agar Gel Assay and Stability to Guanidine Hydrochloride Denaturation

DAVID SHORTLE

Department of Biological Chemistry
The Johns Hopkins University School of Medicine
Baltimore, Maryland 21205

I. INTRODUCTION

As well illustrated by the title and subject matter of this volume, the recent advances in recombinant DNA methodology and *in vitro* mutagenesis have opened up vast new opportunities for investigating the physical and chemical properties of proteins. Perhaps no one area in all of protein chemistry is in greater need of the new experimental approaches provided by genetic engineering than the area of study that has gone by the traditional name of "protein folding." Since the distinction between native and denatured states of proteins was made in the 1930s and 1940s and details of native states of a number of proteins have been determined by X-ray crystallography beginning in the mid-1950s, protein chemists have attempted to understand how the amino acid sequence of a protein encodes the information for a unique native conformation that forms spontaneously and is relatively stable in an aqueous environment. Although a qualitative picture of the relevant forces has been drawn (Creighton, 1983), a quantitative understanding, one that would permit

PROTEIN ENGINEERING
Applications in Science,
Medicine, and Industry

the "engineering" of changes in protein stability, seems rather far in the future.

Basically there are two different strategies for applying the new methods involving recombinant DNA and *in vitro* mutagenesis (Botstein and Shortle, 1985) to the folding/stability problem. First, the very powerful method of oligonucleotide-directed mutagenesis permits the introduction of any mutation, and therefore any amino acid substitution, in a protein for which the gene has been cloned. Once a specific hypothesis has been formulated about the role of a particular amino acid in a folding pathway or a stabilizing the native conformation of a protein, this strategy provides a rigorous, quantitative test of that hypothesis. However, in the absence of sufficient information to identify important amino acid residues, this approach loses much of its strength.

An alternative strategy—one that requires no prior information on the important residues—is to isolate mutations from a randomly mutagenized pool on the basis of phenotype. If a mutant phenotype could be developed that reflects a change in a protein's folding/stability properties, then mutations that specifically perturb these properties could be isolated and characterized in order to identify the revelant amino acid residues and to obtain mutant proteins for subsequent biochemical analysis. With a sufficiently sensitive and specific phenotype, this classical genetic approach has the special merit of focusing on precisely those mutations with the most pronounced effects on folding and stability.

As part of a genetic study of the folding and stability parameters of staphylococcal nuclease, a large collection of phenotypic mutations has been isolated on the basis of an agar plate assay for nuclease activity in individual *Escherichia coli* colonies carrying a recombinant plasmid (Shortle and Lin, 1985). Although this phenotype is by no means specific for mutations that affect folding or stability, it was assumed that mutations which blocked folding or severely destabilized the native conformation would reduce enzyme activity, either directly by lowering the fraction of catalytically active molecules or indirectly by enhancing proteolysis. Not surprisingly, many of the 77 missense mutations in this initial collection altered residues known to play a role in catalysis or substrate binding. However, two-thirds of these mutant proteins have single amino acid substitutions at positions in the native conformation remote from the active site, and of the 12 proteins in this class that have been purified, all show reduced stability to guanidine hydrochloride denaturation (Shortle, 1986; D. Shortle, unpublished data).

To ascertain the mechanism(s) by which this plate assay is able to detect changes in conformational stability, some of which are quite subtle, the experiments described in this chapter were carried out. Evidence is presented that the NUC⁻ phenotype of these stability mutations is due at least in part to reduced levels of nuclease activity in an agar gel matrix, the extent of this reduction correlating fairly well with the reduction in conformational stability. In addition, several mutant proteins displaying greater conformational stability than wild-type staphylococcal nuclease were found to have enhanced levels of activity in this plate assay.

II. MATERIALS AND METHODS

A. Mutant Proteins

The isolation, genetic mapping, and sequence analysis of all of the single and double mutations in the staphylococcal nuclease gene (*nuc*) used in this study are reported in Shortle and Lin (1985). All mutant proteins are designated by their amino acid changes compared to wild-type Foggi strain enzyme, using a three-symbol nomenclature in which the first letter designates the wild-type amino acid, the number the amino acid position, and the second letter the mutant amino acid substituted at that position. To simplify the purification of mutant proteins encoded by the different *nuc* alleles, the overexpression plasmid pFOG405 was constructed. In this plasmid, the nuclease gene is fused to the signal sequence and transcriptional control region of the *E. coli* alkaline phosphatase gene. On starvation for inorganic phosphate, this hybrid gene is very actively expressed, leading to the synthesis and accumulation of large amounts of correctly processed staphylococcal nuclease in the periplasm of the host bacterial cell. By using a simple extraction protocol plus a single column chromatography step, approximately 2–20 mg of protein can be isolated 95–99+% pure from 1 liter of a mid-log culture (Serpersu *et al.*, 1986). [Prior to the purification of a specific mutant nuclease, the corresponding *nuc*⁻ mutation (isolated on plasmid pFOG301) was genetically recombined onto plasmid pFOG405 by a strategy outlined in the first figure in Shortle and Lin (1985)].

B. Enzyme Assays

Protein concentration was measured by absorbance at 280 nm, using the extinction coefficient of 1.16 for a 1 mg/ml solution (Cuatrecasas *et al.*, 1967), and all dilutions used in enzyme assays were made in buffered solutions containing 1mg/ml heat-treated bovine serum albumin (BSA) to prevent loss by nonspecific absorption. Specific nuclease activity was determined at 37° C, using the quantitative assay of Cuatrecasas *et al.*

(1967). This spectrophotometric assay uses saturating concentrations of the substrate single-stranded DNA and calcium ion cofactor and gives an increase in absorbance at 260 nm, the initial rate of which is proportional to enzyme concentration over a 70-fold range.

To quantitate relative activity on the metachromatic dye agar diffusion assay (Lachica *et al.*, 1971), 2 μl of a dilution containing 13×10^{-3} unit/μl (as determined by the spectrophotometric assay) was spotted onto a freshly prepared Petri plate containing 10 ml of agar gel consisting of 50 m*M* Tris · HCl, 1% NaCl, 1% Difco agar, 300 μg/ml boiled salmon sperm DNA, and 0.2 mg/ml toluidine blue O. Plates were then incubated for 48 hr at 37°C, and the diameter of the pink halo formed by the diffusing enzyme was measured with a transparent ruler. In some experiments, the assay conditions were modified by changing the concentration of either toluidine blue O or agar in the mixture, or by adding urea or 1,2-propanediol directly to the standard mixture. All assays were done in duplicate.

C. Guanidine Hydrochloride Denaturation Analysis

The progress of the unfolding reaction at 20.0°C as a function of guanidine hydrochloride concentration was followed by the intrinsic fluorescence of tryptophan 140 at 325 nm (excitation at 295 nm), using a SPEX Fluorolog II spectrophotofluorimeter, as previously described by Shortle (1986).

III. RESULTS

A. Initial Observations

On purification of a number of mutant proteins from *nuc*$^-$ mutant alleles [identified on the basis of reduced enzyme activity in *E. coli* colonies replica-plated to TBD agar plates (Shortle, 1983)], it was found that almost every protein with an amino acid substitution remote from the active site had essentially the same specific activity as wild-type enzyme when assayed at 37°C by the standard DNA hydrolysis assay. To determine the reason for this apparent discrepancy between the *in vivo* and *in vitro* activities, equal numbers of units of different mutant enzymes were spotted on TBD agar plates. For the first 15–60 min of incubation at 37°C, there were no obvious differences in rates of formation of the pink zone of DNA hydrolysis as the enzyme diffused out into the blue agar gel. However, as illustrated in Fig. 1, after incubation for 48 hr different mutant proteins generated halos with very different diameters.

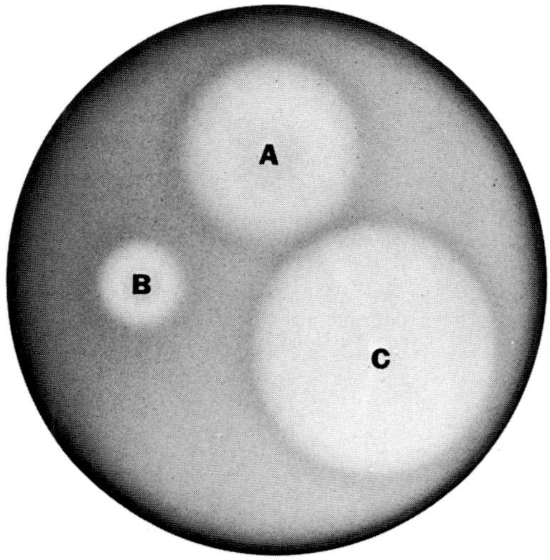

Fig. 1. Toluidine blue O–DNA (TBD) agar gel assay for staphylococcal nuclease. Spots A, B, and C represent zones of DNA hydrolysis formed by 26 milliunits of wild-type Foggi strain enzyme, mutant F76V, and double mutant H124L + V66L, respectively, after 48 hr of incubation at 37°C.

B. Quantitation of the TBD Agar Assay

In their initial characterization of this very sensitive assay for staphylococcal nuclease, Lachica *et al.* (1972) reported that, for 3-hr incubations at 37°C, the diameter of the zone of DNA hydrolysis was proportional to the logarithm of the total enzyme activity in the initial spot. In order to obtain a semiquantitative measure of the variation in enzyme activity displayed by these different mutant proteins, a series of dilutions of wild-type enzyme in a constant volume of 2 μl was spotted on separate plates. As can be seen in Fig. 2, this relationship between diameter and log activity holds remarkably well over a range of four $\log_{10}$ even for much longer incubation times. (Still longer incubation times of 96 hr result in an additional 5–15% increase in diameter.)

C. Activities of Mutant Proteins

To quantitate the activity of purified mutant forms of nuclease by this agar gel method, serial dilutions of each protein were first assayed for their rates of DNA hydrolysis at 37°C spectrophotometrically and then, from the same set of dilutions, 2 μl of enzyme containing 26 milliunits was immediately spotted (in duplicate) onto fresh TBD agar plates and incu-

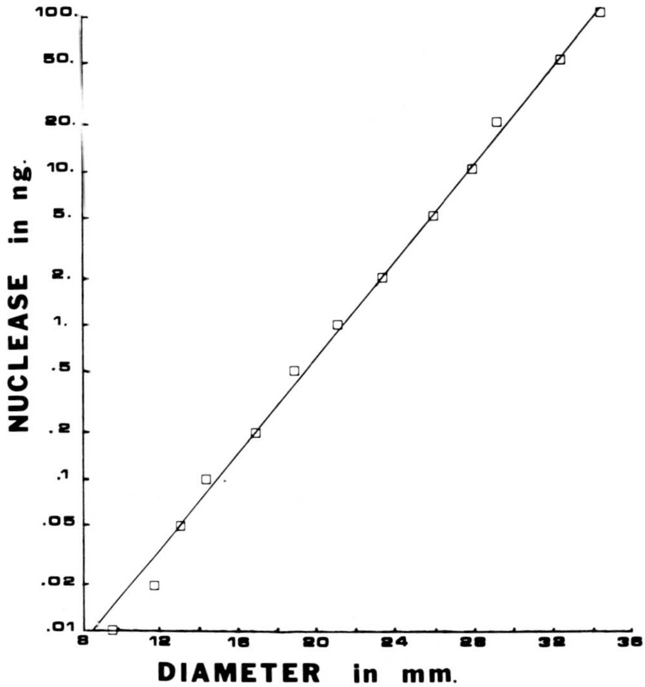

Fig. 2. Plot of $\log_{10}$ of the amount of wild-type enzyme (in nanograms) spotted in 2 μl on TBD agar gel plates versus the diameter of the resulting zone of hydrolysis after 48 hr of incubation at 37°C.

bated for 48 hr. The results are listed in Table I as both the diameter of the zone of hydrolysis and the relative activity compared to wild type on the basis of the standard curve in Fig. 2. Also listed are the enzyme activity levels scored in *E. coli* colonies synthesizing the same enzyme. As can be seen, these two values are in rough agreement. In addition, the concentration of guanidine hydrochloride, C_m, that gives an equilibrium constant for denaturation of one is listed. [Although the estimated free energy of denaturation in the absence of denaturant is a much more appropriate measure of the stability of a protein to reversible denaturation, as described by Shortle (1985), obtaining such estimates for these particular mutant proteins is problematic.]

An even better correlation is apparent between the relative activity in the TBD agar assay and the C_m for guanidine denaturation. All mutants that unfold at lower concentrations of guanidine hydrochloride than the wild type give rise to smaller zones of hydrolysis than the wild type and all mutants that unfold at higher concentrations give larger zones, with rank

TABLE I

Comparison of Nuclease Activity on TBD Agar and Sensitivity to Denaturation by Guanidine Hydrochloride for Several Mutant Forms of Staphylococcal Nuclease

Mutant nuclease	Halo diameter[a] (mm)	Relative activity[b]	Activity from *E. coli* colonies	Guanidine HCl C_m[c] (M)
F76V	14	0.001	+/−	0.39
G88D	15	0.001	+	0.50
G79S	16.3	0.002	+	0.53
F76V + H124L	16.3	0.002	++	0.60
G88D + V66L	19.3	0.005	++	0.58
G79S + V66L	20.5	0.01	++	0.56
Wild type	28	1.0	4+	0.82
V66L	32.8	6	4+	0.96
H124L + V66L	36.5	25	4+	1.25
H124L	37	30	4+	1.08

[a] Zone of DNA hydrolysis formed by 26 milliunits of enzyme in 48 hr at 37°C.
[b] Enzyme activity compared to wild type, obtained from Fig. 1.
[c] C_m is the concentration of guanidine hydrochloride at the midpoint of denaturation at 20.0°C, not at the enzyme assay temperature of 37°C. Most of these values are from Shortle (1985).

orders of diameter and C_m that are very similar. This raises the possibility that the loss in activity is a consequence of spontaneous denaturation of nuclease molecules as they diffuse through the agar, with the molecules that spend more time in the denatured state losing activity at a higher rate.

D. Modifications of the Assay Conditions

To obtain further information on the role of denaturation in loss of activity, TBD agar plates were made with 0.5, 1.0, 1.5, and 2.0 *M* urea and amounts of three different proteins (F76V, wild type, and H124L + V66L) that formed halos of approximately equal size (13 units, 13 milliunits, and 0.7 milliunits, respectively) were spotted on each plate. For all three proteins there were comparable losses in activity that increased with urea concentration, with approximately 0.1% of the activity remaining at 2.0 *M*. In contrast, the wild-type enzyme retained more than 50% of its specific activity in 2 *M* urea in the spectrophotometric assay. However, addition of the protein-stabilizing compound 1,2-propanediol to TBD agar plates in amounts of 5, 10, and 15% (v/v) did not lead to an increase in activity, but rather to slight reductions in the activity of all three enzymes.

Agar gel and spectrophotometric assays both use the same substrates, buffer, and pH; they differ primarily with respect to the duration of the assay and the presence of the dye toluidine blue O and agar. Therefore the very large discrepancies between the activities the two assays measure is likely to involve a time-dependent interaction between the enzyme and either the dye or agar component of the gel assay. To test these possibilities, the range of toluidine blue O was varied from 0.1 to 0.4 mg/ml and agar concentration was tested at 1, 3, and 5%. No significant differences were seen in halo size for wild type, F76V, or H124L + V66L at any of the three dye concentrations, but major reductions in the diameter of the hydrolysis zones were seen with all three enzymes as the percentage of agar increased. Even on incubation of the plates for 72 hr to permit more time for diffusion of enzyme through the tighter gel matrix formed by the more concentrated agar, activity appeared approximately 90% reduced in 3% agar and more than 99% reduced in 5% agar.

IV. DISCUSSION AND CONCLUSIONS

From this initial characterization of purified mutant forms of staphylococcal nuclease, it appears that many proteins encoded by *nuc*$^-$ mutations which alter amino acid residues remote from the active site have essentially normal enzyme activity as measured in a short-term solution assay but have greatly reduced activities in a longer-term agar gel assay. The loss in activity in the agar gel assay correlates well with a reduced stability of the native state of the mutant protein relative to the denatured state, as measured by guanidine denaturation analysis. Given this correlation plus the observed effects of relatively low concentrations of urea in further reducing enzyme activity, it seems likely that activity is lost by a side reaction involving the denatured state. One plausible reaction that could explain these observations plus the pronounced inhibitory effects of higher agar concentrations on nuclease activity is the physical adsorption of denatured molecules to hydrophobic sites in the agar gel matrix.

The presence of a major competing reaction that progressively removes enzymatically active species from the population of protein molecules diffusing through the agar is also suggested by the linear relationship between halo diameter and the logarithm of enzyme activity. If DNA hydrolysis ceased only when diffusion had diluted the concentration of enzyme to some threshold level, then the *volume* of the zone of hydrolysis would be expected to correlate with the nuclease activity. In other words, the square of the diameter instead of simply the diameter should vary with log enzyme activity. This is precisely what is observed with radial immunodiffusion assays, in which the square of the diameter of the precipitin ring formed as antigen diffuses into a gel containing a constant concentration of antibody varies with log antigen (Mancini *et al.,* 1965).

For other proteins found to exhibit a similar reduction in their activities within agar gel matrix, an agar-based activity assay might provide a simple, though not entirely specific, phenotype for identifying mutations that modify conformational stability. Particularly when combined with alterations in temperature, denaturant concentration, or percentage of agar in the assay, this agar–protein interaction might prove useful for "amplifying" the effects of small increases in conformational stability, allowing for the direct identification of amino acid substitutions that stabilize the protein to denaturation. Experiments are currently under way to search for point mutations that increase the thermal stability and guanidine resistance of staphylococcal nuclease.

ACKNOWLEDGMENTS

The author thanks Alan Meeker for his extensive technical assistance and Teresa Dunn for her reading of the manuscript. This work was supported by National Institutes of Health grant GM34171.

REFERENCES

Botstein, D., and Shortle, D. (1985). *Science* **229,** 1193.
Creighton, T. E. (1983). "Proteins: Structures and Molecular Principles," pp. 122–157. Freeman, New York.
Cuatrecasas, P., Fuchs, S., and Anfinsen, C. B. (1967). *J. Biol. Chem.* **242,** 1541.
Lachica, R. V. F., Genigeorgis, C., and Hoeprich, P. D. (1971). *Appl. Microbiol.* **21,** 585.
Lachica, R. V. F., Hoeprich, P. D., and Franti, C. D. (1972). *Appl. Microbiol.* **24,** 920.
Mancini, G., Carbonara, A. O., and Heremans, J. F. (1965). *Immunochemistry* **2,** 235.
Serpersu, E. H. Shortle, D., and Mildvan, A. S. (1986). *Biochemistry* **25,** 68.
Shortle, D. (1983). *Gene* **22,** 181.
Shortle, D. (1986). *J. Cellular Biochem.* **30,** 281.
Shortle, D., and Lin, B. (1985). *Genetics* **110,** 539.

15

Mutagenesis of the Arc Repressor Using Synthetic Primers with Random Nucleotide Substitutions

ANDREW K. VERSHON, KAREN BLACKMER, AND ROBERT T. SAUER

Department of Biology
Massachusetts Institute of Technology
Cambridge, Massachusetts 02139

I. INTRODUCTION

The classic work of Anfinsen and his colleagues (1973) showed that amino acid sequence determines the three-dimensional structure and thus the function of a protein. However, it is also clear from studies of species variation in the hemoglobin and cytochrome families that many different sequences can determine essentially the same structure and function. For example, the sequences of 60 different hemoglobin α chains are known and comparison reveals that only 23 of the 141 residue positions are conserved among these proteins (Dickerson and Geis, 1983). This finding could be interpreted to mean that only a few essential residues are required to determine structure or, more reasonably, to suggest that different residues or groups of residues can serve the same basic function with respect to protein structure and activity. One way to approach this question is to study protein missense mutants in a genetically tractable system. One can then ask whether any residue is a potential site of mutation or whether substitutions at only a few residue positions result in a defective protein. Moreover, once a mutant protein is isolated, reversion analysis can be used to ask if other residues or pairs of residues can restore function without restoring a wild-type amino acid sequence. Studies of

PROTEIN ENGINEERING
Applications in Science,
Medicine, and Industry

this type are currently in progress in the staphylococcal nuclease (Shortle and Lin, 1985) and lambda repressor systems (Hecht and Sauer, 1985; Nelson and Sauer, 1985).

In this chapter we describe a system designed to allow rapid selection and characterization of mutations in the Arc repressor of bacteriophage P22. Arc is a 53-residue protein that binds to operator DNA as a dimer (Susskind, 1980; Sauer *et al.*, 1983; Vershon *et al.*, 1985). The small size of the *arc* gene makes it an ideal subject for genetic manipulation. Moreover, correlation of sequence and structure should soon be possible since determination of the Arc structure by X-ray diffraction is in progress (Jordan *et al.*, 1985; S. Jordon and C. Pabo, personal communication).

In any attempt at genetic saturation of mutant sites, it is desirable that the method of mutagenesis give as wide a range of transitions and transversions as possible and that each site have the same probability of being mutated. One possible means of achieving this goal is to use oligonucleotide synthesis to introduce random mutations into a single-stranded primer and then to use these mutagenic primers to introduce mismatches into the target gene. In this chapter we describe the application of this method in a preliminary study of mutations in the *arc* gene.

II. MATERIALS AND METHODS

A. Microbiology and Cloning

General methods for microbiology and recombinant DNA manipulations followed the procedures in Miller (1972) and Maniatis *et al.* (1982). Restriction fragments smaller than 800 base pairs (bp) were generally purified by electrophoresis on 6% TBE polyacrylamide gels cross-linked with 0.1% bisacrylcystamine (BioRad). After electrophoresis, gel slices were excised and solubilized in a low-salt buffer containing 20 m*M* Tris-HCl (pH 7.5), 200 m*M* NaCl, 0.1 m*M* EDTA, and 10% 2-mercaptoethanol. The DNA was then loaded on a Schleicher & Scheull Elutip, washed with buffer without 2-mercaptoethanol, and eluted with a high-salt buffer containing 20 m*M* Tris-HCl (pH 7.5), 1 *M* NaCl, and 0.1 m*M* EDTA. Larger restriction fragments were purified by electrophoresis on TBE low-melting-point agarose gels (American Laboratory Supply). Gel slices were excised, solubilized by heating to 65°C in low-salt buffer (without 2-mercaptoethanol), and purified on Elutips as described above. The DNA purified by either method was recovered from the high-salt buffer by ethanol precipitation.

B. Construction of pTA200

Plasmid pTA200 is a pBR322 derivative containing an M13 origin of replication and the bacteriophage P22 *arc* gene under control of the hybrid *trp–lac* promoter P_{tac}. The construction of pTA200 required several steps, as described below.

Plasmid pMS385 (M. M. Susskind, unpublished) contains a fusion of the *lac*UV5 promoter and the *arc* gene, cloned into the *Eco*RI/*Pvu*II backbone of pBR322. This plasmid was digested with *Msp*I and the 388-bp fragment containing the −10 region of the *lac* promoter and the *arc* gene was gel-purified. Plasmid pEA300, which contains the −35 region of the *trp* promoter (Amann *et al.*, 1983), was digested with *Cla*I and the 5′ phosphates were removed by using calf intestinal phosphatase (Boehringer Mannheim). The 388-bp *Msp*I fragment was ligated with the dephosphorylated pEA300 and the mixture was transformed into *Escherichia coli* strain X90 [F′*lac*I^{Q1}] (Amman *et al.*, 1983). Transformants were selected by resistance to ampicillin and the structure of the resulting P_{tac}–*arc* fusion plasmid, pTA100, was verified by restriction mapping. To construct a P_{tac}–*arc* fusion plasmid containing a phage M13 origion of replication, the 3400-bp *Pst*I–*Pvu*II fragment containing the P_{tac}–*arc* fusion from the pTA100 and the 2043-bp *Pst*I–*Pvu*II fragment containing the M13 and plasmid origins from pZ150 (Zagursky and Berman, 1984) were gel-purified and ligated together. The ligation mix was transformed into strain X90 and transformants resistant to ampicillin were selected. The resulting strain, XTA200, contained plasmid pTA200 whose structure is shown in Fig. 1.

The level of Arc protein expressed from plasmid pTA200 was assayed by sodium dodecyl sulfate (SDS)–polyacrylamide gel electrophoresis. An overnight culture of XTA200 was diluted into LB broth supplemented with 100 μg/ml ampicillin and was grown to late log phase. At this time isopropyl-β-D-thiogalactoside (IPTG) was added to a final concentration of 100 μg/ml and growth was continued for another 2.5 hr. A 1.0-ml portion of the IPTG-induced culture was centrifuged in a microfuge for 5 min, and the pellet was suspended in 200 μl SDS gel sample buffer [3% SDS, 65 m*M* Tris-HCl (pH 7.5), 10% glycerol, 5% 2-mercaptoethanol, and 0.002% bromphenol blue dye] and heated at 90°C for 2 min. This lysate was then centrifuged in a microfuge for 15 min, and the viscous DNA pellet and cellular debris were removed with a toothpick and discarded. A 20-μl portion of the remaining lysate was electrophoresed on a 13.5% SDS–polyacrylamide gel, using the buffer system of Laemmli (1970).

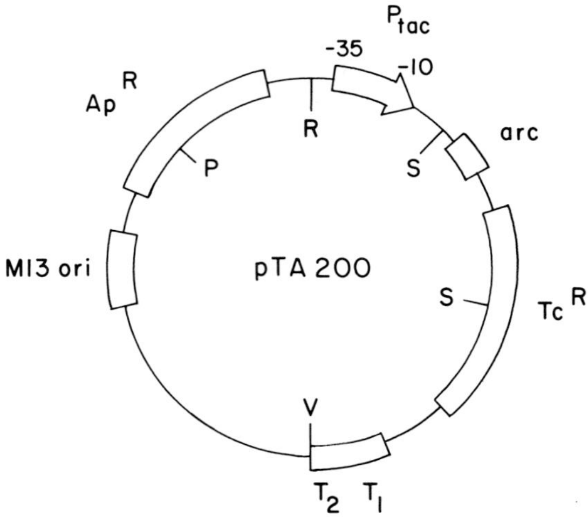

Fig. 1. Structure of plasmid pTA200. P_{tac}, Hybrid promoter containing the -35 region of the *trp* promoter and the −10 region of the *lac*UV5 promoter; Ap^R and Tc^R, genes coding for resistance to ampicillin and tetracycline, respectively; T_1 and T_2, transcription terminators of the rrnB operon and 5S gene (Amman *et al.*, 1983); M13 ori, the phage M13 origin of replication and packaging (Zagursky and Berman, 1984); R, *Eco*RI; S, *Sal*I; V, *Pvu*II; P, *Pst*I.

C. Construction of Selection Strain UA2F

Strain UA2F is a *rec*A, *su*⁻ *E. coli* strain containing a λ*imm*21 prophage bearing a fusion of the P22 *ant* promoter to the selectable chloramphenicol resistance gene *cat*. Since Arc represses P_{ant}, this strain can be used to select for Arc-defective mutations. UA2F was constructed in several steps as follows.

Plasmid pMS172 (M. M. Susskind, unpublished) contains the 818-bp *Sau*3A fragment of the P22 *imm*I *Vy*delA2 mutant cloned into the *Eco*RI/*Pvu*II backbone of pBR322. The *Vy*delA2 mutation deletes a small (10 bp) *Ava*II fragment just downstream from P_{ant} and creates a new *Sal*I restriction site at the deletion junction. Plasmid pMS172 was digested with *Sal*I and dephosphorylated by treatment with calf intestinal phosphatase. Plasmid pCM1 contains a *Sal*I-linkered *cat* gene cloned into the *Sal*I site of pBR322 (Close and Rodriguez, 1982). The 780-bp *Sal*I fragment of pCM1 containing the *cat* gene was gel-purified and ligated with the dephosphorylated pMS172. After ligation, ampicillin-resistant transformants of strain HB101 were selected and screened for resistance to 50 μg/ml chloramphenicol. Plasmid pAC200, which contains a fusion of the P22 *ant* promoter to *cat*, was obtained from one such candidate and its structure

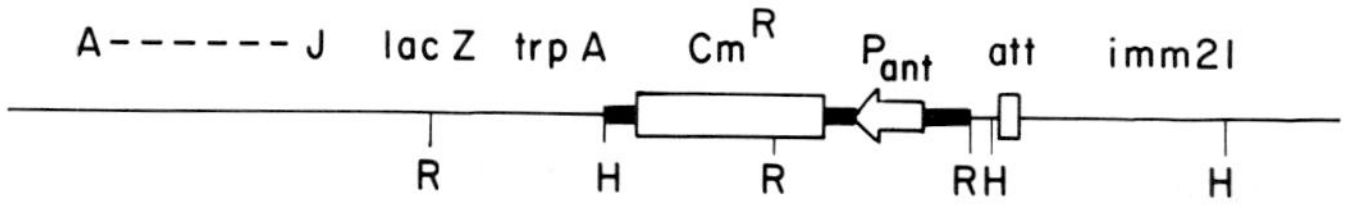

Fig. 2. Partial restriction map of phage λAC201. The λ genes A through J along with the *imm*21 immunity region and the *att* site are from phage λ112 (Maurer *et al.*, 1980). The thick line represents the region cloned from plasmid pAC201. The arrow shows the direction of transcription from the promoter P_{ant}. Chloramphenicol resistance (Cm^R) is coded by the *cat* gene, which is expressed from p_{ant}. R, *Eco*RI; H, *Hin*dIII.

was verified by restriction mapping. The pAC200 DNA was then digested with *Eco*RI and the 304-bp fragment containing P_{ant} and a portion of *cat* was gel purified. Plasmid pCM7 (Close and Rodriguez, 1982), which contains the *cat* gene cloned into the *Hin*dIII site of pBR322, was digested with *Eco*RI, dephosphorylated, and the 3650-bp backbone fragment gel purified. The backbone and 304-bp fragments were ligated, transformed into HB101, and ampicillin-resistant colonies were selected and screened for resistance to chloramphenicol. The structure of the resulting plasmid, pAC201, was verified by restriction mapping.

DNA from phage λ112 (*imm*^{21}, λP_{RM}–*c*I–*trpA*–*lacZ* fusion; Maurer *et al.*, 1980) was cleaved with *Hin*dIII to generate a 25-kb fragment containing the left arm of the phage and the *lacZ* and *trpA* genes. The right arm of λ112 was generated by cleaving with *Eco*RI. Both arms were gel purified on a 0.5% low-melting agarose gel. pAC201 was digested to completion with *Hin*dIII and then partially digested with *Eco*RI. The 854-bp *Eco*RI/*Hin*dIII fragment containing the P_{ant}–*cat* fusion was gel purified from a 0.7% low-melting agarose gel and ligated together with the purified right and left arms of λ112. The ligation mix was packaged into phage as described in Maniatis *et al.* (1982) and plated on strain YMC9 (strain 294, Δ*lac–pro,* L. Guarente) on LB plates supplemented with 20 μg/ml Xgal. Blue plaques were struck out on Xgal/LB plates to obtain single-colony lysogen candidates, and phage stocks from the same plaques were prepared on strain YMC9. Blue colonies were screened for resistance to chloramphenicol and λ*imm*21*c*I$^-$. A phage that formed *lacZ*$^+$, Cm^R, λ*imm*21*c*I$^-$-resistant lysogens was picked and designated λAC201. The structure of this phage is shown in Fig. 2. This phage was plated on strain US3 (*thi*$^-$, *his*$^-$, *lacZ*$^-$, *lacY*$^+$, *sup*^{0}, *recA*$^-$, U. Schmeissner) and lysogens, designated UA2, were screened to confirm the expected *lacZ*$^+$, Cm^R, λ*imm*21*c*I$^-$-resistant phenotype. Strain UA2 was then mated with SF1148 (Δ*lac–pro, spc*^{R}, *F'lacI*^{Q}, *lacZ* : Tn*5*, *pro*$^+$, obtained from S. Froshauer), and strain UA2F was selected as a chloramphenicol- and kanamycin-resistant survivor of the mating. Plasmid pTA200 was transformed into UA2F to give strain UTA200. When Arc synthesis in strain UTA200 was induced with 0.2 μg/ml IPTG, the strain failed to grow on plates supplemented with 20 μg/ml chloramphenicol.

D. Oligonucleotide Synthesis of Mutagenic Primers

All primers were synthesized on a Systec Microsyn 1450A, using American BioNuclear CED phosphoramidites. Mutagenic primers were synthesized by adding 1% each of the three incorrect nucleotides to each of the phosphoramidite stocks. Oligonucleotides were deprotected and cleaved from the solid support by suspending in 1.0 ml of concentrated ammonium hydroxide and shaking overnight at 55°C. The suspension was centrifuged for 2 min in a microfuge and the supernatant was saved. The solid support was washed with 0.5 ml of ammonium hydroxide, and the supernatant was pooled with the first supernatant and lyophilized. The oligonucleotides were then suspended in 300 μl of 50 m*M* ammonium acetate pH 7.0 and chromatographed on a C18 high-performance liquid chromatography (HPLC) column (Waters, 10 μm), using a 90% A : 10% B to 40% A : 60% B linear gradient over 10 min at 1.0 ml/min (solution A, 50 mM ammonium acetate, pH 7.0; solution B, 50 m*M* ammonium acetate, pH 7.0, and 50% acetonitrile). The main peak, eluting after 15 min, was collected and lyophilized overnight. The dimethyltrityl group was removed by suspending the oligonucleotide in 0.1 ml of 5% acetic acid for 30 min at room temperature. A 200-μl portion of 5% triethylamine was then added, and two ether extractions were performed to remove the cleaved trityl groups. The oligonucleotide was then rechromatographed on HPLC, using the conditions described above. The main peak, eluting at about 12 min, was collected and lyophilized overnight. The oligonucletides were suspended in 500 μl of 10 m*M* Tris-HCl (pH 7.5), 0.1 m*M* EDTA and stored at −20°C. Seven oligonucletides were synthesized by this technique and designated MP1 (bases 394–417), MP2 (466–489), MP3 (490–513), MP4 (372–396), MP5 (418–444), MP6 (445–468), and MP7 (508–531).

E. Oligonucleotide Mutagenesis

Single-strand template was prepared by diluting an overnight culture of UTA200 (strain UA2F carrying the M13 origin plasmid pTA200) 50-fold into 2xYT broth and infecting the culture with 2 × 10^7, plaque-forming units (pfu) of M13 phage derivative RV1 (Levinson *et al.*, 1984). After 6 hr of growth at 37°C, single-stranded DNA was purified by the procedure described in Zagursky and Berman (1984). The single-stranded DNA was then diluted with an equal volume of formamide and chromatographed on a Sephacryl-S200 column equilibrated in 10 m*M* Tris-HCl (pH 8.0), 1 m*M* EDTA. Fractions containing DNA were pooled and stored at −20°C. A 100-pmol portion of each primer was treated with T4 polynucleotide kinase and ATP (Maniatis *et al.*, 1982). Annealing reactions contained 5 pmol each of the mutagenic and secondary pBR322 *Hin*dIII primer and 0.25 pmol of single-strand template in 20 μl of Klenow buffer (10 m*M* Tris-HCl, ph 7.5), 10 m*M* $MgCl_2$, 1 m*M* DTT, and 50 m*M* NaCl). The mixture was heated to 65°C for 20 min and then slowly cooled to room temperature over 30 min. A 2.5-μl portion of extension buffer (400 m*M* Tris-HCl, pH 7.5, 250 m*M* $MgCl_2$, 250 m*M* DTT, 10 m*M* ATP, 10 m*M* dNTPs), 1 unit of DNA polymerase large fragment (Boehringer Mannheim), and 40 units of T4 DNA ligase (New England BioLabs) were added to the DNA and allowed to incubate for 90 min at room temperature. At this time, an additional 1 unit of DNA polymerase large fragment and 40 units of ligase were added and the reaction was continued at 14°C overnight.

The reaction mixtures were diluted about 10-fold in 50 m*M* $CaCl_2$ and divided into five aliquots. Each aliquot was transformed into MH90 (strain AB1157 with an *mutS* allele and F′*lacI*^{Q} *pro*$^+$; M. Hecht), and the entire transformation mixture was diluted into 5 ml of LB broth supplemented with 200 μg/ml ampicillin and grown overnight at 37°C. Plasmid DNA from each transformation was purified and transformed into the selection strain UA2F. Transformants were selected on LB plates supplemented with 100 μg/ml ampicillin, 25 μg/ml kanamycin, and 0.4 μg/ml IPTG. Arc mutants were selected by the ability of their host strains to grow on the above plates supplemented with 100 μg/ml chloramphenicol. Single-stranded plasmid DNA was purified as described above, and a small sample of each mutant preparation was electrophoresed on a 0.7% agarose gel to check for large deletions in the plasmids. The mutants were then sequenced by the dideoxy procedure (Sanger *et al.*, 1977), using a primer complementary to a region past the 3′ end of the *arc* gene.

III. RESULTS

A. Mutant Selection Strategy

By cloning the *arc* gene under the control of the P_{tac} promoter in the M13 origin plasmid pTA200, we obtained a vector that has several desirable features for mutant studies. First, the intracellular level of Arc can be controlled since P_{tac}-mediated expression of *arc* is regulated by Lac repressor, which in turn can be regulated by the amount of IPTG in the

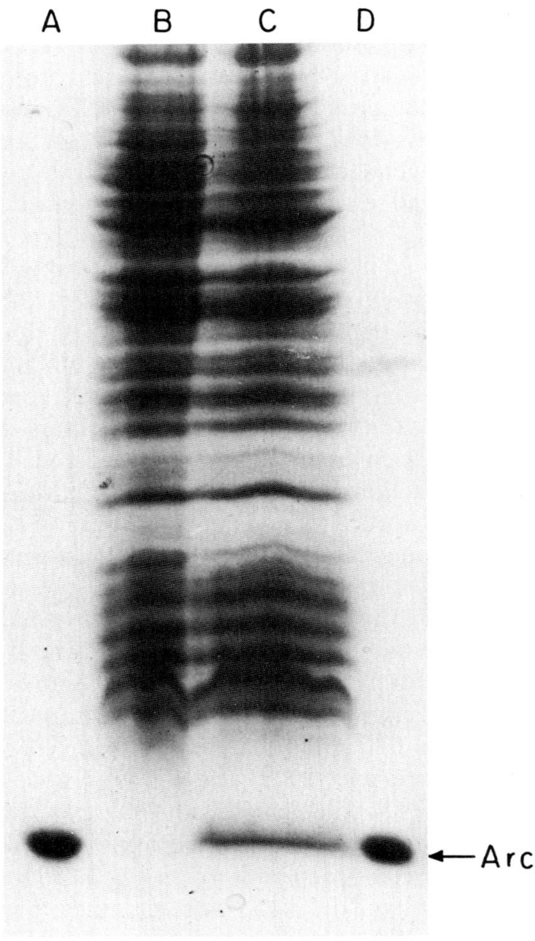

Fig. 3. Induction of Arc repressor production from plasmid pTA200 in strain X90. Lysates of induced and noninduced cultures were run on 13.5% SDS–polyacrylamide gels as described in Section II. Lanes A and D, 1 μg of purified Arc protein; lane B, noninduced culture; lane C, IPTG-induced culture.

medium. It is important to be able to limit Arc production since high-level expression is lethal, and at the same time it is useful to be able to produce large quantities of the wild-type or mutant proteins for biochemical studies. The P_{tac}–*arc* fusion satisfies both requirements. Figure 3 shows lysates of strain X90/pTA200 under noninducing and fully inducing conditions. Under the inducing conditions, approximately 2% of the cellular protein is Arc repressor. Under the noninducing conditions, Arc is not visible in the lysate. Second, the M13 origin of pTA200 allows single-stranded plasmid DNA to be packaged following helper phage infection. This single-stranded DNA can then be used either as a template for mutagenesis or for dideoxy sequencing of potential *arc* mutations.

The strategy we used to isolate *arc* mutations was similar to that used to isolate mutations in the λ *c*I and Cro repressors (Hecht *et al.,* 1983; Nelson *et al.,* 1983; Pakula *et al.,* 1987), except that we employed a selection for defective mutations whereas the *c*I and Cro mutations were identified by a screening procedure. A schematic representation of our selection strategy for *arc* mutations is shown in Fig. 4. Plasmid pTA200 is mutagenized and transformed into strain UA2F, and ampicillin- and chloramphenicol-resistant transformants are selected in the presence of IPTG. In the presence of IPTG, cells containing wild-type Arc are chloramphenicol-sensitive since Arc is expressed, binds to O_{arc}, and represses *cat* transcription from P_{ant}. In cells containing mutant Arc repressors, which fail to repress P_{ant}, the *cat* gene is expressed and the cell is chloramphenicol-resistant. The stringency of the selection can be varied by changing the IPTG concentration and thus the amount of Arc produced, or by changing the chloramphenicol concentration.

B. Random Primer Mutagenesis

Conventional techniques or random mutagenesis involve treatment with mutagens *in vitro* or *in vivo* or growth in mutator strains of *E. coli.* Such methods often result in isolation of only a limited spectrum of transitions or transversions. Moreover, mutations that occur at "hot spots" are often isolated repeatedly. Specific oligonucleotide-directed mutagenesis allows any desired mutational change to be introduced at any given site, but in general some prior knowledge about the structure or activity of a protein is needed before it becomes sensible to make a particular residue change. Some of the limitations of both methods of mutagenesis can be removed by combining the two techniques. Random mutations over a limited region of a gene can be generated by using oligonucleotides that have a low level of non-wild-type bases incorporated at each position. When these oligonucleotides are used to prime complementary strand synthesis on single-stranded templates, mismatches in the primer region should result. In the experiments reported here, we synthesized oligonucleotides by using phosphoramidate stocks containing 1% of each of the incorrect bases (for example, the C stock contains 97% C, 1:% A, 1% G, 1% T). Under these conditions, we expect that synthesis of a 24-base oligonucleotide will result in a population in which approximately 50% of the primers have one or more mutations.

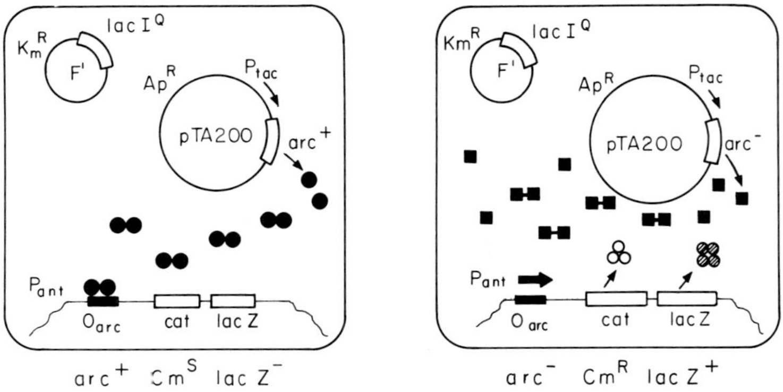

Fig. 4. Representation of the system used to select mutations in the Arc repressor. The left cell contains a plasmid with a wild-type *arc* gene on pTA200. Lac repressor is used to regulate the amount of synthesis of Arc from P_{tac}. Wild-type Arc repressor dimerizes and binds to O_{arc}, shutting off transcription from P_{ant} and thereby making the cell phenotypically chloramphenicol-sensitive and *lacZ*$^-$. The right cell contains a pTA200 plasmid with a mutation in the *arc* gene. Mutant Arc protein is synthesized, but it can no longer bind to O_{arc}. Transcription occurs from P_{ant} and the cell becomes chloramphenicol-resistant and *lacZ*$^+$.

To apply this method of mutagenesis, seven mutagenic primers, which span the *arc* gene, were synthesized. These primers were separately annealed to single-stranded pTA200 DNA along with a secondary primer, and second-strand synthesis was carried out with DNA polymerase and ligase (Zoller and Smith, 1984). To avoid mismatch repair, the heteroduplex DNA was transformed into a *mutS* strain defective in repair. After several generations, plasmid DNA was prepared and transformed into strain UA2F to select Arc-defective mutants as described above. From four independent rounds of primer mutagenesis, 77 independent *arc*$^-$ mutations were isolated. In the first three experiments only three mutagenic primers were employed, and thus each primer is not equally represented in the collection. Mutations were isolated at a frequency of 10^{-3} to 10^{-2}, but this high frequency may be misleading since cells bearing mutant Arc repressors may have a growth advantage and thus be amplified relative to wild type during passage through the *mutS* strain.

The positions of some of the *arc*$^-$mutations are shown in Fig. 5. Of the 77 mutant candidates, 36 contained single point mutations in the *arc* gene. These include 23 different missense and 4 different nonsense mutations. One double mutation (Val 18→Glu : Leu 19→Ser) and one triple mutation (Gly 17→Gly : Asp 20→Val : plus a silent change at residue 32) were also isolated. Fourteen of the *arc*$^-$ mutations were either single or tandem base deletions in the gene; these are represented by triangles below the DNA sequence in Fig. 5. One single base insertion was isolated; this is

```
                                  AMB VAL      PHE
                CYS               LEU SER      AMB TRP ARG

MET-LYS-GLY-MET-SER-LYS-MET-PRO-GLN-PHE-ASN-LEU-ARG-TRP PRO
 1                                                       15

                T                 TT  GC        AT T    C

ATG AAA GGA ATG AGC AAA ATG CCG CAG TTC AAT TTG CGG TGG CCT
|                                    ▼   ▼ ▼              |
370                                                     414


                VAL
                ALA                                 THR
    GLY GLU SER TYR PHE                             TYR

ARG-GLU-VAL-LEU-ASP-LEU-VAL-ARG-LYS-VAL-ALA-GLU-GLU-ASN-GLY
 16                                                      30
                 T
    G   A   C   TC   T                              TC

AGA GAA GTA TTG GAT TTG GTA CGC AAG GTA GCG GAA GAG AAT GGT
|                                                         |
415                                                     459


                        GLY         PRO             GLY
TRP SER GLY LYS         LYS     CYS AMB     ALA     OCH

ARG-SER-VAL-ASN-SER-GLU-ILE-TYR-GLN-ARG-VAL-MET-GLU-SER-PHE
 31                                                      45

T     A G    A      AG   ▼    G  TC         C     TG

CGG TCT GTT AAT TCT GAG ATT TAT CAG CGA GTA ATG GAA AGC TTT
|                    ▼▼   ▼                      ▼        |
460                 ▼    ▼                              504


THR             PRO

LYS-LYS-GLU-GLY-ARG-ILE-GLY-ALA
 46                         53

 C              C

AAG AAG GAA GGG CGC ATT GGC GCG TAA
|    ▼                           |
505                             531
```

Fig. 5. Mutations in the *arc* gene generated by random primer-directed mutagenesis. Mutations are shown above the wild-type DNA sequence along with the corresponding changes in the amino acid sequence. Small triangles below the DNA sequence represent a deletion of that base pair. Large triangles represent deletions of two adjacent base pairs. The triangle above the DNA sequence represents insertion of a base pair between the base pairs indicated. The numbering of bases is from Sauer *et al.* (1983). AMB and OCH represent TAG and TAA termination codons, respectively.

represented by a triangle above the DNA sequence. Sixteen of the arc^- plasmids contained large deletions presumably affecting the *arc* gene and/or the P_{tac}promoter. In eight cases, the site of the presumed arc^- mutations was not identified. This last category could include plasmid copy number mutations or unlinked cellular mutations.

The large deletion mutations were identified by an increase in electrophoretic mobility of single-stranded plasmid DNA on an agarose gel. In several of these cases we attempted to sequence the plasmids, but they had all lost the sequencing primer annealing site. The 16 large deletions arose from a single mutagenic experiment. We do not know how these deletions were generated.

Ten of the 14 single or double base deletions fell within the region covered by the mutagenic primer employed. Of the four mutants that fell outside the primer employed, two were in the promoter region and one was 5′ and the other 3′ to the mutagenic primer. The number of small deletions isolated in each round of mutagensis varied considerably. Most of these deletions were isolated in the first mutagenic experiment, in which 100 units of ligase was used; however, reducing the ligase to 40 units in subsequent experiments diminished but did not eliminate these deletion mutations. We have no direct evidence, however, that the deletions arise as a consequence of aberrant ligation events. It is possible that these mutations arise from primers containing deletions or by other, unspecified mechanisms.

The lone insertion mutation adds a thymine between bases 478 and 479. This position occurs after a run of three thymines and thus might occur as a result of strand slippage during replication. It is also possible that this change was incorporated during the synthesis of the mutagenic primer.

Surprisingly, only 16 of the 36 point mutations occurred within the region covered by the mutagenic primer employed. The remaining 20 mutations either were "external" to the primer or were isolated from control reactions in which no mutagenic primer was used. The "external" mutations occurred on the 5′ and 3′ sides of the mutagenic primer with approximately equal frequency. It is not clear whether this class of mutations arises as a consequence of a high mutation frequency in single-stranded DNA, unfaithful replication during the *in vitro* annealing and extension reactions, or passage through the *mutS* strain. We have not studied any of these possibilities systematically.

The four rounds of mutagenesis produced approximately equal numbers of transitions and transversions. A large portion of the mutations occurred in wild-type A : T and T : A base pairs, but this bias may not be statistically significant. The mutagenic spectra were also similar for those mutations that fell within the mutagenic primers and those outside the primer regions. Kunkel *et al.* (1983) observed that adenine is often incorporated during Pol I synthesis across depurinated sites on the template strand. However, the majority of our mutations do not contain adenine on the synthesize strand, and thus depurination of single-stranded template is unlikely to account for the mutations that occur outside the primers.

IV. DISCUSSION

We have described a simple genetic system that permits rapid isolation and sequencing of defective mutations in the *arc* repressor gene. In this system, Arc represses transcription of the gene encoding chrolamphenicol acetyltransferase (*cat*) and thus Arc-defective mutants can be selected by antibiotic resistance. Moreover, since the *arc* gene is present on a multicopy plasmid and is expressed from the highly efficient P_{tac} promoter, overproduction and purification of *arc* missense mutants should be straightforward.

We have also investigated the use of oligonucleotide primers with random nucleotide substitutions as a method of directing mutagenesis to specific regions of the *arc* gene. By using this technique, we isolated a spectrum of different transition and transversion mutations in the *arc* gene. However, two problems were encountered that may limit the general utility of this mutagenic technique. First, deletion mutations accounted for nearly 40% of the total mutations that were obtained, and such mutations are of little use for studies of protein structure and activity. We anticipate that the frequency of these deletion mutations can be reduced by appropriate alteration of the experimental conditions used for second-strand synthesis, but we have not studied this problem in detail. Second, over half of the missense mutations occurred at positions 3′ or 5′ to the mutagenic primer employed. Since the *arc* gene is small, such mutations were easily detected. However, for larger genes, the advantages of directed mutagenesis would be lost if a substantial number of mutations were not located within the mutagenized region. We assume that this "external" class of mutations arises either during second-strand synthesis *in vitro* or during propagation of the plasmid as single-stranded DNA *in vivo*. What is not clear, however, is whether the high frequency of such mutations was a consequence of the particular reaction conditions or strains that we used or is a general problem associated with the basic method of oligonucleotide mutagenesis. In most oligonucleotide-directed mutagenesis experiments that have been reported, hybridization has been used to identify the desired product and random mutations occurring at frequencies of 0.1–1% would not have been detected.

For the experiments presented here, approximately 20% of the mutations were single base substitution mutations within the region covered by the mutagenic primer. This frequency may be acceptable for certain applications. Furthermore, if the method can be optimized to limit deletion mutations and possibly to limit mutations outside the primer sequence, its usefulness should increase substantially.

ACKNOWLEDGMENTS

We thank Andy Pakula for suggesting the use of the M13 origin plasmid, Arlene Wyman for the lambda phage packaging extract, Mimi Susskind for plasmids pMS385 and pMS172, Theresa Karplus for pEA300 DNA, and Michael Hecht for strain MH90. This work was supported by National Institutes of Health grant #AI-16892. A.K.V. was supported by NIH training grant GM-07287.

REFERENCES

Amman, E., Brosius, J., and Ptashne, M. (1983). *Gene* **25,** 167–178.

Anfinsen, C. B., (1973). *Science* **181,** 223–230.

Close, T, J., and Rodriguez, R. L. (1982). *Gene* **20,** 305–316.

Dickerson, R. E., and Geis, I. (1983). "Hemoglobin." Benjamin/Cummings Publ Co., Menlo Park, California.

Hecht, M. H., and Sauer, R. T. (1985). *J. Mol. Biol.* **185,** 53–63.

Hecht, M. H., Nelson, H. C. M., and Sauer, R. T. (1983). *Proc. Natl. Acad. Sci. U.S.A.* **80,** 2676–2680.

Jordan, S. R., Pabo, C. O., Vershon, A. K., and Sauer, R. T. (1985). *J. Mol. Biol.* **185,** 445–446.

Kunkel, T. A., Schaaper, R. M., and Loeb, L. A. (1983). *Biochemistry* **22,** 2378–2384.

Laemmli, U. K. (1970). *Nature (London)* **227,** 680–685.

Levinson, A., Silver, D., and Seed, B. (1984). *J. Mol. Appl. Genet.* **2,** 507–517.

Maniatis, T., Fritsch, E. F., and Sambrook, J. (1982). "Molecular Cloning: A Laboratory Manual." Cold Spring Harbor Lab., Cold Spring Harbor, New York.

Maurer, R., Meyer, B. J., and Ptashne, M. (1980). *J. Mol. Biol.* **139,** 147–161.

Miller, J. H. (1972). "Experiments in Molecular Genetics." Cold Spring Harbor Lab., Cold Spring Harbor, New York.

Nelson, H. C. M., and Sauer, R. T. (1985). *Cell* **42,** 549–558.

Nelson, H. C. M., Hecht, M. H., and Sauer, R. T. (1982). *Cold Spring Harbor Symp. Quant. Biol.* **47,** 441–449.

Pakula, A., Young, V., and Sauer, R. T. (1987). *Proc. Natl. Acad. Sci. U.S.A.* (in press).

Sanger, F., Nicklens, S., and Coulson, A. R. (1977). *Proc. Natl. Acad. Sci. U.S.A.* **74,** 5463–5467.

Sauer, R. T., Krovatin, W., DeAnda, J., Youderian, P., and Susskind, M. M. (1983). *J. Mol. Biol.* **168,** 699–713.

Shortle, D., and Lin, B. (1985). *Genetics* **110,** 539–555.

Susskind, M. M. (1980). *J. Mol. Biol.* **138,** 685–713.

Vershon, A. K., Youderian, P., Susskind, M. M., and Sauer, R. T. (1985). *J. Biol. Chem.* **260,** 12124–12129.

Zagursky, R. J., and Berman, M. L. (1984). *Gene* **27,** 183–191.

Zoller, M. J., and Smith, M. (1984) *DNA* **3,** 479–488.

16

Investigation of the Structural Roles of Disulfides by Protein Engineering: A Study with T4 Lysozyme

RONALD WETZEL

Biocatalysis Department
Genentech, Inc.
South San Francisco, California 94080

I. INTRODUCTION

The disulfide bond is a common element in the structures of many extracellular globular proteins. Anfinsen and his colleagues (Anfinsen, 1973) utilized these bonds as structural probes for protein folding, and disulfides continue to be exploited in such studies (for reviews, see Creighton, 1978; Freedman and Hillson, 1980). Nature's purpose, however, in inventing and utilizing disulfides has been little explored and remains for the most part obscure. This chapter describes experiments designed to contribute to the elucidation of the roles disulfides play in protein structure. These experiments involve introduction of disulfides, via protein engineering techniques, into proteins that do not normally contain them. The chapter begins with a review of some features of disulfides which make these bonds especially attractive subjects for study. There follow a description of the rationale for choosing T4 lysozyme as a model for these studies and a review of the cloning of the lysozyme gene for expression in *Escherichia coli*. The bulk of the chapter describes the introduction of a disulfide bond into lysozyme, the comparative properties of various lysozyme derivatives, and some preliminary interpreta-

PROTEIN ENGINEERING
Applications in Science,
Medicine, and Industry

tions of the results. The chapter concludes with some speculation on other ways in which engineered disulfide bonds might be exploited to better understand the biosynthesis, activity, stability, and folding pathways of proteins.

II. PROPERTIES OF DISULFIDES IN GLOBULAR PROTEINS

Disulfide bonds can be considered elements of both the primary and tertiary structures of proteins. Since the disulfide is covalent and exists even when the protein is denatured, it has aspects of primary structure. At the same time the bond clearly defines a point of close approach in the native folding pattern by two discontinuous elements of the polypeptide chain. When proteins which have been reduced are refolded *in vitro* under sufficiently oxidizing conditions, correct, native disulfides are formed only after most or all of the protein tertiary structure has formed (Anfinsen, 1973; Anfinsen and Scheraga, 1975). *In vivo,* some cytoplasmic proteins contain relatively stable disulfides (Eklund *et al.,* 1984), but most such bonds have been characterized in extracellular proteins, in which they form during (Bergman and Kuehl, 1979) or after (Scheele and Jacoby, 1982) secretion from the cytoplasm.

Protein disulfides may exist in equilibrium with the reduced state in relatively reducing environments such as thiol/disulfide "redox buffers" (Saxena and Wetlaufer, 1970). The degree to which this equilibrium favors the protein disulfide in such buffers depends not only on disulfide chemistry but also on noncovalent interactions of other groups on the protein, with each other and with the oxidized or reduced protein cysteines. It can thus be said that other elements of the protein's structure can stabilize or destabilize the disulfide in a particular redox environment. [This is inherent in the ability of ribonuclease, with its disulfides "scrambled," to reestablish its native disulfide arrangement in a thiol/disulfide exchange buffer (Haber and Anfinsen, 1962)]. Conditions which weaken structure, such as denaturants, thus "weaken" the disulfide bond, that is, make it less stable to a particular redox potential.

The situation is the reverse in a relatively oxidizing environment. In this case the disulfide, as a secure covalent bond, is more stable than noncovalent packing interactions in the protein (for a discussion of the forces which contribute to protein stability, see Schultz and Schirmer, 1979). The disulfide in such conditions can serve to stabilize the protein toward various stresses such as heat and denaturants.

The contribution of disulfides to unfolding/folding pathways of proteins thus depends on whether these bonds are formed, broken, or themselves in transition as conformational processes proceed. Correspondingly, studies of disulfides tend to be framed differently depending on whether investigators approach protein folding as a problem in biosynthesis or in biopolymers.

Since polypeptides emerge from the ribosome with cysteines predominantly in the reduced state, thiol/disulfide interchange to generate disulfides must occur as part of the overall biosynthetic process (which, in addition to chain folding, may include secretion, signal peptide cleavage, and other posttranslational modifications). Folding studies involving disulfide formation mediated by thiol–disulfide interchange thus not only exploit disulfides as probes of the structures of intermediates but also may be relevant to the biosynthetic process.

Protein folding may also be approached as a problem in the biophysics of polymers. How do the primary amino acid sequences of these biopolymers determine their stability, structures, and unfolding and refolding pathways in *in vitro* denaturation/renaturation experiments? Some of these studies may also be relevant to the biosynthetic folding problem. Beyond that, proteins are chemical entities with structure/function relationships that are independent of the mode of synthesis. Studies of proteins whose disulfides remain intact throughout the experiment fall into this category.

Disulfides have been found predominantly in extracellular globular proteins. The simple explanation for this pattern is that it is more difficult for a stable protein disulfide to evolve in the highly reducing (Freedman and Hillson, 1980) environment of the cytoplasm. Are disulfides of general utility in protein structure, but simply not in evolution's tool kit for intracellular molecules because of this restriction? The fact that many intracellular proteins function quite nicely without disulfides suggests that these bonds play no indispensable role in protein stability; but might they not serve special roles for particular classes of proteins?

The role of disulfides in determining the physical properties of proteins is probably more complex than is currently believed. Polymer theory predicts that cross-links should stabilize a protein toward *reversible* unfolding by heat or denaturants (Anfinsen and Scheraga, 1975); this predicted contribution to the free energy of stabilization is based on the difference in entropies between cross-linked and non-cross-linked unfolded polypeptide chains. In the few cases where the thermodynamics have been investigated, cross-links have, in fact, been shown to influence the free energy of stabilization, largely through the entropy term (Johnson *et al.*, 1978; Lin *et al.*, 1984). While it seems possible that changes in packing forces on cross-linking might make additional contributions (posi-

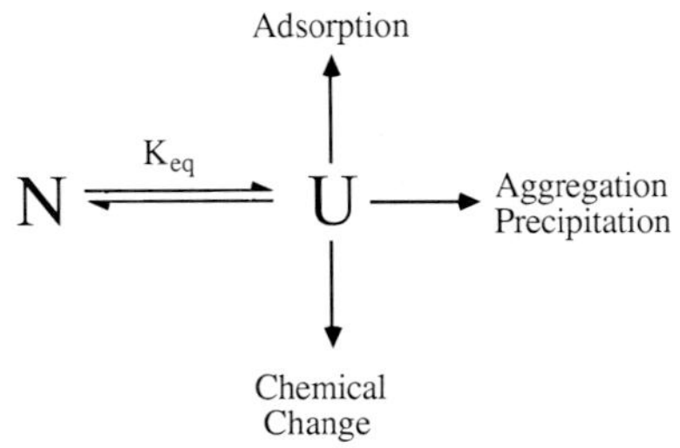

Fig. 1. Simple model for mechanisms of irreversible thermal inactivation. N, Native structure; U, unfolded state in a simple two-state unfolding model.

tive or negative) to thermostability, it is clear that disulfides can add to stability against reversible unfolding by at least the chain entropy effect.

While stability toward reversible inactivation is important, the major stability problems facing nature during evolution, as well as workers in applied and basic biochemistry, are those of irreversible inactivation. The simplest mechanistic model for the latter processes involves the initial reversible unfolding of the native form (Klibanov, 1983) to a species that can, in turn, enter one or more pathways of irreversible decay (Fig. 1). As reviewed above, a disulfide may influence the unfolding equilibrium. As the steady-state concentration of the unfolded form is affected, so may be the observed rate of the subsequent irreversible inactivation, if that step is rate-limiting. However, disulfides might make additional contributions to stability toward irreversible processes. The research described in this chapter is largely directed toward testing this hypothesis.

In previous work we established a stabilizing role for the disulfides of human interferon α (Morehead *et al.,* 1984). Loss of its two disulfides imparts onto interferon a sensitivity to irreversible thermal denaturation at 37°C. While, in principle, this sensitivity could be due to a shift in the unfolding equilibrium (Fig. 1), other explanations are possible. Interferon, however, is not an ideal subject for characterizing this effect: no X-ray structure has been elucidated; neither is there a groundwork of fundamental studies on the solution structure and stability of the molecule.

If there are general roles for disulfides in globular proteins, it should be possible to understand them by studying any of a number of well-characterized model systems. We might choose to remove disulfides selectively from proteins such as hen egg white lysozyme or ribonuclease, the workhorses of protein folding studies for several decades. Another approach is to attempt to introduce disulfides into a protein which does not normally contain them. In doing so, we can explore the advantages and disadvantages of disulfides without encountering any built-in bias in a protein whose structure has coevolved with such bonds.

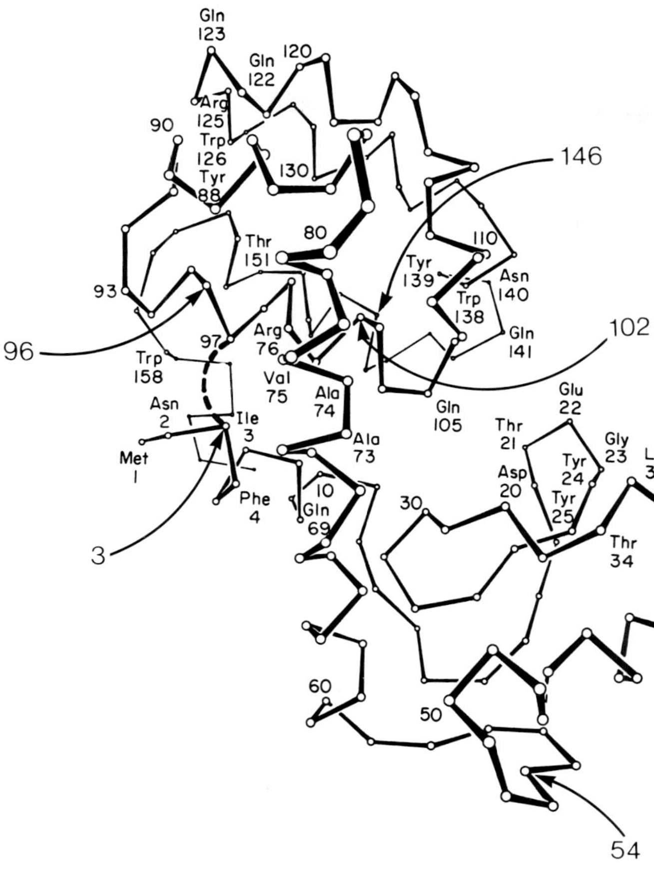

Fig. 2. X-ray crystal structure of T4 lysozyme, highlighting residues of interest discussed in the text (adapted from Remington *et al.*, 1978). An Ile→Cys replacement at position 3 makes possible the formation of a disulfide with Cys 97 of the wild type (shown schematically by the dashed line). A Cys at position 54 interferes with the stabilizing effect of the disulfide. Single variants at positions 96, 102, and 146 introduce temperature sensitivity into the molecule. (From Hawkes *et al.*, 1984; reprinted with permission from Academic Press.)

III. T4 LYSOZYME

We chose as a model T4 lysozyme (Fig. 2). The X-ray structure of the molecule has been determined (Remington *et al.*, 1978), and the thermodynamics of denaturant and thermal unfolding have been studied (Hawkes *et al.*, 1984; Desmadril and Yon, 1984). The existence of a number of well-characterized temperature-sensitive mutants (Hawkes *et al.*, 1984; see Fig. 2) makes the system especially attractive. The molecule (164 amino acids; MW 18,720) contains two free cysteines and no disulfide bonds. While T4 lysozyme's primary sequence bears no significant homology to that of hen egg white lysozyme (which contains four disulfides and is quite stable toward irreversible thermal inactivation), the two molecules share significant homology in their three-dimensional structures (Matthews *et al.*, 1981; Rossman and Argos, 1976), especially around their active sites. Most importantly, T4 lysozyme exhibits only limited stability toward thermal incubation in the neutral pH range (Tsugita *et al.*, 1968), and the loss of activity can be reversed to some extent by denaturation/renaturation (F. R. Dahlquist, personal communication); both properties are characteristic of the inactivation of disulfide-free interferon α (Morehead *et al.*, 1984).

The choice of T4 lysozyme entailed some risks and shortcomings. While there is genetic evidence that the cytolytic effect of T4 lysozyme *in vivo* requires expression of other T4 genes (Tsugita, 1971), we were concerned that high-level expression of this protein in *E. coli* might nonetheless prove toxic. Another disadvantage of this system is the limitations of the enzyme assay. In contrast to hen egg white lysozyme, there are no low-molecular-weight substrates readily available for the T4 enzyme, due to more specialized requirements for substrate binding (Bienkowska and Taylor, 1979; Jensen *et al.*, 1976). The only assays readily available involve the lytic effect of lysozyme on cells (Tsugita *et al.*, 1978) or isolated peptidoglycan (Becktel and Baase, 1985). These assays do not lend themselves to measurement of kinetic parameters. Since our work is focused elsewhere, this was not a serious obstacle; at the same time, it is clear that enzyme kinetics is a powerful technique in evaluating the effects of amino acid replacements (Fersht *et al.*, 1985).

IV. EXPRESSION OF T4 LYSOZYME GENE IN *E. coli*

The *e* gene of the T4 genome which codes for T4 lysozyme has been subcloned into a lambda vector in *E. coli* and its DNA sequence determined (Owen *et al.*, 1983). No efforts to express the protein product were reported. Using the genetic and restriction maps of the T4 genome (O'Farrell *et al.*, 1980) and the restriction map and sequence of the *e* gene (Owen *et al.*, 1983) we devised a strategy for cloning the lysozyme gene for expression behind the strong, inducible *tac* promoter (de Boer *et al.*, 1982). We succeeded in obtaining a vector which directs the synthesis of T4 lysozyme at levels of about 2% of the cellular protein (Perry *et al.*, 1985), which is more than adequate for obtaining useful amounts of protein. The *E. coli* strains harboring this plasmid exhibit no significant toxic effect of the expression of lysozyme, although there does seem to be a slight growth disadvantage to cells producing active lysozyme.

One strategy we devised in anticipation of toxic effects in the expression of lysozyme was the design of a prolysozyme molecule which could be constructed at the DNA level. There is good evidence that Asp 20 of T4 lysozyme is important for enzymatic activity (Matthews *et al.*, 1981). We reasoned that if we converted this Asp to Asn by site-directed mutagenesis, we might produce an inactive lysozyme. We further reasoned that it should be possible to deamidate this residue *in vitro* to generate an Asp and enzyme activity. This variant, T4 lysozyme(D20N),* in fact exhibits little or no lysozyme activity, confirming the critical role played by Asp 20 in catalysis. In addition, it has been possible to partially activate this molecule by incubation at basic pH (R. Wetzel, unpublished results). The amount of activity generated in this prolysozyme is low, but a more activatable zymogen should be attainable by further mutagenesis. Although such an engineered zymogen was not necessary for expression of lysozyme, the technique may be important for expression of other proteins.

V. CRITERIA AND CHOICE OF A CROSS-LINKING SITE

Correlations in proteins whose crystal structures have been deduced show that some kinds of secondary structure tend to harbor disulfides more than other kinds of structure (Thornton, 1981) and that the dihedral angle of the sulfur–sulfur bond tends to fall within well-defined limits (Richardson, 1981). To obtain maximal stability from a cross-link via the chain entropy effect (see above), a new disulfide should span as much of

* Our ability to generate protein variants containing multiple amino acid replacements will require the establishment of a new nomenclature system. An immediate problem encountered in the work described here is the designation of the chemical status of particular cysteine residues. Until guidelines are established, we have adopted the following system: (1) Replacements are designated by using the single-letter code (Dayhoff, 1972), with the wild-type residue followed by the residue number followed by the replacement residue. Thus T4 lysozyme containing an Asp→Asn substitution at position 20 is T4 lysozyme(D2ON). (2) Multiple replacements are listed in numerical order separated by slashes: T4 lysozyme(I3C/C54T). (3) The chemical status of cysteine is designated by superscripts. Lack of a superscript implies that there is no special information about the chemical status of the cysteine. The superscript "SH" designates a reduced thiol; "CM" designates a carboxymethyl group; "−97C" designates an intramolecular disulfide with the cysteine at position 97. Thus T4 lysozyme with an Ile→Cys replacement at position 3, in which the new Cys is disulfide-bonded to Cys 97 and Cys 54 is in the reduced state, is T4 lysozyme(I3C^{-97C}/54C^{SH}).

the primary sequence as possible (Anfinsen and Scheraga, 1975). At the same time, if the three-dimensional structure of the wild-type protein is to be preserved, the cross-linked positions should lie in space at a separation and geometry appropriate for disulfide bond formation. Given these constraints, one might expect a certain amount of difficulty in finding sites in a protein structure suitable for disulfide bond cross-linking. One test is simply whether the disulfide forms in the variant protein.

The interpretation of such an experiment is complicated, however, by the predominance of the chemical driving force of the oxidation reaction in the thermodynamics of protein disulfide formation. As Haber and Anfinsen (1962) showed, it is possible to induce formation of incorrect disulfides in proteins. Thus, we might expect to be able to drive formation of a disulfide bond (intra- or intermolecular) in any protein which contains two free cysteine residues. At this level, the test becomes: What are the minimum conditions necessary for formation of the desired disulfide? In other words, how similar is the inherent redox potential of the new disulfide to potentials of naturally occurring disulfides?

One of the first issues that protein engineering studies can address is the question of the flexibility of protein structures in accommodating substitutions without suffering major changes in the folding pattern. While examination of naturally occurring variants indicates that structures have flexibility at some positions, these data are of limited utility. First, most such variants contain multiple substitutions, leading to the possibility of compensating effects in the structure. Second, genetic selection in the wild and most selections in the laboratory will only deliver viable substitutions. For the first time we are in a position to make single substitutions regardless of the consequences for protein structure/function, and we must be prepared to revisit the question of flexibility at higher resolution. In the present context, having established minimum conditions for disulfide formation, we can ask: By how much, and where, has the structure changed to allow formation of the disulfide? How does the introduction of the cross-link (as opposed to the introduction of the cysteine residues) affect the properties of the molecule? While the emerging results of protein engineering experiments suggest that protein structures have considerable freedom to accommodate the effects of substitutions at some sites, we might expect to encounter more disruptive structural changes in cross-linking experiments, given our ability (see above) to overcome energy barriers to misfolding reactions by utilizing the favorable thermodynamics of the thiol oxidation.

Our initial search for cross-linking sites in T4 lysozyme was made on the molecular model in Brian Matthews' lab at the University of Oregon. Using the refined coordinates for T4 lysozyme (L. Weaver and B. W.

Matthews, unpublished), we then made a more systematic search with the MIDAS software of T. Ferrin and R. Langridge at the University of California, San Francisco. In recent modeling studies we used FRODO adapted for the Evans and Sutherland 330 by J. Pflugrath and F. Quiocho (unpublished). After searching the molecule for positions widely separated in the primary sequence but lying close in space, we investigated these positions manually for their ability to harbor a good disulfide (as measured by structural parameters of well-characterized protein disulfides; see Richardson, 1981; Perry and Wetzel, 1984). There now exist algorithms for carrying out an initial search in the computer (E. Suchanek and C. Pabo, personal communication; R. C. Ladner, personal communication).

Although none of the disulfides we modeled in MIDAS matched any known disulfides in geometry, a few were closer to typical, low energy disulfides than the rest. One of these is a disulfide between residues 3 and 97. Since position 97 is occupied by one of the two wild-type cysteine residues, we chose the 3–97 bond as our first target.

VI. INTRODUCTION OF THE 3–97 DISULFIDE INTO T4 LYSOZYME

To introduce an amino acid replacement we used a modification of the double-priming M13 mutagenesis procedure of Zoller and Smith (1984). The *tac* expression system produced the variant protein at levels comparable to wild-type expression. The product was purified in the presence of reducing agent by ion exchange and gel filtration chromatography (Perry and Wetzel, 1986) and was homogeneous by polyacrylamide gel electrophoresis. The purified protein, T4 lysozyme(I3C), contained three free cysteines by Ellman's titration and a Cys at position 3 by N-terminal amino acid sequencing analysis (Perry and Wetzel, 1984, 1986).

Using mobility on a sodium dodecyl sulfate (SDS)–polyacrylamide gel to assess cross-linking, we exposed the new protein to a variety of oxidizing conditions. Air oxidation and sodium tetrathionate both proved to be effective agents. More physiologically relevant glutathione redox buffers also successfully introduced a cross-link into the molecule; this system is now used routinely for preparative synthesis of molecules containing the 3–97 bond (Perry and Wetzel, 1986). While we have not yet rigorously determined the redox potential of the new disulfide, we are encouraged that no harsher conditions, such as molecular oxygen, tetrathionate, or Ellman's reagent, are required to drive its formation; this is some indication of the compatibility of the disulfide with the structure.

We determined the connectivity of the disulfide by looking for mercaptoethanol-sensitive peaks in a high-performance liquid chromatography (HPLC) separation of fragments of a trypsin digestion (Perry and Wetzel, 1986). Of the three disulfides possible in a protein with three cysteines, we found evidence for only one, the 3–97 disulfide. That a viable disulfide can be made between residues 3 and 97 was unequivocally demonstrated when we subsequently produced cross-links in variants of lysozyme with Cys 54 replaced (see below).

VII. PROPERTIES OF T4 LYSOZYME(I3C)

Although our original data suggested that cross-linked T4 lysozyme(I3C^{-97C})* exhibited enzymatic activity in the cell-clearing assay indistinguishable from that of the wild type (Perry and Wetzel, 1984), subsequent experiments indicated that the molecule has about 80% of the specific activity of the parent (Perry and Wetzel, 1986). Our latest results suggest that this activity change is entirely due to the Ile 3→Cys substitution, not to the cross-linking.

Experiments directed toward solving the three-dimensional structure of this lysozyme variant are in progress in collaboration with Brian Matthews and co-workers. The lack of effect of cross-linking on both specific activity (Perry and Wetzel, 1986) and ultraviolet–visible and circular dichroism (CD) spectra (Mulkerrin *et al.*, 1985) suggest an essentially unaltered structure. The CD spectra of the cross-linked variant show small changes from the wild type, all of which may be due to contributions from the disulfide itself rather than perturbations of the structure by the disulfide. Clearly, however, there may be small changes in the structure which will show up only in a highly refined crystal structure.

VIII. STABILITY TOWARD IRREVERSIBLE THERMAL INACTIVATION

Preliminary experiments showed disulfide cross-linked T4 lysozyme(I3C^{-97C}) to be significantly more stable toward irreversible thermal inactivation than the wild-type molecule at 67°C (Perry and Wetzel, 1984). The stabilization is lost when the disulfide is broken by reducing agent, indicating that it is the cross-link and not the amino acid replacement which provides stability (see Fig. 3).

* See footnote on p. 263.

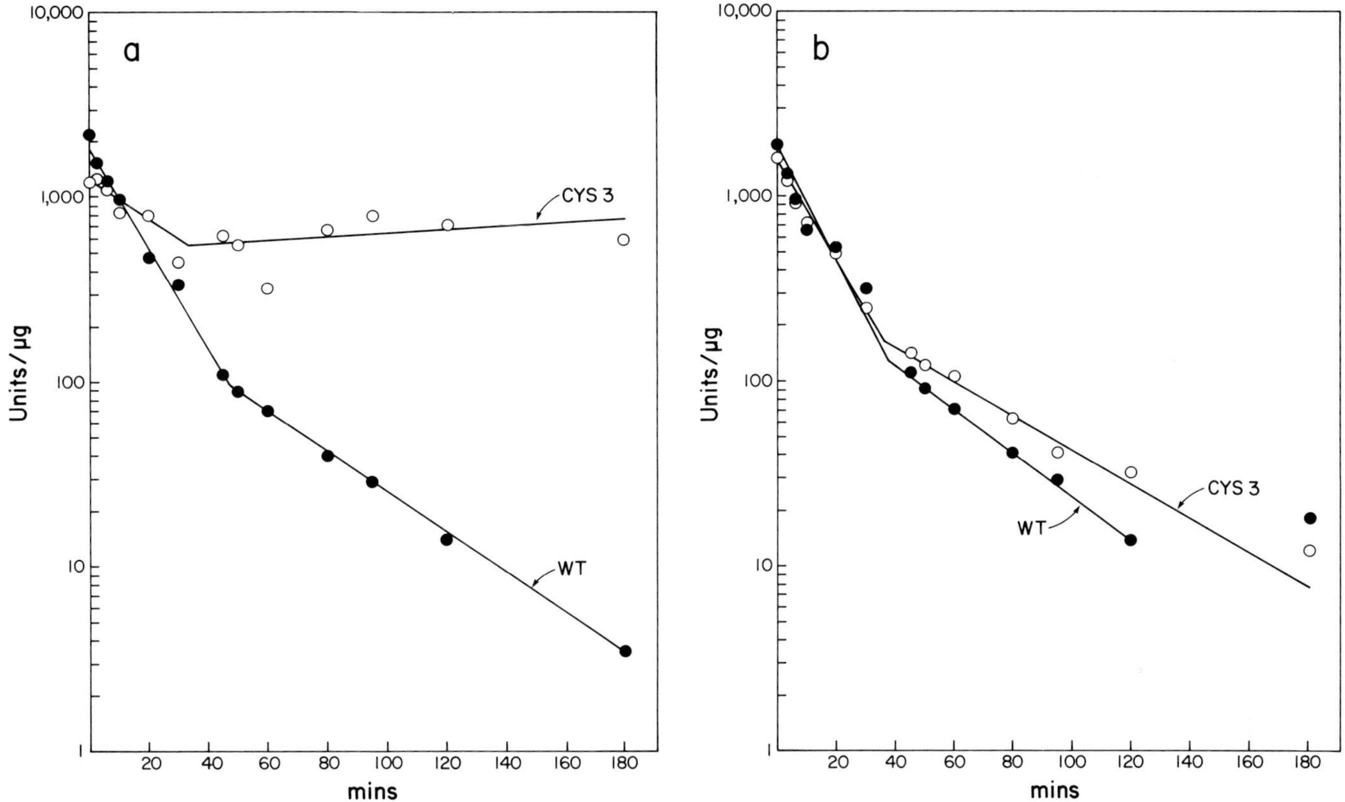

Fig. 3. Kinetics of inactivation of lysozyme derivatives at pH 6.5 and 67°C. At various times, samples were plunged into ice, then diluted and assayed for residual enzymatic activity at 20°C. (a) Wild-type T4 lysozyme (●) and T4 lysozyme(I3C^{-97C}) (○); (b) wild-type T4 lysozyme (●) and T4 lysozyme (I3C^{SH}) (○) in the presence of 10 m*M* β-mercaptoethanol. (From Perry and Wetzel, 1984; copyright 1984 by the AAAS.)

In our initial studies T4 lysozyme(I3C^{-97C}) behaved as if it were composed of variable proportions of stable and unstable molecules. For example, Fig. 3a shows that 50% of the preparation decays quickly, while the other 50% is quite stable. By studying the stability of lysozymes modified by reagents highly selective for thiol groups, we were able to show that a disulfide cross-linked lysozyme with a free thiol at Cys 54 is much less stable than one in which that thiol is blocked. We also showed that one mechanism by which the thiol of Cys 54 compromises the disulfide's ability to function involves attack by Cys 54 on the disulfide of another molecule to generate an intermolecular disulfide; the final inactive product in this case consists of a series of disulfide-linked oligomers (Perry and Wetzel, 1986).

Given this evidence for a negative effect of Cys 54, we decided to replace that residue with another amino acid. Modeling studies suggested that Thr and Val would both be reasonable replacements, and these derivatives were constructed in both the wild-type and disulfide backgrounds (Perry and Wetzel, 1986). As single replacements these lowered the specific activity of lysozyme to about 90% wild type; the cross-linked double variants were 50–60% wild-type activity. Figure 4 shows the enhanced stability of T4 lysozyme(I3C^{-97C}/C54T) compared to wild type at 75°C (Mulkerrin *et al.*, 1985).

Proteins which contain both a disulfide and a free cysteine are often unstable due to thiol–disulfide interchange (Torchinsky, 1981; Sogami *et al.*, 1969; Nikkel and Foster, 1971; McKenzie *et al.*, 1972; Huggins *et al.*, 1951; Frensdorff *et al.*, 1953). Site-directed mutagenesis was used previously to stabilize two such structures by replacing the free cysteine (Mark *et al.*, 1984; Wang *et al.*, 1984); the mechanisms by which the wild type is destabilized by the free thiol in these proteins have not been reported. Since free cysteines such as Cys 54 are likely to confuse our studies of conformational lability, our intention is to work in the future in a background free of unpaired cysteines.

Figure 4 also shows that T4 lysozyme(I3C^{-97C}/C54T) is essentially as stable as hen egg white lysozyme at pH 6.5 and 75°C. While T4 and hen egg white lysozyme do have some structural homology, it is not so great that we might expect to convert the T4 enzyme to a molecule with the stability of the hen egg white lysozyme simply by transferring a structural feature of the latter (a disulfide) to the former. The observed effect probably occurs not because of any structural homology, but because of some general features in the energy requirements for various mechanisms of irreversible inactivation of proteins (Feeney, 1980; Klibanov, 1983; Ahern and Klibanov, 1985).

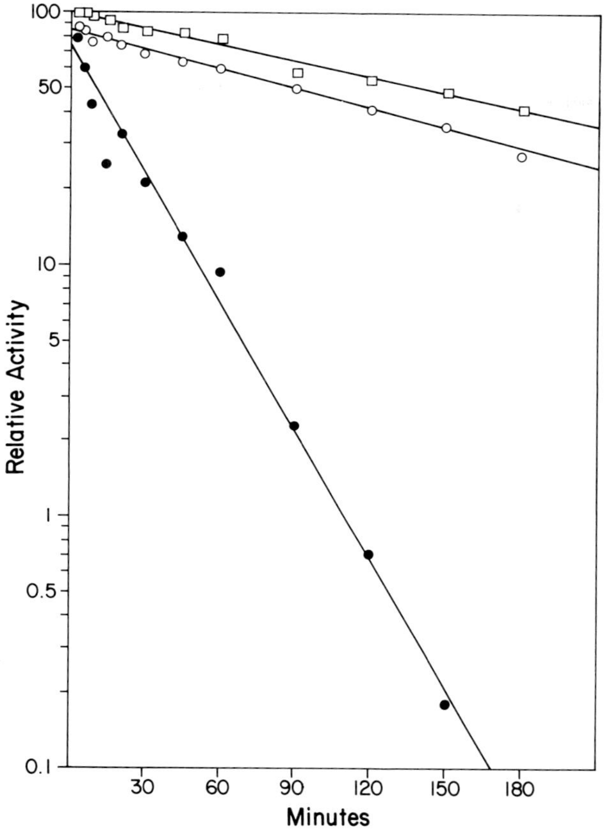

Fig. 4. Kinetics of inactivation of lysozyme derivatives at pH 6.5 and 75°C. Hen egg white lysozyme (□); wild-type T4 lysozyme (●); T4 lysozyme(I3C^{-97C}/C54T) (○).

IX. HOW DOES THE 3–97 DISULFIDE STABILIZE T4 LYSOZYME?

There are at least three general ways in which a disulfide might stabilize T4 lysozyme in these experiments. Since unpaired cysteines are amino acids especially susceptible to oxidation, we may have achieved stability in the cross-linked double mutants simply by manufacturing a molecule lacking unpaired cysteines. A second possibility is that stability to thermal inactivation is related to the stabilization toward reversible unfolding expected from an intramolecular cross-link (see above). Finally, it seems possible that a cross-link might significantly alter conformational mechanisms of inactivation in other ways, such as affecting folding and unfolding pathways and the structures of unfolded states.

Chemical modification of the wild-type cysteines has been shown to reduce the rate of thermal inactivation of the wild-type protein (Perry and Wetzel, 1986). Recently we have confirmed this by preparing a cysteine-free, non-cross-linked protein by directed mutagenesis (L. J. Perry and R. Wetzel, unpublished). Although previous work (Perry and Wetzel, 1984, 1986) suggests that oxidation of cysteines is not a major inactivation pathway for the wild-type, these experiments were conducted in buffers containing EDTA, which is known to inhibit air oxidation of thiols by scavenging trace metal ions. In fact, EDTA was employed in these experiments precisely for this reason.

We have recently shown that air oxidation of cysteines becomes more pronounced in the thermal inactivation of wild-type T4 lysozyme in buffers lacking EDTA (R. Wetzel and L. J. Perry, unpublished). As expected, this reaction increases in rate as the pH becomes higher. Thus, one way in which the disulfide stabilizes T4 lysozyme is by simply "blocking" the free thiols of T4 lysozyme. Of course, this is only possible when the other wild-type cysteine at position 54 is also blocked by alkylation or mutagenesis (see above).

There is, however, a significant nonoxidative pathway of inactivation open to non-cross-linked T4 lysozymes, and the 3-97 disulfide also seems to inhibit this reaction. The wild-type enzyme in EDTA and in reducing agent (Perry and Wetzel, 1984), alkylated wild type (Perry and Wetzel, 1986), and a cysteine-free non-cross-linked mutant (L. J. Perry and R. Wetzel, unpublished) all inactivate much more rapidly than cross-linked T4 lysozyme(I3C/C54T). This inactivation process seems to be related to alterations in the protein's folding pattern and/or solubility, and not in its chemical state, since exposure of inactivated protein to a denaturation/renaturation cycle in guanidine hydrochloride is sufficient to restore almost all of the enzymatic activity (Perry and Wetzel, 1986; R. Wetzel and L. J. Perry, unpublished).

Figure 1 shows with a simple model how proteins might undergo this kind of inactivation process. According to this model, the protein can enter various irreversible pathways of inactivation from an unfolded state which is in equilibrium with the native state; accordingly, if one of the irreversible steps is rate-limiting and identical for two proteins, the observed rates will depend on the steady state concentrations of the unfolded forms of the two proteins at the incubation temperature (Lumry and Eyring, 1954; Klibanov, 1983).

A disulfide might influence the observed rate by virtue of its expected effect (see above) on the equilibrium constant between N and U; a more thermodynamically stable molecule will be less unfolded at any given temperature and therefore will decay more slowly. In fact, the observed T_m for T4 lysozyme(I3C^{-97C}/C54T) is shifted a few degrees higher than that of wild type between pH 2 and 7 (W. Baase and W. Becktel, unpublished). Similarly, free energies of stabilization extrapolated to 67°C indicate that the wild type is more unfolded at that temperature than the cross-linked lysozyme.

It appears, however, that the disulfide has a more radical effect on the conformational inactivation mechanism. Even at temperatures above the T_m found for cross-linked T4 lysozyme(I3C/C54T) in reversible melting experiments, this molecule inactivates much more slowly than, for example, the cysteine-free non-cross-linked mutant. At temperatures where the cross-linked T4 lysozyme does undergo significant thermal inactivation, the inactivated product is only poorly reactivated by denaturation/renaturation. These and other observations suggest that the primary mechanisms by which the cross-linked T4 lysozyme undergoes thermal inactivation are chemical in nature; the conformational modes of inactivation found for non-cross-linked T4 lysozymes are effectively blocked by the 3-97 disulfide (R. Wetzel and L. J. Perry, unpublished).

Recently we obtained other evidence that the 3–97 disulfide makes more significant contributions to stability than via a purely thermodynamic effect on the unfolding equilibrium. We constructed triple variants which incorporate several known temperature-sensitive loci (Hawkes *et al.,* 1984) into the 3–97 disulfide background (L. J. Perry and R. Wetzel, unpublished). These cross-linked molecules exhibit T_m values lower than those of the wild-type molecule, at least in the acid pH range (W. Baase and W. Becktel, unpublished). However, they continue to be much more stable than the wild-type or cys-free molecules toward irreversible thermal inactivation (R. Wetzel, unpublished).

How, then, does the disulfide stabilize lysozyme against conformational modes of decay? It might slow the initial unfolding reaction to the extent that unfolding becomes rate-limiting. It might change the solution structure of the unfolded form, or alter the forward or backward rate constants for the "irreversible" step(s) following unfolding. Or the mechanism of conformational rates of inactivation for the wild type might be more complex than that pictured in Fig. 1. While the disulfide stabilization mechanism(s) is yet to be elucidated, we have shown that this stability is based on more than a change in the free energy of the initial unfolding equilibrium. These results suggest that although refined X-ray structures of such cross-linked derivatives may have much to tell us about thermodynamic stability, they will not necessarily be helpful in understanding stability toward irreversible inactivation. At least in the lysozyme example described above, the structures of unfolding intermediates may be more relevant than that of the native molecule.

If some disulfides contribute more to protein stability than by incremental free energy effects on unfolding equilibria, what does this tell us about the roles of these bonds in the evolution of protein structure and function? It seems possible that a new and distinct mode of stabilization might have a greater impact on protein evolution than would an additional structural feature capable of incrementally adding to the free energy of stabilization. From this point of view, one might visualize the cross-linked, triple variants alluded to above as molecules on one path of a simulated evolution of T4 lysozyme to a structure with some properties more like those of hen egg white lysozyme. Specifically, introduction of more ts loci into cross-linked T4 lysozyme should, in principle, generate structures which continue to be relatively stable to irreversible inactivation but which have such weakened packing interactions that they, like hen egg white lysozyme, depend on the free energy contributions of disulfides for net conformational stability under physiological conditions. Is this possible in practice? What will the enzymatic properties and stability of such molecules be? What new features will arise in these molecules which might give us an insight into some of the purposes for the wide existence of disulfides in extracellular globular proteins?

X. USES OF ENGINEERED DISULFIDES

Although the structural limits to the installation of disulfide bonds have not been formally addressed, it is already apparent that the synthesis of the disulfide cross-linked lysozyme described here is not a special case; Kraut and co-workers have described the introduction of a disulfide into dihydrofolate reductase (Villafranca *et al.*, 1983), and intramolecular disulfides have been placed at a number of sites, which vary considerably in their predicted compatibility with known disulfides, in the subtilisin molecule (P. N. Bryan, personal communication; J. A. Wells, personal communication). It thus seems likely that introduction of disulfides will become routine. This ability opens up a variety of studies directed to both fundamental and practical aspects of the fields of protein folding, biosynthesis, stability, and structure/function.

For example, one might use the redox potential of a disulfide as a measure of packing energies and of changes induced in these energies with further amino acid replacements. It may be possible to test molecular mechanisms of protein folding by looking for changes in folding kinetics on introduction of a carefully placed cross-link (Lin *et al.*, 1985). Variants of secretory proteins, designed to form disulfides in the native folding pattern, may give new information on mechanisms of protein secretion. If it is possible to prepare variants in which clear distortions in chain folding occur on disulfide cross-linking, such derivatives may be useful in protein structure/function studies, since the distortions may be of a different character than those obtained by amino acid replacement.

In our studies, we have found it possible to stabilize a protein toward both reversible and irreversible denaturation by the introduction of a disulfide. Clearly, this ability can be of commercial importance. At the same time, examination of the mechanisms by which stability toward both processes is achieved, as well as the relationship between them, is providing new insight into the ways in which disulfides contribute to the structure, stability, and evolution of proteins.

ACKNOWLEDGMENTS

H. L. Heyneker (Genencor), L. J. Perry and M. G. Mulkerrin (Genentech), and W. J. Becktel and W. A. Baase (U. Oregon) made valuable contributions to the work described in this chapter. I gratefully acknowledge discussions with R. Stroud, A. Fink, R. Baldwin, and I. Kuntz. My thanks to A. Gray and W. Anstine for help with the illustrations and preparation of the manuscript and to L. J. Perry for critical reading of this chapter. This work was supported by Genentech and Genencor.

REFERENCES

Ahern, T. J., and Klibanov, A. M. (1985). *Science* **228,** 1280–1284.

Anfinsen, C. B. (1973). *Science* **181,** 223–230.

Anfinsen, C. B., and Scheraga, H. A. (1975). *Adv. Protein Chem.* **29,** 205–300.

Becktel, W. J., and Baase, W. A. (1985). *Anal. Biochem.* **150,** 258–263.

Bergman, L. W., and Kuehl, W. M. (1979). *J. Biol. Chem.* **254,** 5690–5694, 8869–8876.

Bienkowska, K., and Taylor, A. (1979). *Eur. J. Biochem.* **96,** 581–584.

Creighton, T. E. (1978). *Prog. Biophys. Mol. Biol.* **33,** 231–297.

Dayhoff, M. O., ed. (1972). *In* "Atlas of Protein Sequence and Structure," p. D-2. National Biomedical Research Foundation, Silver Spring, Maryland.

de Boer, H., Heyneker, H., Comstock, L., Wieland, A., Vasser, M., and Horn, T. (1982). *Miami Winter Symp.* **19,** 309–327.

Desmadril, M., and Yon, J. M. (1984). *Biochemistry* **23,** 11–19.

Eklund, H., Cambillau, C., Sjöberg, B. M., Holmgren, A., Jornvall, H., Hoog, J.-O., and Branden, C.-I. (1984). *EMBO J.* **3,** 1443–1449.

Feeney, R. E. (1980). *ACS Symp. Ser.* **123,** 1–47.

Fersht, A. R., Shi, J. P., Knill-Jones, J., Lowe, D. M., Wilkinson, A. J., Blow, D. M., Brick, P., Carter, P., Waye, M. M. Y., and Winter, G. (1985). *Nature (London)* **314,** 235–238.

Freedman, R. B., and Hillson, D. A. (1980). *In* "Enzymology of Post-translational Modification of Proteins" (R. B. Freedman and H. C. Hawkins, eds.), Vol. 1, pp. 157–210. Academic Press, New York.

Frensdorff, H. K., Watson, M. T., and Kauzmann, W. (1953). *J. Amer. Chem. Soc.* **75,** 5157–5166.
Haber, E., and Anfinsen, C. B. (1962). *J. Biol. Chem.* **237,** 1839.
Hawkes, R., Gruetter, M. G., and Schellman, J. (1984). *J. Mol. Biol.* **175,** 195–212.
Huggins, C., Tapley, D. F., and Jensen, E. V. (1951). *Nature (London)* **167,** 592–593.
Jensen, H. B., Kleppe, G., Schindler, M., and Mirelman, D. (1976). *Eur. J. Biochem.* **66,** 319–325.
Johnson, R. E., Adams, P., and Rupley, J. A. (1978). *Biochemistry* **17,** 1479–1484.
Klibanov, A. M. (1983). *Adv. Appl. Microbiol.* **29,** 1–28.
Lin, S. H., Konishi, Y., Denton, M. E., and Scheraga, H. A. (1984). *Biochemistry* **23,** 5504–5512.
Lin, S. H., Konishi, Y., Nall, B. T., and Scheraga, H. A. (1985). *Biochemistry* **24,** 2680–2686.
Lumry, R., and Eyring, H. (1954). *J. Amer. Chem. Soc.* **58,** 110–120.
McKenzie, H. A., Ralston, G. B., and Shaw, D. C. (1972). *Biochemistry* **11,** 4539–4547.
Mark, D. F., Lu, S. D., Creasey, A. A., Yamamoto, R., and Lin, L. S. (1984). *Proc. Natl. Acad. Sci. U.S.A.* **81,** 5662–5666.
Matthews, B. W., Remington, S. J., Gruetter, M. G., and Anderson, W. F. (1981). *J. Mol. Biol.* **157,** 545–558.
Morehead, H., Johnston, P. D., and Wetzel, R. (1984). *Biochemistry* **23,** 2500–2507.
Mulkerrin, M. G., Perry, L. J., and Wetzel, R. (1985). *In* "Protein Structure, Folding and Design" (C. F. Fox and D. Oxender, eds.). Alan R. Liss, Inc., New York (in press).
Nikkel, H. J., and Foster, J. F. (1971). *Biochemistry* **10,** 4479–4486.
O'Farrell, P. H., Kutter, E., and Nakanishi, M. (1980). *Mol. Gen. Genet.* **179,** pp. 421–435.
Owen, J. E., Schultz, D. W., Taylor, A., and Smith, G. R. (1983). *J. Mol. Biol.* **165,** 229–248.
Perry, L. J., and Wetzel, R. (1984). *Science* **226,** 555–557.
Perry, L. J., and Wetzel, R. (1986). *Biochemistry* **25,** 733–739.
Perry, L. J., Heyneker, H. L., and Wetzel, R. (1985). *Gene* **38,** 259–264.
Remington, S. J., Anderson, W. F., Owen, J., Ten Eyck, L. F., Grainger, C. T., and Matthews, B. W. (1978). *J. Mol. Biol.* **118,** 81–98.
Richardson, J. (1981). *Adv. Protein Chem.* **34,** 167–339.
Rossman, M. G., and Argos, P. (1976). *J. Mol. Biol.* **105,** 75–95.
Saxena, V. P., and Wetlaufer, D. B. (1970). *Biochemistry* **9,** 5015–5022.
Scheele, G., and Jacoby, R. (1982). *J. Biol. Chem.* **257,** 12277–12282.
Schultz, G. E., and Schirmer, R. H. (1979). "Principles of Protein Structure." Springer-Verlag, Berlin and New York.
Sogami, M., Peterson, H. A., and Foster, J. F. (1969). *Biochemistry* **8,** 49–58.
Thornton, J. M. (1981). *J. Mol. Biol.* **151,** 261–287.
Torchinsky, Y. M. (1981). "Sulfur in Proteins." Pergamon, Oxford.
Tsugita, A. (1971). *In* "The Enzymes" (P. D. Boyer, ed.), 3rd ed., Vol. 5, pp. 344–411. Academic Press, New York.
Tsugita, A., Inouye, M., Terzaghi, E., and Streisinger, G. (1968). *J. Biol. Chem.* **243,** 391–397.
Villafranca, J. E., Howell, E. E., Voet, D. H., Strobel, M. S., Ogden, R. C., Abelsen, J. N., and Kraut, J. (1983). *Science* **222,** 782–788.
Wang, A., Lu, S. D., and Mark, D. F. (1984). *Science* **224,** 1431–1433.
Zoller, M. J., and Smith, M. (1984). *DNA* **3,** 479–488.

17

Genetic Identification of Amino Acid Sequences Influencing Protein Folding

JONATHAN KING, MYEONG-HEE YU,[1] JAVED SIDDIQI,[2] AND CAMERON HAASE

Department of Biology
Massachusetts Institute of Technology
Cambridge, Massachusetts 02139

I. THE CODING ASPECT OF THE FOLDING PROBLEM

A fundamental aspect of gene expression is the folding of the newly synthesized polypeptide chain into the precisely organized spatial conformation required for biological activity. Over 20 years have passed since Christian Anfinsen (1973) demonstrated the refolding of purified denatured ribonuclease to the fully active form in the absence of any cellular components. These results established that all the information for determining the spatial organization of a protein could reside in the amino acid sequence.

At present, the three-dimensional structures of well over 100 native proteins are known to atomic resolution. The amino acid sequence of each of these proteins, and of thousands more, is also precisely known. Yet it is still not possible to deduce or predict from inspection of the sequence of a protein of unknown structure and function precisely how the polypeptide backbone and side chains will be arranged in space. Nor do we understand, for those proteins whose structure has been solved, why the chain takes the conformation it does—why the hemoglobin se-

[1] Present address: Center for Genetic Engineering, Korean Advanced Institute for Science and Technology, Cheong Ryang Ri, Dong Dae Moon-ku, Seoul, Korea.

[2] Present address: Dana Farber Institute, Boston, Massachusetts 02115.

PROTEIN ENGINEERING
Applications in Science,
Medicine, and Industry

quence forms hemoglobin molecules and the influenza hemagglutinin sequence forms hemagglutinin spikes.

Clearly there is some program or code or set of rules embedded in the amino acid sequences that determines the outcomes of the folding and subunit assembly processes. In a sense only half of the genetic code has been solved—the code that relates the linear sequence of nucleotides in DNA to the linear sequence of amino acids in proteins. This code is very simple, and context-independent—UUU means phenylalanine independent of its position in the gene and independent of when in the cell cycle the gene is transcribed and translated. However, an immensely complex cellular apparatus is required to read out the code.

The second half of the code, which relates amino acid sequence to final conformation, is not only unsolved, but not well formulated. Its readout apparatus is much less complex, involving only interactions between segments of the polypeptide chain with each other and with solvent. However, the code or grammar is probably much more complex and is context-dependent.

Why has the problem eluded solution for so long? A number of experimental and conceptual barriers have impeded progress in this direction. Protein chemists have often viewed the sequence as directly determining the final structure of the protein. This is probably an inadequate formulation, since it leaves out the role of the sequence in determining the kinetic steps and conformational intermediates in the folding pathway (King, 1986).

A. Importance of Intermediates

Experimental analysis of refolding pathways has clearly established the existence of intermediates in folding pathways (Kim and Baldwin, 1982). The work of Creighton with bovine pancreatic trypsin inhibitor (BPTI) (Creighton, 1980; Creighton and Goldenberg, 1984) and our work on the P22 tail spike (Goldenberg and King, 1982; Goldenberg *et al.*, 1983) indicate that some of the intermediates are obligatory. Creighton has identified sequential intermediates in BPTI and shown that at least one of these intermediates has a conformation not found in the native protein (Creighton, 1980). This means that one cannot deduce folding pathways by inspection of the native structure. The pathways will be revealed only by direct investigation.

The formation of folding intermediates must be specified in the information encoded by the amino acid sequence. To decipher this information requires knowledge of the conformation of the intermediate. Such information has been experimentally difficult to obtain, given the instability of folding intermediates, and only recently have techniques been developed

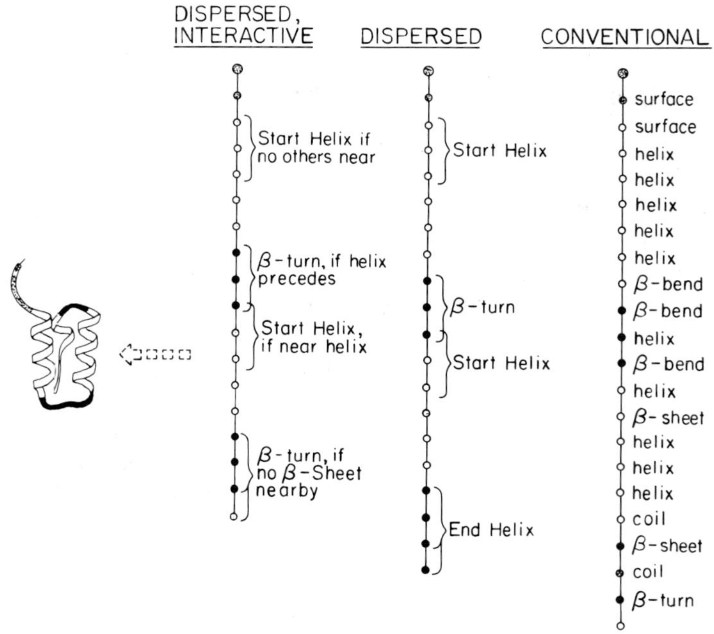

Fig. 1. The strings of beads represent an amino acid sequence that will form the helix–turn–helix conformation shown at the left. The "conventional" model at the right assumes that all residues in the sequence contribute to specifying chain conformation. The model in the center assumes that specialized sequences direct the folding pathway while the intervening residues are passive, presumably awaiting their later role in the activity of the native protein. The model at the left, which we prefer, assumes that dispersed sequences direct the folding pathway, but that these sequences are sensitive to their local environment so that identical sequences would specify different conformations, depending on context.

to elucidate the conformation of short-lived intermediates (Bierzynski *et al.*, 1982).

B. Linear Density of Folding Information

Another impediment, both conceptual and empirical, has been in the assessment of the density of the coding information in the amino acid sequence. Almost all algorithms which attempt to relate structure to sequence implicitly assume that every residue carries information for the folding process, as shown in the right-hand part of Fig. 1. Each residue in the chain is assigned some propensity to be in various secondary structures, and these values are multiplied or added to get a final prediction.

In fact, experimental data indicate that many residues are not involved in determining structure. Comparison of sequences from homologous chains from different species, such as globins, shows clearly that many residues can be varied with no effect on the native conformation. Second, the widespread existence of amber mutations, where multiple amino acids can be inserted and still yield a functional protein, indicates that some residues do not contribute to native conformation (Miller *et al.,* 1979). These results suggest that the folding information is, at a minimum, dispersed through the polypeptide chain, as shown in the center of Fig. 1. In this formulation certain regions of the chain would drive the folding pathway, while other regions have their conformation determined by those sequences or residues.

Unfortunately there have been very few direct experimental efforts to identify the meaningful residues. Recent studies of signal sequences and of ribonuclease, T4 lysozyme, and the P22 tail spike have begun to fill this gap in our knowledge (Lehnhardt *et al.,* Chapter 10, this volume; Bierzynski *et al.,* 1983; Hawkes *et al.,* 1984; Goldenberg *et al.,* 1983).

II. GENETIC ANALYSIS OF PROTEIN FOLDING

We have taken a genetic approach to questions of how the folding and conformational information is distributed through a polypeptide chain. These questions include:

How densely is the folding information distributed through the sequence; how many residues are important?

Which amino acid residues in a polypeptide chain drive the folding pathway?

How are these residues organized? As words, for example, "start helix", "stop helix"?

What are the size and general character of these words, how many residues are in them; are they contiguous or dispersed?

Our approach has been to isolate and characterize mutants which specifically alter the folding and subunit association pathway of a polypeptide chain, without altering the native protein. Such mutants distinguish residues involved in the kinetic control of conformation from residues involved in the stability and activity of the native protein. This approach is complementary to the efforts to characterize mutations which alter the stability of the mature protein (Shortle, Chapter 14, this volume; Hawkes *et al.,* 1984). It is likely that many residues will have roles in both aspects of the functioning of the polypeptide chain. We thought it likely, however, that at least with large proteins these aspects might be segregated in different local sequences.

In order to distinguish effects of amino acid substitutions on intermediates from effects on the native structure, we have focused on a protein in which folding intermediates and end products are kinetically and energetically well separated. The protein utilized in these studies is the thermostable cell attachment protein of *Salmonella* phage P22 (Berget and Poteete, 1980).

A. The Thermostable Tailspike of P22

The tailspike is a trimer of polypeptide chains coded for by gene 9 of P22 (Berget and Poteete, 1980; Goldenberg *et al.,* 1982). The gene 9 polypeptide chain is 666 amino acids long, without unusual features of amino acid composition or sequence (Sauer *et al.,* 1982).

Figure 2 summarizes the essential properties of the protein. One end binds to phage heads *in vivo,* providing an activity assay (Israel *et al.,* 1967). The other end is catalytic and is an endorhamnosidase, cleaving the O antigen of *Salmonella* (Iwashita and Kanegasaki, 1976). The protein is a strong antigen and is the target of rabbit anti-phage antibodies.

The native protein is very stable, consistent with its role as an external structural protein of the phage. It is stable to heat up to 80°C (Goldenberg and King, 1981). It is resistant to proteases and is not denatured by sodium dodecyl sulfate up to 60°C, though, if heat-denatured, the chains form normal SDS–polypeptide chain complexes (Goldenberg *et al.,* 1982).

The tailspike maturation pathway is purely conformational, with no proteolytic cleavage, glycosylation, phosphorylation, or other covalent modification known.

B. Folding and Subunit Assembly Pathway for the Tailspike

Since the native protein is SDS- and trypsin-resistant, partially folded intermediates can be easily distinguished from the mature spikes by SDS sensitivity or protease sensitivity (Goldenberg *et al.,* 1983). Studies of the wild-type pathway have revealed the following sequence: the newly synthesized polypeptide chains are released from the ribosome as single chains. They partially fold to a structure that has the specificity for chain–chain recognition. These chains associate into a complex called the protrimer. This kinetic intermediate in the folding and assembly pathway can be trapped by rapid cooling (Goldenberg and King, 1982). The protri-

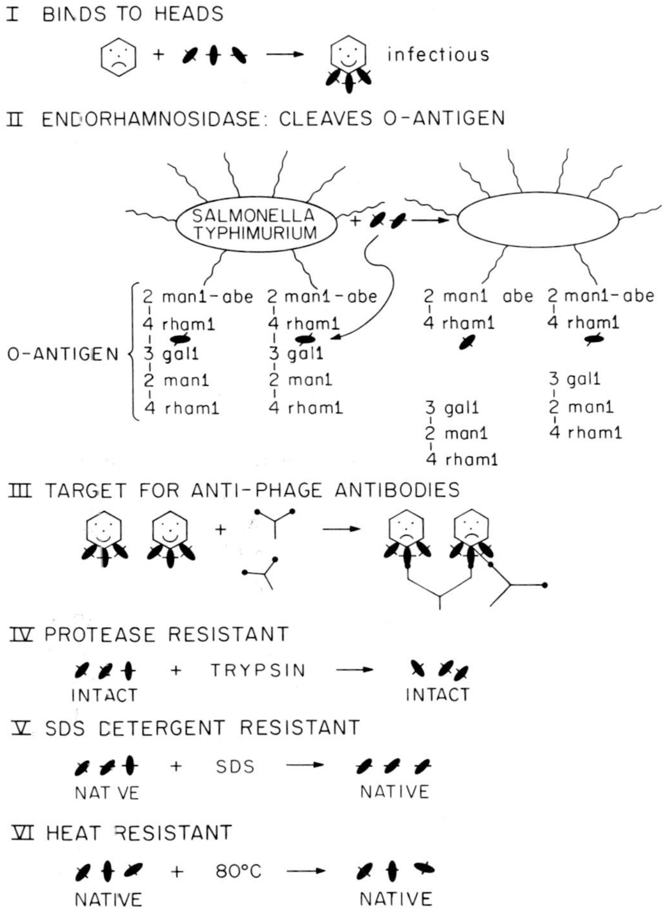

Fig. 2. Properties of the native P22 tailspike.

mer intermediate has the three chains associated but not fully folded (Fig. 3). It migrates more slowly in a native gel and is sensitive to proteases and detergent. On warming to room temperature, this species converts to the thermostable native spike *in vitro*.

Note that in this pathway, as with myosin or collagen, there is no species corresponding to a native monomer. The individual polypeptide chains require interaction with their sibs in order to achieve the native conformation (Goldenberg and King, 1982).

The early steps in this pathway are thermolabile. As the temperature is increased, the fraction of chains successfully passing through the pathway decreases (Goldenberg *et al.*, 1982; Yu and King, 1984). The last step, the

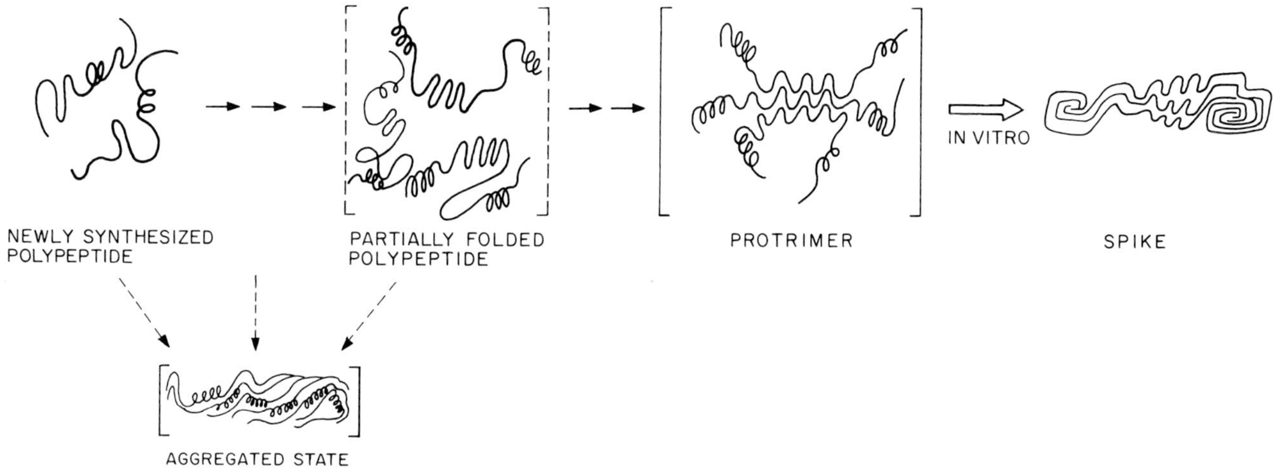

Fig. 3. Pathway of tailspike maturation. Newly released polypeptide chains fold partially into a conformation that has enough structure for specific chain–chain recognition. These species associate into the protrimer, in which the chains are associated but not fully folded. Both the unassociated and protrimer forms are sensitive to proteases and SDS. The protrimer folds further to form the highly stable native spike, which is SDS- and protease-resistant and is stable to 80°C. This last reaction proceeds *in vitro* (Goldenberg and King, 1982). Although the native spike is heat-stable, an early intermediate is heat-labile (Goldenberg and King, 1981; Goldenberg *et al.*, 1982).

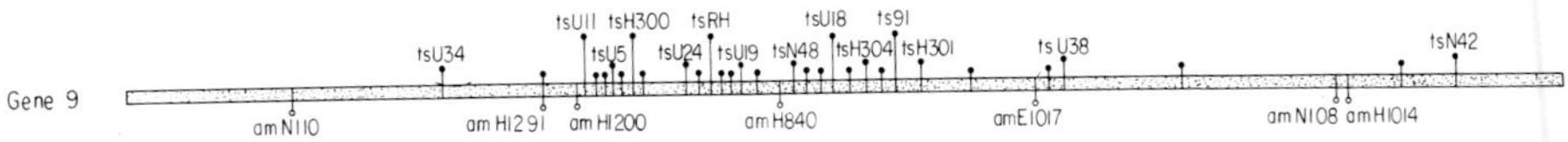

Fig. 4. Map of temperature-sensitive folding mutations in gene 9. The 31 tags above the line represent sites of TSF mutations (Smith *et al.*, 1980; Smith and King, 1981; Goldenberg *et al.*, 1983), some of which are labeled for further reference. The sites of seven amber mutations are shown below the line.

protrimer-to-trimer transition, has the opposite character. It is inhibited in the cold, permitting trapping, but proceeds at optimal efficiency in the 37–40°C temperature range (Goldenberg and King, 1982).

C. Temperature-Sensitive Folding (TSF) Mutations

To isolate mutants blocked in the folding pathway, we focused on temperature-sensitive mutations, since a number of lines of experimentation indicate that many TSS (temperature-sensitive synthesis) mutants were indeed folding mutations (Smith *et al.*, 1980). We isolated over 100 temperature-sensitive mutations in the gene for this thermostable protein. Figure 4 shows a map of 31 sites of such mutations. The mapping is not yet complete, but the mutations are distributed over a minimum of 55 sites. They are concentrated in the central region.

These are conventional ts mutants—they form a plaque at low temperature but not at high temperature (Edgar and Lielausis, 1964). To study the mutant proteins, we introduce an additional mutation blocking capsid shell assembly, so that tailspikes accumulate as soluble precursors and also are overproduced.

In considering the temperature-sensitive mutants it is important to bear in mind that two protein preparations must be characterized for each mutant protein, polypeptide chains released from the ribosome at restrictive temperature, and the protein synthesized at permissive temperature. The character of the TSF mutants is such that these have very different spatial conformations.

At permissive temperature the polypeptide chains carrying TSF substitutions proceed through the folding and subunit assembly pathway and form native tailspikes. This can be seen in the pulse/chase experiment shown in the top part of Fig. 5. Newly synthesized but incompletely folded or associated polypeptide chains are dissociated by SDS and migrate to the position expected of 71,000-dalton chains. The mature spikes do not bind SDS and migrate more slowly (Goldenberg *et al.*, 1982).

Although these native proteins carry the TSF amino acid substitution, they are essentially indistinguishable from the wild type. In particular, they are as thermostable as the wild type, requiring temperatures above 80°C for heat denaturation (Goldenberg and King, 1981). The substitutions have little effect on the mature state of the protein. In a sense, they represent residues that are not involved in the activities of the mature protein.

Some of the native forms of the TSF proteins are distinguishable from the wild type in having an altered electrophoretic mobility, for example, tsH302 in Fig. 5.

At a restrictive temperature, the results are very different. The polypeptide chains carrying the mutant amino acid are synthesized at normal rates and are stable within the cell. However, they are blocked relatively early in the folding pathway and accumulate as SDS-sensitive polypeptide chains (Fig. 5, lower gel). These chains display none of the activities of the native protein—head binding, endorhamnosidase, absorption, or antigenicity—and do not form the protrimer (Smith and King, 1981; Goldenberg *et al.*, 1983). In fact, they are precipitated by antibody made against guanidine-denatured tailspike chains indicating that their conformation is closer to the unfolded (Goldenberg *et al.*, 1983). On cell lysis, and possibly before, these incompletely folded chains associate into large aggregates.

The high-temperature state of the tailspike polypeptide chain is not simply a misfolded species. On shifting infected cells to a permissive temperature, active tailspikes are recovered (Smith and King, 1981). Thus the mutant polypeptide chains accumulating are reversibly related to intermediates in the folding pathway.

The mutant proteins coded by TSF mutations at 33 sites have been studied in some detail. All of them have the same general character; they do not affect the native protein formed at the permissive temperature, but they block the formation of the native protein if the polypeptide chain was released from the ribosome at the restrictive temperature. Thus, this genetic map is a map of sites in the polypeptide chain which play a critical role in the folding pathway, at least at elevated temperatures. (Fig. 6).

D. The Role of Aggregation

Characterization of the physical state of the inactive TSF mutant polypeptide chains synthesized at a restrictive temperature indicates that they are aggregated. After lysis of infected cells, the mutant gene 9 polypeptide chains are in a rapidly sedimenting form. The complex bands with a density of 1.3 g/ml in a cesium chloride density gradient, indicating that it is aggregated polypeptide chains.

These aggregates represent dead ends formed from incompletely folded

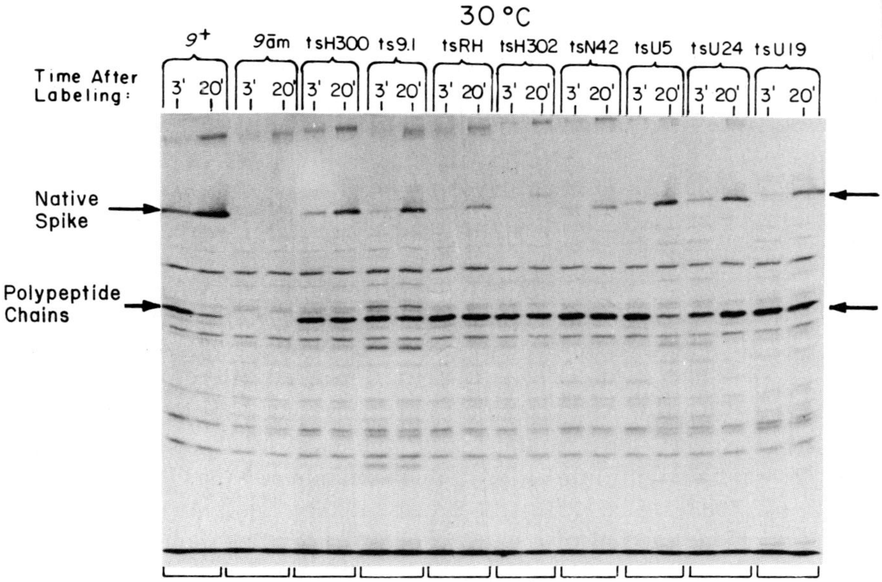

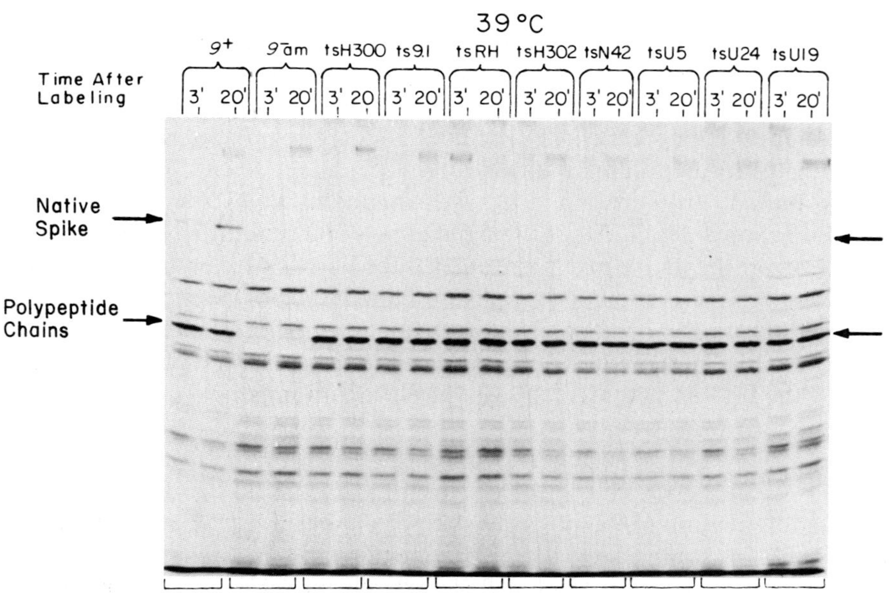

Fig. 5. The state of newly synthesized TSF gene 9 polypeptide chains at different temperatures. Incompletely folded and native forms of the tailspike are resolved in these gels. Both categories are missing from cells infected with a nonsense mutant of gene 9, as can be seen in the second set of lanes. The native spike is formed only by the mutants at

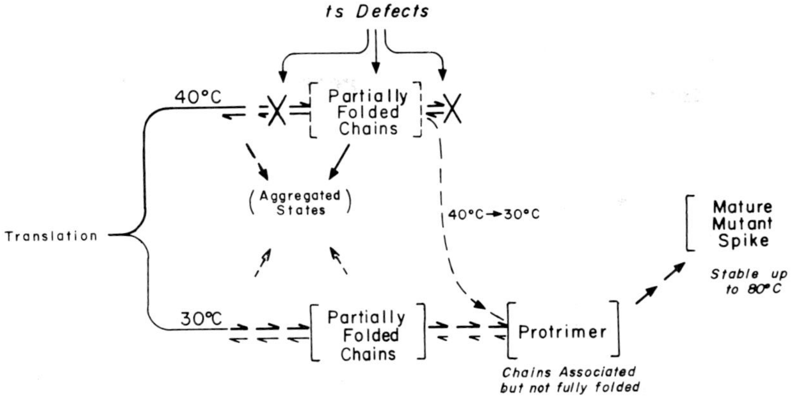

Fig. 6. Locus of the TSF defects in the tailspike folding and subunit association pathway. Most of the mutants appear to act by blocking an early step in chain folding, prior to the stage of chain association (Goldenberg *et al.*, 1983). This may involve destabilizing an already thermolabile intermediate in the wild-type pathway or speeding up an off-pathway aggregation reaction (Yu and King, 1984). Presumably the mutant chains at high temperature are unable to achieve the conformation required for chain–chain association. Since the mutations map throughout the gene, it is unlikely that they affect a specific domain required for chain association.

polypeptide chains and are also present in wild-type infected cells at high temperatures. However, the aggregation step could be a primary effect of the TSF mutations. Increasing the rate of an off-pathway step would have the same overall effect as decreasing the stability of an on-pathway intermediate.

Measurements of the overall rate of the tailspike maturation pathway as a function of temperature reveal that the mutations do not alter the rates, but rather reduce the yield with increasing temperature (Yu and King, 1984). This is consistent with an increase in the efficiency of an off-pathway step such as aggregation.

30°C; at 39°C the mutant polypeptide chains accumulate as SDS-sensitive forms. Cell cultures infected with reference and various TSF mutants of the tailspike were divided and incubated at permissive and restrictive temperatures (Smith and King, 1981). The cultures were exposed to a short pulse of ^{14}C-labeled amino acids. Samples were withdrawn at 3 and at 20 min and the reaction stopped by freezing. Samples were thawed and heated in SDS prior to electrophoresis through an acrylamide gel.

TABLE I

Amino Acid Substitutions of Gene 9 TS Folding Mutants

Mutation	Residue	Substitution	Local sequence
tsU9	177	Gly > Arg	Phe · Ile · Gly · Asp · Gly · Asn · Leu · Ile · Phe
tsH304	244	Gly > Arg	Val · Lys · Phe · Pro · Gly · Ile · Glu · Thr · Leu
tsH302	323	Gly > Asp	Asn · Tyr · Val · Ile · Gly · Gly · Arg · Thr · Ser
tsU38	435	Gly > Glu	Leu · Leu · Val · Arg · Gly · Ala · Leu · Gly · Val
tsH300	235	Thr > Ile	Gly · Tyr · Gln · Pro · Thr · Val · Ser · Asp · Tyr
tsU18	307	Thr > Ala	Asp · Gly · Ile · Ile · Thr · Phe · Glu · Asn · Leu
tsH301	368	Thr > Ile	Thr · Trp · Gln · Gly · Thr · Val · Gly · Ser · Thr
tsU5	227	Ser > Phe	Thr · Leu · Lys · Gln · Ser · Lys · Thr · Asp · Gly
tsN48	333	Ser > Asn	Gly · Ser · Val · Ser · Ser · Ala · Gln · Phe · Leu
tsU19	285	Arg > Lys	Gly · Phe · Leu · Phe · Arg · Gly · Cys · His · Phe
tsU53	382	Arg > Ser	Asn · Leu · Gln · Phe · Arg · Asp · Ser · Val · Val
tsH303	250	Pro > Ser	Glu · Thr · Leu · Leu · Pro · Pro · Asn · Ala · Lys
tsU11	250	Pro > Leu	Glu · Thr · Leu · Leu · Pro · Pro · Asn · Ala · Lys
tsU24	258	Ile > Leu	Lys · Gly · Gln · Asn · Ile · Thr · Ser · Thr · Leu
tsRAF tsRH	270	Val > Gly	Glu · Cys · Ile · Gly · Val · Glu · Val · His · Arg
ts9.1	334	Ala > Val	Ser · Val · Ser · Ser · Ala · Gln · Phe · Leu · Arg

E. Amino Acid Substitutions in the TSF Mutants

We have determined the amino acid substitutions of a number of these mutant proteins by isolating restriction fragments from the phage DNA and sequencing the fragments. The genetic map was used to select the fragments to be isolated (Yu and King, 1984). A number of mutations have also been sequenced by cloning restriction fragments into M13 and sequencing the insert. In all cases where we found nucleotide substitutions, they represented a single amino acid substitution. These are shown in Table I.

III. DISCUSSION

A. Sites of TSF Mutations

We initially thought that many of these mutations would occur at the sites of prolines. The restricted rotation around the peptide bond would be more important at high temperature than at low temperature in controlling the correct trajectory during the folding process. In fact, only one proline residue has been identified as a site of TSF mutations, though we have a number of independent occurrences at that site.

The substitutions are quite diverse, as shown in Table I. However, two classes stand out: replacements of glycines and replacements of threonines.

1. Glycine Replacements

Four glycine residues have been identified as sites of TSF mutations. For all four the substitutions are by bulky charged amino acids. Although it was not surprising that these substitutions would cause problems, it was difficult to understand how they could be accommodated in the native protein at a permissive temperature. In fact, the native mutant proteins specified by these mutants have altered electrophoretic mobilities and isoelectric points, indicating that they are on the protein surface. This allows the charge residues to be accommodated at sites normally occupied by glycines (M. H. Yu and J. King, unpublished results.)

The low-temperature result shows that there is no steric hindrance in the protein. Therefore, the high-temperature defect must be due to a kinetic effect on the folding pathway, as indicated by the mutant phenotypes (Smith and King, 1981).

Since turns are concentrated at the surface of proteins and glycines are common at turns, we interpret these sites as representing glycines at surface turns. The charged residue would alter the interaction with the solvent, or perhaps introduce interactions with other parts of the chain, which at an elevated temperature interferes with the folding pathway.

2. Threonine Replacements

Three of the sequenced TSF mutations are replacements of threonines. The mutant amino acids lack the side-chain hydroxyl group. The side-chain hydroxyl groups of threonines are most frequently involved in internal hydrogen bonds with an amido or carbonyl group of a nearby backbone. If we imagine a bend or turn in which the threonine hydroxyl is hydrogen-bonded to the backbone of the opposite side of the turn, it is easy to understand the temperature effect. At a permissive temperature a turn could be stable enough to continue through the pathway. However, at an elevated temperature the turn would be thermolabile without the stabilizing bond.

Examination of the local sequence surrounding the two threonine → isoleucine replacements (shown in Table II) reveals distinctive homology. At three positions the residues are identical. Tyrosine and tryptophan are related, and at the II position of a β turn, glycine and proline are equiva-

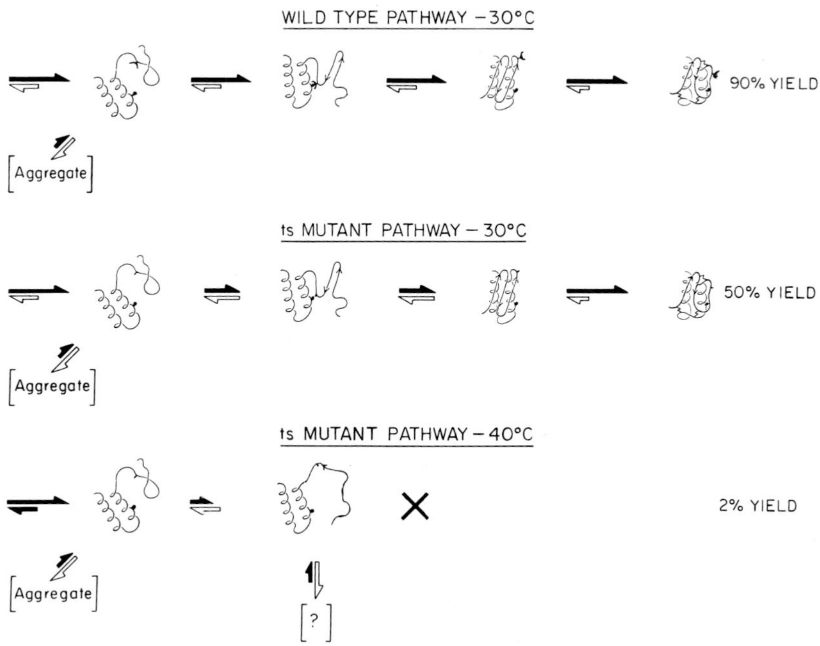

Fig. 7. Model for the molecular basis of a TSF defect. This model shows an interaction between a segment of polypeptide chain forming β strands and a nucleating pair of α helices. The interaction stabilizes the complex, ensuring the correct pathway at elevated temperatures. At lower temperatures the reaction still proceeds with lower efficiency. This interaction is present only in the intermediate and is not found in the final structure, as indicated in the pathway. Thus the substitution does not destabilize the native structure, but only the intermediate. The intermediates are from drawings of Jane Richardson.

TABLE II

Potential Foldon for a Stable Turn in the Tailspike[a]

Residues	Allele	Sequence
231–239	Wild type	Gly · **Tyr** · **Gln** · **Pro** · **Thr** · **Val** · Ser · Asp · Tyr
231–239	tsH300	Ile
364–372	Wild type	Thr · **Trp** · **Gln** · **Gly** · **Thr** · **Val** · Gly · Ser · Thr
364–372	tsH301	Ile

[a] Homologous sequences are shown in boldface.

lent amino acids. Since the replacements of threonine by isoleucine cause similar functional defects, the local sequences may well specify similar steps in the folding pathway. For example, in light of the previous discussion, these sequences might be specifying turns in the folding pathway. We term such a sequence a "foldon." We were able to identify the sequence relationship because contiguous amino acids were homologous. However, if the relationship was between residues that were somewhat dispersed, the pattern would probably go unrecognized.

B. Are TSF Sites Involved in Chain Folding or Chain Association?

At restrictive temperatures, all of the TSF mutant polypeptide chains we studied are blocked prior to the formation of the protrimer intermediate (Goldenberg *et al.,* 1983). We have interpreted these and other results as indicating that the mutant amino acid substitutions prevent the chains from achieving the conformation needed for specific chain–chain recognition and association. This kinetic explanation shows why the mutant proteins, once native, are as thermostable as the wild type.

An alternative model is that the mutations do not act kinetically, but destabilize a specific domain needed for association. However, the distribution of the mutations throughout the polypeptide chain indicates that a very large fraction of the polypeptide chain participates in the "domain"—in fact, the entire chain. Furthermore, since amino acid substitutions at over 30 different sites destabilize this special domain, each one cannot make a very large contribution to its stability. Yet the different mutants all sharply reduce protrimer formation at restrictive temperatures.

The observations are better accounted for by a model in which the mutations interfere with kinetic steps in the pathway, such that the chain never achieves the correct conformation for association. The substitutions could also affect kinetic steps by speeding up off-pathway interactions.

C. Density of the Folding Information

We have established that the TSF mutations mark sites in the protein important for folding and subunit assembly at restrictive temperatures. These sites could be essential sites for folding, regardless of temperature. In that case, our substitutions are a selected subset which interfere only at elevated temperatures. Presumably other amino acid substitutions would prevent correct folding at all temperatures. Alternatively, they could mark residues which are needed only at higher temperatures, for example, the model for the threonine turns.

Peter Berget and colleagues at the University of Texas have isolated absolute mutations in gene 9. Many of these appear to be defective in some step in folding or subunit association. Determination of the precise position of these mutations should answer the question of the nature of the TSF sites.

D. Silent or Spacer Sites

A second category of sites relevant to folding consists of those which do not carry critical information for these processes. The existence of such sites can be deduced in certain protein families by the prevalence of considerable variability between related species. In phage proteins some of these sites are identified by amber mutations. For example, substitution of the wild-type residue by serine, tyrosine, glutamic acid, or leucine, at a number of amber sites in gene 9, produces a fully functional protein. We interpret these results as meaning that these sites do not carry critical information for the folding pathway (King, 1986).

Although we have begun to characterize the defects at amber sites in gene 9, we do not yet have a large enough collection to compare their distribution with the distribution of TSF sites.

E. Is the Folding Code Context-Dependent?

The fact that similar sequences can appear in different conformations establishes that the signals for folding are context-dependent (Kabsch and Sander, 1984). The information assumed by any local region is determined by the conformation nearby. Thus one cannot generalize from sequence to conformation, since this procedure is context-independent. Determining the context effects will require working out the actual kinetic pathways, since the context develops kinetically as the protein is synthesized and soon after release from the ribosome.

ACKNOWLEDGMENTS

This work was supported by grant DMB-8402546 from the National Science Foundation and GM17,980 from the National Institutes of Health. We thank David Goldenberg and Donna Smith for valuable discussions. Ruth Martin provided expert assistance in preparation of the manuscript.

REFERENCES

Anfinsen, C. B. (1973). *Science* **181,** 223–230.

Berget, P. B., and Poteete, A. R. (1980). *J. Virol.* **34,** 234–243.

Bierzynski, A., Kim, P. S., and Baldwin, R. L. (1980). *Proc. Natl. Acad. Sci. U.S.A.* **79,** 2470–2474.

Creighton, T. E. (1980). *In* "Protein Folding" (R. Jaenicke, ed.), pp. 427-446. Elsevier/North-Holland Biomedical Press, Amsterdam.

Creighton, T. E., and Goldenberg, D. P. (1984). *J. Mol. Biol.* **179,** 497–526.

Edgar, R. S., and Lielausis, I. (1964). *Genetics* **49,** 649–662.

Goldenberg, D. P., and King, J. (1981). *J. Mol. Biol.* **145,** 633–651.

Goldenberg, D. P., and King, J. (1982). *Proc. Natl. Acad. Sci. U.S.A.* **79,** 3403–3407.

Goldenberg, D. P., Berget, P. B., and King, J. (1982). *J. Biol. Chem.* **257,** 7864–7871.

Goldenberg, D. P., Smith, D. H., and King, J. (1983). *Proc. Natl. Acad. Sci. U.S.A.* **80,** 7060–7064.

Hawkes, R., Grutter, M. G., and Schellman, J. (1984). *J. Mol. Biol.* **175,** 195–212.

Israel, J. V., Anderson, T. F., and Levine, M. (1967). *Proc. Natl. Acad. Sci. U.S.A.* **57,** 284–291.

Iwashita, S., and Kanegasaki, S. (1976). *Eur. J. Biochem.* **65,** 87–94.

Kabsch, W., and Sander, C. (1984). *Proc. Natl. Acad. Sci. U.S.A.* **81,** 1075–1078.

Kim, P. S., and Baldwin, R. L. (1982). *Annu. Rev. Biochem.* **51,** 459–489.

King, J. (1986). *Biotechnology* **4,** 297–303.

Miller, J. H., Coulondre, C., Hofer, M., and Schmeissner, U., Sommer, H., and Schmitz, A. (1979). *J. Mol. Biol.* **131,** 191–222.

Sauer, R. T., Krovatin, W., Poteete, A. R., and Berget, P. B. (1982). *Biochemistry* **21,** 5811–5815.

Smith, D. H., and King, J. (1981). *J. Mol. Biol.* **145,** 653–676.

Smith, D. H., Berget, P. B., and King, J. (1980). *Genetics* **96,** 331–352.

Yu, M.-H., and King, J. (1984). *Proc. Natl. Acad. Sci. U.S.A.* **81,** 6584–6588.

III

Complex Systems

18

Structural Basis for Acetylcholine Receptor Function

MASAYOSHI MISHINA AND SHOSAKU NUMA

Departments of Medical Chemistry and Molecular Genetics
Kyoto University Faculty of Medicine
Kyoto 606, Japan

I. INTRODUCTION

The nicotinic acetylcholine receptor (AChR) is a ligand-gated ionic channel present in the postsynaptic membrane at the vertebrate neuromuscular junction and is the best characterized neurotransmitter receptor from both structural and functional viewpoints. The AChR from the electric organ of the ray *Torpedo californica* is composed of four kinds of subunits assembled in a molar stoichiometry of $\alpha_2\beta\gamma\delta$ (for reviews, see Karlin, 1980; Changeux, 1973; Conti-Tronconi and Raftery, 1982). This pentameric transmembrane protein complex contains both the binding site for acetylcholine (ACh) and the ionic channel. The primary structures of all four subunits of fish and mammalian AChRs have been deduced from the cDNA or genomic DNA sequences (Noda *et al.*, 1982, 1983a,b,c; Sumikawa *et al.*, 1982; Claudio *et al.*, 1983; Devillers-Thiery *et al.*, 1983; Tanabe *et al.*, 1984; Takai *et al.*, 1984, 1985; LaPolla *et al.*, 1984; Nef *et al.*, 1984; Kubo *et al.*, 1985; Shibahara *et al.*, 1985).

In an attempt to understand the molecular mechanism underlying AChR function, we have developed an expression system in which the functional receptor can be formed from the cloned cDNAs (Mishina *et al.*, 1984, 1985). Furthermore, functional analysis of mutant AChRs produced by site-directed mutagenesis of the cDNA has allowed us to locate functional regions of the AChR (Mishina *et al.*, 1985).

PROTEIN ENGINEERING
Applications in Science,
Medicine, and Industry

II. PRIMARY STRUCTURE

The amino acid sequences of the four subunits of the *T. californica* AChR deduced from the cDNA sequences (Noda *et al.*, 1982, 1983a,b) are aligned in Fig. 1. The α-, β-, γ-, and δ-subunit precursors are composed of 461, 493, 506, and 522 amino acids, including a hydrophobic prepeptide of 24, 24, 17, and 21 amino acids, respectively. The prepeptides resemble the signal peptide of secretory proteins (Blobel and Dobberstein, 1975; Steiner *et al.*, 1980), being involved presumably in the translocation of the AChR subunits into the membrane. The molecular weights of the mature α, β, γ, and δ subunits are calculated to be 50,116, 53,681, 56,279, and 57,565, respectively. The four subunits show marked sequence homology. This suggests that the AChR subunit genes have been generated from a single common ancestor by gene duplications. The hydropathy profiles of the four subunits along the aligned amino acid sequences (Fig. 2) as well as the predicted secondary structures are similar for all the subunits (Noda *et al.*, 1982, 1983a,b). Thus, the four subunits are probably oriented in a pseudosymmetric fashion across the membrane. In all four subunits, there are four strongly hydrophobic regions at equivalent positions (Noda *et al.*, 1982, 1983a,b; Claudio *et al.*, 1983; Devillers-Thiery *et al.*, 1983). Each of these four hydrophobic regions, designated as M1, M2, M3, and M4, corresponds to a continuous stretch of 19–27 uncharged amino acids, most of which are nonpolar residues (Fig. 1). These segments are bounded on both sides by charged residues. Because such a sequence is characteristic of transmembrane segments, segments M1–M4 of each AChR subunit most likely traverse the membrane. The α-helical structure requires about 24 residues to span the hydrocarbon area of the lipid bilayer, and the extended β configuration requires about 11 residues. In view of the lengths of segments M1–

Fig. 1. Alignment of the amino acid sequences of the *T. californica* AChR subunit precursors. The one-letter amino acid notation is used. Sets of identical residues and residues considered to be favored amino acid substitutions (Dayhoff *et al.*, 1978) are represented by large capital letters and the remaining residues by small capital letters. Sets of three or four identical residues at one aligned position are enclosed within continuous lines, and the fourth residues regarded as favored substitutions at the same position are enclosed with dotted lines. The positions in the aligned sequences including gaps (—) are numbered beginning with that of the initiative methionine. The putative transmembrane segments M1–M4 and MA and the putative disulfide bridges (S–S) are indicated. The asparagine residues as potential *N*-glycosylation sites are marked with asterisks; only those preceding segment M1 are shown. [From Noda *et al.*, 1983b. Reprinted with permission from *Nature (London)* **302,** 528–532. Copyright © 1983 Macmillan Journals Ltd.]

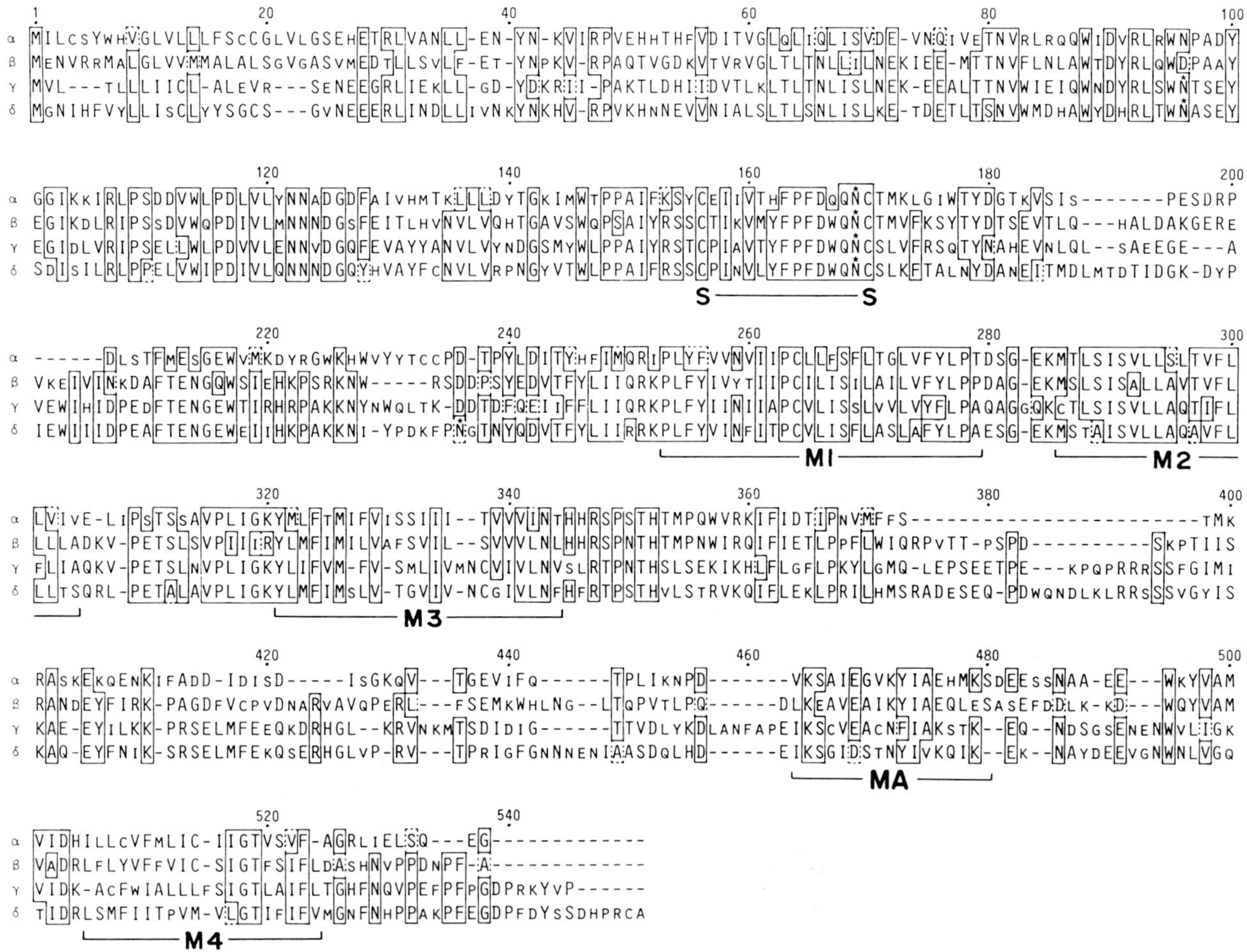

S
S
M1
M2
M3
MA
M4

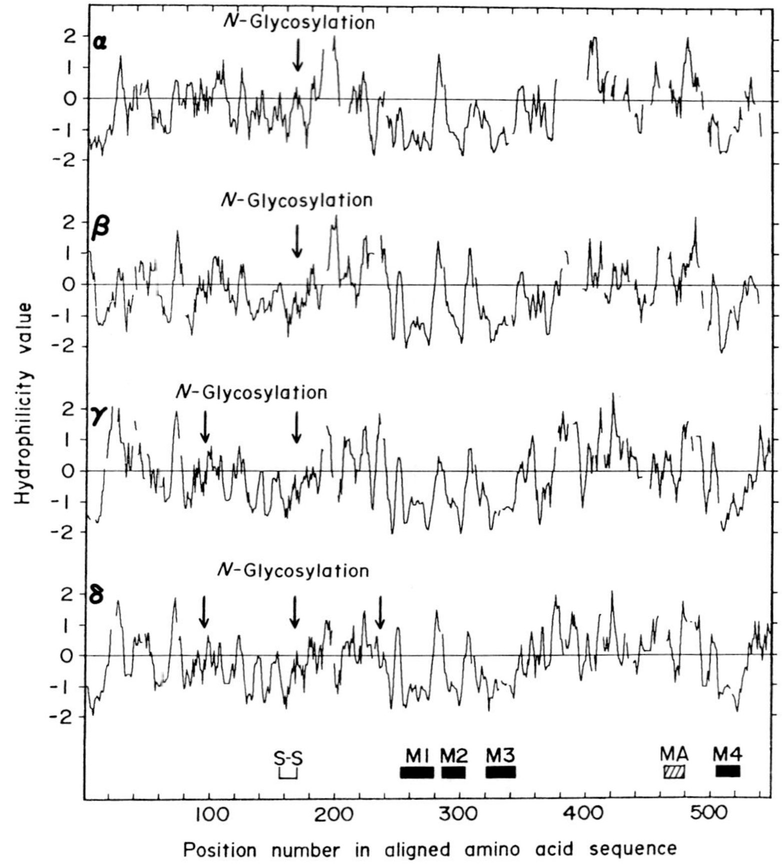

Fig. 2. Hydropathy profiles of the *T. californica* AChR subunit precursors. The average hydrophilicity value of a hexapeptide composed of residues at $i-2$ to $i+3$ is plotted against i, where i represents the position number in the aligned sequences (see Fig. 1); no plots are given for the positions corresponding to gaps. The locations of the putative transmembrane segments (M1–M4 and MA) and the putative disulfide bridge are shown. The potential *N*-glycosylation sites preceding segment M1 are indicated by arrows. (From Numa *et al.*, 1983. Reprinted with permission from *Cold Spring Harbor Symp. Quant. Biol.* **48**, 57–69.)

M4, it seems reasonable to assume that segments M1–M4 form α-helical structures.

It has been shown that the amino-terminal two-thirds of the *Torpedo* AChR subunits are resistant to proteolysis and are glycosylated (Wennogle and Changeux, 1980; Wennogle *et al.*, 1981; Strader and Raftery, 1980). In fact, this region contains the single potential *N*-glycosylation site (Marshall, 1974) in the α and β subunits, all three potential *N*-glycosylation sites in the δ subunit, and two of the four potential *N*-glycosylation sites in the γ subunit (Figs. 1 and 2). Therefore, the amino-terminal half of the mature subunit molecules preceding the putative transmembrane segment M1 is probably exposed on the extracellular surface of the membrane, forming a rigid conformation.

III. EXPRESSION OF CLONED cDNAs

The entire protein-coding sequence of the cloned cDNA for each AChR subunit was inserted into a simian virus 40 (SV40) vector containing the SV40 early gene promoter, donor and acceptor splice sites derived from the rabbit β-globin gene, and polyadenylation sites derived from the rabbit β-globin gene and the SV40 early gene (Mishina *et al.,* 1984). These recombinant plasmids contain the replication origin of SV40 and are thus expected to propagate to high copy numbers in COS monkey cells (Gluzman, 1981), which produce SV40 T-antigen constitutively. Each recombinant plasmid was introduced into COS cells by transfection. S1 nuclease mapping of poly $(A)^+$ RNA isolated from the transfected cells indicated that the transcripts were formed under the control of the SV40 early gene promoter and that the intronic sequence of the rabbit β-globin gene was eliminated (Mishina *et al.,* 1984). Thus, the AChR subunit-specific mRNAs produced in the COS cells contained no AUG triplet upstream of the translational initiation site for the subunit precursors. The content of each subunit-specific mRNA in poly$(A)^+$ RNA from the COS cells, as quantified by S1 nuclease analysis, was comparable to that in *T. californica* electroplax poly$(A)^+$ RNA. The α-, β-, γ-, and δ-subunit-specific mRNAs were combined and injected into *Xenopus* oocytes. The four AChR subunit polypeptides synthesized in the oocytes were indistinguishable in electrophoretic mobility on sodium dodecyl sulfate (SDS)–polyacrylamide gel from those synthesized in oocytes injected with *T. californica* electroplax poly$(A)^+$ RNA (Mishina *et al.,* 1984). The oocytes injected with the four subunit-specific mRNAs were studied electrophysiologically by recording the intracellular potentials or the membrane currents under voltage clamp; recordings were made in the presence of atropine, which blocks the native muscarinic AChR known to exist in the membrane of *Xenopus* oocytes (Kusano *et al.,* 1977, 1982). As shown in Fig. 3, an inward current with a short latency was observed in response to ACh applied ionophoretically, and this response was sensitive to (+)-tubocurarine. The reversal potential of the ACh response was -5 to -1 mV and was not altered in a Cl^--free solution. When ACh was applied repeatedly at short intervals (10–20 sec), the response was progressively diminished because of desensitization of the receptor. These results indi-

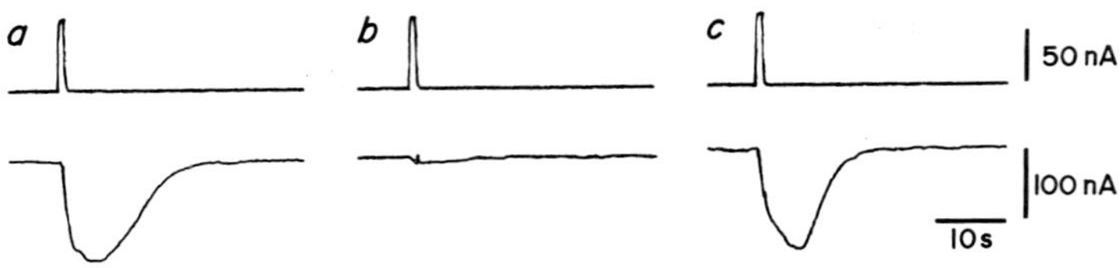

Fig. 3. Acetylcholine responses recorded from a *Xenopus* oocyte injected with the four subunit-specific mRNAs produced from the recombinant plasmids in COS cells. (Upper traces) Currents used for ionophoresis of ACh (backing current, 12 nA); (lower traces) ACh currents recorded under voltage clamp (downward deflection indicating inward current). The responses to ACh before (*a*) and 2.5 min after perfusion with 2.5 μM (+)-tubocurarine (*b*) and 15 min after washing out the antagonist (*c*) are shown. [From Mishina *et al.*, 1984. Reprinted with permission from *Nature* (*London*) **307,** 604–608. Copyright © 1984 Macmillan Journals Ltd.]

cate that the four subunit-specific mRNAs formed by transcription of the cloned cDNAs can direct the synthesis of the functional AChR of the nicotinic type.

IV. SUBUNIT REQUIREMENT

In view of the marked amino acid sequence homology found among the four subunits of the AChR, the question arises of whether all four subunits are indispensable for the receptor function. The expression system described above enabled us to examine this issue. Oocytes were injected with various combinations of three of the subunit-specific mRNAs. Electrophoretic analysis of the [^{35}S]methionine-labeled translation products revealed that the three mRNAs introduced into oocytes directed the synthesis of the corresponding subunits, although the amounts of the polypeptides found were generally lower than when all four mRNAs were injected (Mishina *et al.*, 1984).

The results in Table I show that the cell extracts containing all but the β, γ, or δ subunit exhibited 3, 4, and 16%, respectively, of the α-bungarotoxin binding activity of the complete receptor synthesized in oocytes, whereas the cell extract containing all but the α subunit had no detectable activity. This is consistent with the finding that the isolated α subunit alone is capable of binding α-bungarotoxin (Haggerty and Froehner, 1981; Tzartos and Changeux, 1983). Table I also reveals that most of the oocytes injected with all four subunit-specific mRNAs responded to ACh when tested electrophysiologically. In contrast, the oocytes injected with three of the mRNAs exhibited no ACh potential, except that a few of the oocytes injected with all but the γ- or δ-subunit-specific mRNA showed a

TABLE I

Effect of Deletion of Each Subunit on AChR Function[a,b]

AChR subunit-specific mRNAs injected	^{125}I-Labeled α-bungarotoxin binding (fmol/oocyte)	Response to ACh: Number of responsive oocytes/number of oocytes tested	Response to ACh: Mean ACh sensitivity (mV/μC)
α, β, γ, δ	2.06	126/170	37 (0.4–460)
β, γ, δ	<0.01	0/82	—
α, γ, δ	0.06	0/82	—
α, β, δ	0.09	2/82	1 (0.9–1)
α, β, γ	0.33	3/105	2 (0.2–4)
None	<0.01	0/80	—

[a] The ^{125}I-labeled α-bungarotoxin binding activities were determined by the antibody precipitation method (Weinberg and Hall, 1979). The mean ACh sensitivities were calculated only for the responsive oocytes; the ranges of ACh sensitivity are given in parentheses, and our detectable limit was ~0.2 mV/μC.

[b] From Mishina *et al.*, (1984). Reprinted by permission from *Nature (London)* **307,** 604–608. Copyright © 1984 Macmillan Journals Ltd.

weak (+)-tubocurarine-sensitive response. In view of the close amino acid sequence homology between the γ and δ subunits (Fig. 1), it is possible that these two subunits replace each other to some extent for the receptor function.

V. DELETION MAPPING OF FUNCTIONAL REGIONS

In the experiments described below, the AChR subunit-specific mRNAs were synthesized *in vitro* (Mishina *et al.*, 1985). Each of the subunit cDNAs including the poly(A) tail, linked with the bacteriophage SP6 promoter, was transcribed with SP6 polymerase, and the resulting transcript was capped with vaccinia virus mRNA quanylyltransferase. Thus, pure preparations of the subunit-specific mRNAs were obtained in relatively large quantities. Internal deletions were introduced at various loci of the cDNA corresponding to the carboxy-terminal half of the AChR α subunit, where the putative transmembrane segments are located (Fig. 4). Each of the α-subunit-specific mRNAs synthesized with the cDNA templates with internal deletions, combined with the wild-type β-, γ-, and δ-subunit-specific mRNAs, was injected into *Xenopus* oocytes. Electrophoretic analysis of [^{35}S]methionine-labeled translation products indi-

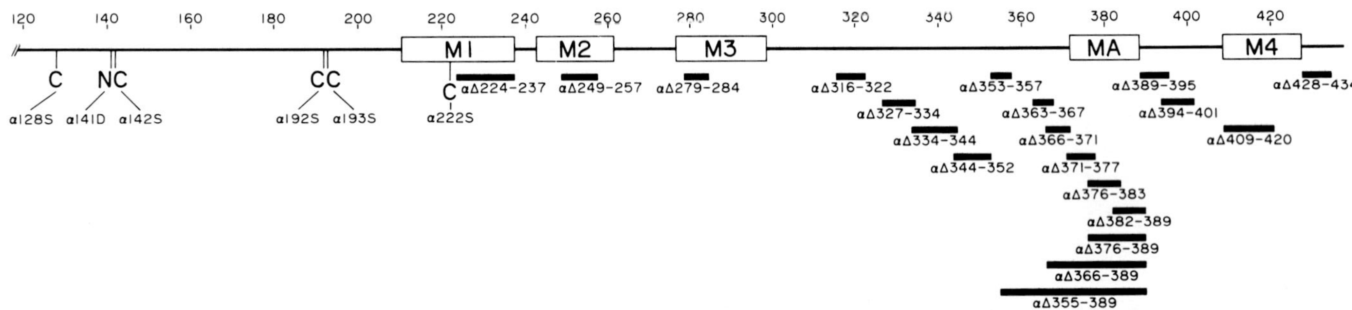

Fig. 4. Mutations introduced into the *T. californica* AChR α-subunit polypeptide. The putative transmembrane segments M1–M4 and MA are represented by open boxes, and the amino acid numbers of the AChR α subunit (Noda *et al.*, 1982) are given above the diagram. The positions of deletions are indicated by bars, and those of the cysteine and asparagine residues subjected to point mutations by the one-letter code; the mutant numbers indicate the number(s) of the amino acid residue(s) deleted or altered. [From Mishina *et al.*, 1985. Reprinted with permission from *Nature* (*London*) **313,** 364–369. Copyright © 1985 Macmillan Journals Ltd.]

cated that there were sufficient concentrations of the four subunit polypeptides (at least 10% of the wild type) to permit assay of the functional parameters for the AChR (Mishina *et al.*, 1985).

The effects of the various deletions in the AChR α subunit on the ^{125}I-labeled α-bungarotoxin binding activity in cell extracts and on the membrane potential response to ACh are shown in Table II. The toxin binding activities of most of the AChR mutants with a deletion in the α subunit were comparable to the wild-type activity, whereas the AChR mutant with the α subunit having a deletion in segment M1 ($\alpha\Delta$224–237) exhibited a markedly reduced activity. Because segment M1 is adjacent to the extracellularly located amino-terminal half of the subunit molecule, this mutation may result in a conformational change in the region encompassing the toxin binding site. The relatively low toxin binding activities of some other AChR mutants (for example, $\alpha\Delta$279–284) can roughly be accounted for by the observed decreases in the amount of the α subunit, which is responsible for α-bungarotoxin binding (see above).

All the oocytes injected with the four wild-type subunit-specific mRNAs responded to ACh. In contrast, the AChR mutants with the α subunit having a deletion in putative transmembrane segment M1, M2, M3, or M4 failed to respond to ACh. Of the AChR mutants with the α subunit having a deletion between M3 and M4, some exhibited little or no response to ACh, whereas the others showed an essentially normal response. The dispensable segment corresponds to most of the putative cytoplasmic region, but it remains possible that this segment is responsible for other features of the AChR, such as clustering of receptor molecules or phosphorylation of the receptor. The positions of the deletions in the α subunit of the AChR mutants showing little or no ACh response ($\alpha\Delta$355–389, $\alpha\Delta$366-389, $\alpha\Delta$371–377, $\alpha\Delta$376–383, $\alpha\Delta$376–389, and $\alpha\Delta$382–389) and of the responsive mutants with $\alpha\Delta$366–371 or $\alpha\Delta$389–395 indicate that the segment composed of residues 372–388 (designated as segment MA) is required for forming the functional receptor. This segment coincides nearly with the amphipathic segment proposed by Guy (1984) and Finer-Moore and Stroud (1984). All the unresponsive AChR mutants exhibited essentially normal apparent binding affinity for the agonist, as estimated by inhibition of α-bungarotoxin binding by carbamylcholine (Table II). (The effect of the deletion in segment M1 on agonist binding could not be assessed because the AChR mutant with this deletion showed too little toxin binding to allow measurement of competition by the agonist.) Our results thus suggest that the hydrophobic segments M1, M2, M3, and M4 and the amphipathic segment MA are involved directly or indirectly in the formation of the ionic channel. (The possibility that the deletion in M1 also affects agonist binding cannot be excluded.) This is

TABLE II

Functional Properties of AChR Mutants with Altered α Subunit[a,b]

α Subunit	Residue change		^{125}I-Labeled α-bungarotoxin binding			Response to acetylcholine	
	From	To	Activity (%)	Percent inhibition by carbamylcholine: 0.5 mM	Percent inhibition by carbamylcholine: 10 mM	No. of responsive oocytes/no. of oocytes tested	Mean acetylcholine sensitivity (%)
Wild type			100	38	80	82/82	100
αΔ224–237	LFSFLTGLVF YLPT	PSS	0.5	NT	NT	0/15	ND
αΔ249–257	VLLSLTVFL	SS	17	47	87	0/15	ND
αΔ279–284	LFTMIF	PSS	8	50	80	0/15	ND
αΔ316–322	FIDTIPN	RAR	60	39	86	28/29	31
αΔ327–334	STMKRASK	RAR	80	44	82	14/14	70
αΔ334–344	KEKQENKIFA D	A	71	44	82	29/29	55
αΔ344–352	DDIDISDIS	ELA	101	36	79	14/14	122
αΔ353–357	GKQVT	R	58	37	82	29/29	142
αΔ355–389	QVTGEVIFQT PLIKNPDVKS AIEGVKYIAE HMKSD	PSS	45	38	77	14/14	3

$\alpha\Delta$363–367	QTPLI	PSS	83	39	77	14/14	101
$\alpha\Delta$366–371	LIKNPD	SS	81	34	81	29/29	31
$\alpha\Delta$366–389	LIKNPDVKSA IEGVKYIAEH MKSD	SS	37	38	78	2/14	0.1
$\alpha\Delta$371–377	DVKSAIE	A	44	38	81	4/14	0.2
$\alpha\Delta$376–383	IEGVKYIA	EL	43	46	84	0/14	ND
$\alpha\Delta$376–389	IEGVKYIAEH MKSD	SS	36	37	78	0/14	ND
$\alpha\Delta$382–389	IAEHMKSD	TSS	29	40	81	0/14	ND
$\alpha\Delta$389–395	DEESSNA	EL	48	38	80	29/29	65
$\alpha\Delta$394–401	NAAEEWKY	TSS	40	54	89	0/42	ND
$\alpha\Delta$409–420	ILLCVFMLIC II	EL	23	38	77	0/29	ND
$\alpha\Delta$428–434	GRLIELS	R	80	39	83	27/27	15
α128S	C	S	0.08	NT	NT	0/14	ND
α142S	C	S	ND	NT	NT	0/14	ND
α192S	C	S	39	8	27	0/14	ND
α193S	C	S	28	13	53	0/14	ND
α222S	C	S	79	36	83	28/28	107
α141D	N	D	0.09	NT	NT	0/14	ND

[a] The mutant numbers indicate the number of the amino acid residue altered. ND, Not detectable; NT, not tested. The toxin binding activity of the wild-type AChR was 14–114 fmol/oocyte and our detectable limit was ~0.01 fmol/oocyte. The mean ACh sensitivity of the wild-type AChR ranged from 150 to 1100 mV/μC and our detectable limit was ~0.2 mV/μC.

[b] From Mishina *et al.*, (1985). Reprinted by permission from *Nature (London)* **313,** 364–369. Copyrignt © 1985 Macmillan Journals, Ltd.

apparently consistent with the view that these five segments traverse the membrane (Guy, 1984; Finer-Moore and Stroud, 1984). It remains possible, however, that these segments are required for the correct folding, assembly, or transport of the subunit molecule. In this context, it is to be noted that the AChR mutant with $\alpha\Delta355$–389, in which the entire segment MA is deleted, shows a slight but unambiguous response to ACh. This finding seems to contradict the hypothesis that the amphipathic segment represents the transmembrane segment forming the inner wall of the channel (Guy, 1984; Finer-Moore and Stroud, 1984), although the possibility cannot be excluded that the segment of residues 340–354, which has an amphipathic α-helical character to some extent (Kosower, 1983; Finer-Moore and Stroud, 1984), takes over the original position of segment MA in this mutant and may thus be substituted for segment MA to show some function. Further site-directed mutagenesis studies as well as analysis of channel properties by single-channel current measurements will be needed for elucidating the roles of the putative transmembrane segments in the formation of the ionic channel.

VI. MUTATIONS IN THE ACETYLCHOLINE BINDING REGION

Evidence has been provided to indicate that the α subunit contains part or all of the ACh binding site and that a disulfide bridge is present in the vicinity of the ACh binding site on the α subunit (Karlin, 1969, 1980). There are seven cysteine residues in the mature α subunit of the *T. californica* AChR (Noda *et al.*, 1982). Two of them, which are located within segment M4, can be deleted without affecting apparent agonist binding affinity (see above). Each of the remaining five cysteine residues (residues 128, 142, 192, 193, and 222) was converted into serine by means of oligonucleotide-directed mutagenesis (Fig. 4) because this amino acid is similar in structure to cysteine but cannot form a disulfide linkage. For the AChR mutants with the altered α subunit α192S, α193S, or α222S, the contents of the subunit polypeptides were comparable to those of the wild-type AChR, whereas the AChR mutants with α128S or α142S contained substantially smaller amounts of the α, γ, and δ subunits (Mishina *et al.*, 1985). Thus, the two latter mutations may affect the conformation, assembly, or transport of the α subunit.

The AChR mutants with α128S or α142S showed hardly detectable or no α-bungarotoxin binding activity and failed to respond to ACh (Table II), even though the contents of the subunit polypeptides should have been sufficient to elicit a readily measurable response. The AChR mutants with α192S or α193S had nearly normal α-bungarotoxin binding activity, but were not responsive to ACh. Furthermore, the apparent agonist binding affinities of the mutants with α192S or α193S were ~30- and ~10-fold lower than that of the wild type, respectively (Mishina *et al.*, 1985). Neither the ACh response nor the agonist binding was impaired in the AChR mutant with α222S.

Thus, mutations of cysteine residues 128 and 142 resulted in similar functional alterations, as was the case for mutations of cysteine residues 192 and 193. Because the mutation of either of two cysteine residues forming a disulfide bridge is expected to affect the functional properties of the AChR in an analogous manner, our results are consistent with the presence of a disulfide linkage between cysteine residues 128 and 142 as well as between cysteine residues 192 and 193. The disulfide bridge between adjacent cysteine residues seems rather unlikely (Dayhoff, 1976), but there are precedents for it (Ovchinnikov *et al.*, 1985). Cysteine residues 192 and 193 are unique to the α subunit, being conserved in all the fish and mammalian AChR α subunits analyzed (Noda *et al.*, 1982, 1983c; Sumikawa *et al.*, 1982; Devillers-Thiery *et al.*, 1983), whereas cysteine residues 128 and 142 are conserved in all kinds of AChR subunits (Noda *et al.*, 1982, 1983a,b,c; Sumikawa *et al.*, 1982; Claudio *et al.*, 1983; Devillers-Thiery *et al.*, 1983; Tanabe *et al.*, 1984; Takai *et al.*, 1984, 1985; LaPolla *et al.*, 1984; Nef *et al.*, 1984; Kubo *et al.*, 1985; Shibahara *et al.*, 1985). Thus, we suggest that cysteine residues 192 and 193 of the α subunit, presumably forming a disulfide bridge near the ACh binding site, play a specific role in agonist binding and possibly also in signal transduction, while cysteine residues 128 and 142 of all subunits, which probably form a disulfide bridge in each subunit, are essential for maintaining the proper conformation of the extracellular region of the AChR molecule. Of interest in this context is the recent finding of Karlin's group that cysteine residue 192 and possibly cysteine residue 193 are affinity-labeled by the cholinomimetic alkylating agent MBTA [4-(*N*-maleimido)benzyltri[^{3}H]methylammonium iodide] (Kao *et al.*, 1984).

The α subunits of fish and mamalian AChRs contain a single potential site of *N*-glycosylation, asparagine residue 141, conserved in all kinds of AChR subunits (for references, see above). This asparagine residue of the α subunit was converted to aspartic acid (Fig. 4). SDS–polyacrylamide gel electrophoresis of the AChR mutant with the altered α subunit α141D revealed that this subunit migrated faster than the wild type, suggesting the absence of carbohydrate attachment (Mishina *et al.*, 1985). A marked decrease in the amount of the δ subunit was also observed, suggesting that the lack of *N*-glycosylation of the α subunit may affect the assembly of the δ subunit. This AChR mutant showed hardly detectable α-bungarotoxin binding activity and no ACh response (Table II) despite the presence of sufficient amounts of the subunit polypeptides. This implies that the *N*-glycosylation at asparagine residue 141 is essential for the correct folding, assembly, or transport of the α subunit, and thus is required for AChR function.

VII. CONCLUDING REMARKS

The application of recombinant DNA techniques has provided considerable information concerning the structure and function of the nicotinic AChR. The expression system and site-directed mutagenesis of the cloned cDNAs, in conjuction with single-channel current analysis, will be useful for further studies on the question of what roles subunits, domains, and specific amino acids play in the operation of the AChR.

REFERENCES

Blobel, G., and Dobberstein, B. (1975). *J. Cell Biol.* **67**, 852–862.

Changeux, J.-P. (1973). *FEBS Lett.* **33,** 109–113.

Claudio, T., Ballivet, M., Patrick, J., and Heinemann, S. (1983). *Proc. Natl. Acad. Sci. U. S. A.* **80,** 1111–1115.

Conti-Tronconi, B. M., and Raftery, M. A. (1982). *Annu. Rev. Biochem.* **51,** 491–530.

Dayhoff, M. O. (1976). *In* "Atlas of Protein Sequence and Structure" (M. O. Dayhoff, ed.), Vol. 5, Suppl. 2, pp. 77-268. Natl. Biomed. Res. Found., Silver Spring, Maryland.

Dayhoff, M.O., Schwartz, R. M., and Orcutt, B. C. (1978). *In* "Atlas of Protein Sequence and Structure" (M. O. Dayhoff, ed.),Vol. 5, Suppl. 3, pp. 345–352. Natl. Biomed. Res. Found., Silver Spring, Maryland.

Devillers-Thiery, A., Giraudat, J., Bentaboulet, M., and Changeux, J.-P. (1983). *Proc. Natl. Acad. Sci. U. S. A.* **80,** 2067–2071.

Finer-Moore, J., and Stroud, R. M. (1984). *Proc. Natl. Acad. Sci. U. S. A.* **81,** 155–159.

Gluzman, Y. (1981). *Cell* **23,** 175–182.

Guy, H. (1984). *Biophys. J.* **45,** 249–261.

Haggerty, J. G., and Froehner, S. C. (1981). *J. Biol. Chem.* **256**, 8294–8297.

Kao, P. N., Dwork, A. J., Kaldany, R.-R. J., Silver, M. L., Wideman, J., Stein, S., and Karlin, A. (1984). *J. Biol. Chem.* **259,** 11662–11665.

Karlin, A. (1969). *J. Gen. Physiol.* **54,** 245s-264s.

Karlin, A. (1980). *Cell Surf. Rev.* **6,** 191–260.

Kosower, E. M. (1983). *Biochem. Biophys. Res. Commun.* **111**, 1022-1026.

Kubo, T., Noda, M., Takai, T., Tanabe, T., Kayano, T., Shimizu, S., Tanaka, K., Takahashi, H., Hirose, T., Inayama, S., Kikuno, R., Miyata, T., and Numa, S. (1985). *Eur. J. Biochem.* **149,** 5–13.

Kusano, K., Miledi, R., and Stinnakre, J. (1977). *Nature (London)* **270,** 739–741.

Kusano, K., Miledi, R., and Stinnakre, J. (1982). *J. Physiol. (London)* **328,** 143-170.

LaPolla, R. J., Mayne, K. M., and Davidson, N. (1984). *Proc. Natl. Acad. Sci. U. S. A.* **81,** 7970–7974.

Marshall, R. D. (1974). *Biochem. Soc. Symp.* **40,** 17–26.

Mishina, M., Kurosaki, T., Tobimatsu, T., Morimoto, Y., Noda, M.,Yamamoto, T., Terao, M., Lindstrom, J., Takahashi, T., Kuno, M., and Numa, S. (1984). *Nature (London)* **307,** 604–608.

Mishina, M., Tobimatsu, T., Imoto, K., Tanaka, K., Fujita, Y., Fukuda, K., Kurasaki, M., Takahashi, H., Morimoto, Y., Hirose, T., Inayama, S., Takahashi, T., Kuno, M., and Numa, S. (1985). *Nature (London)* **313,** 364–369.

Nef, P., Mauron, A., Stalder, R., Alliod, C., and Ballivet, M. (1984). *Proc. Natl. Acad. Sci. U. S. A.* **81,** 7975–7979.

Noda, M., Takahashi, H., Tanabe, T., Toyosato, M., Furutani, Y., Hirose, T., Asai, M., Inayama, S., Miyata, T., and Numa, S. (1982). *Nature (London)* **299,** 793–797.

Noda, M., Takahashi, H., Tanabe, T., Toyosato, M., Kikyotani, S., Hirose, T., Asai, M., Takashima, H., Inayama, S., Miyata, T., and Numa, S. (1983a). *Nature (London)* **301,** 251–255.

Noda, M., Takahashi, H., Tanabe, T., Toyosato, M., Kikyotani, S., Furutani, Y., Hirose, T., Takashima, H., Inayama, S., Miyata, T., and Numa, S. (1983b). *Nature (London)* **302,** 528-532.

Noda, M., Furutani, Y., Takahashi, H., Toyosato, M., Tanabe, T., Shimizu, S., Kikyotani, S., Kayano, T., Hirose, T., Inayama, S., and Numa, S. (1983c). *Nature (London)* **305,** 818–823.

Numa, S., Noda, M., Takahashi, H., Tanabe, T., Toyosato, M., Furutani, Y., and Kikyotani, S. (1983). *Cold Spring Harbor Symp. Quant. Biol.* **48,** 57–69.

Ovchinnikov, Y. A., Lipkin, V. M., Shuvaeva, T.M., Bogachuk, A. P., and Shemyakin, V. V. (1985). *FEBS Lett.* **179,** 107–110.

Shibahara, S., Kubo, T., Perski, H. J., Takahashi, H., Noda, M., and Numa, S. (1985). *Eur. J. Biochem.* **146,** 15–22.

Steiner, D. F., Quinn, P. S. , Chan, S. J., Marsh, J., and Tager, H. S. (1980). *Ann. N. Y. Acad. Sci.* **343.** 1–16.

Strader, C. D., and Raftery, M. A. (1980). *Proc. Natl. Acad. Sci. U. S. A.* **77,** 5807–5811.

Sumikawa, K., Houghton, M., Smith, J. C., Bell, L., Richards, B. M., and Barnard, E. A. (1982). *Nucleic Acids Res.* **10,** 5809–5822.

Takai, T., Noda, M., Furutani, Y., Takahashi, H., Notake, M., Shimizu, S., Kayano, T., Tanabe, T., Tanaka, K., Hirose, T., Inayama, S., and Numa, S. (1984). *Eur. J. Biochem.* **143,**109–115.

Takai, T., Noda, M., Mishina, M., Shimizu, S., Furutani, Y., Kayano, T., Ikeda, T., Kubo, T., Takahashi, H., Takahashi, T., Kuno, M., and Numa, S. (1985). *Nature (London)* **316,** 761–764.

Tanabe, T., Noda, M., Furutani, Y., Takai, T., Takahashi, H., Tanaka, K., Hirose, T., Inayama, S., and Numa, S. (1984). *Eur. J. Biochem.* **144,** 11–17.

Tzartos, S. J., and Changeux, J.-P. (1983). *EMBO J.* **2,** 381–387.

Weinberg, C. B., and Hall, Z. W. (1979). *Proc. Natl. Acad. Sci. U. S. A.* **76,** 504–508.

Wennogle, L. P., and Changeux, J.-P. (1980). *Eur. J. Biochem.* **106,** 381–393.

Wennogle, L. P., Oswald, R., Saitoh, T., and Changeux, J.-P. (1981). *Biochemistry* **20,** 2492–2497.

19

Protein Engineering of Antibody Molecules

MICHAEL S. NEUBERGER
AND GARETH T. WILLIAMS

Medical Research Council
Laboratory of Molecular Biology
Cambridge CB2 2QH, United Kingdom

I. INTRODUCTION

Antibodies are bifunctional molecules: the amino-terminal portion of the protein consists of the variable (V) regions, which determine the antigen specificity, whereas the carboxy-terminal portion (the Fc region) is responsible for antibody effector functions. Oligonucleotide-directed, site-specific mutagenesis studies should yield important information about the contacts that antibodies make both with their antigens and with the effector proteins such as Fc receptors and complement component C1q. Such studies clearly require an expression system for immunoglobulin gene DNA.

In vitro manipulation of antibody genes should also yield novel molecules which could be employed in both biology and medicine. At present, antibodies are used for affinity purification procedures and an enormous range of immunoassays as well as being employed for therapeutic and prophylactic purposes in medicine. Several expression systems for antibody genes or cDNA molecules have been tested (e.g., see Falkner and Zachau, 1982; Cabilly *et al.*, 1983; Neuberger, 1983; Ochi *et al.*, 1983a,b; Oi *et al.*, 1983; Boss *et al.*, 1984; Deans *et al.*, 1984; Kenten *et al.*, 1984; Wood *et al.*, 1985). However, stable tranfection of myeloma cells at present appears to be the best system for the production of milligram

PROTEIN ENGINEERING
Applications in Science,
Medicine, and Industry

quantities of antibody. Here we discuss in some detail the myeloma transfection procedure that we currently use.

II. STABLE TRANSFECTION OF MYELOMA CELLS

A. Cell Lines

The first papers on the transfection of myeloma cells appeared in 1983 (Neuberger, 1983; Ochi *et al.*, 1983a,b; Oi *et al.*, 1983), and since then several cell lines, selectable markers, and transfection procedures have been tested. Many myeloma-derived cell lines (e.g., X63Ag8, NSO, Sp2/0, MOPC 315.26, J558L, Y3, HMy2) as well as several hybridomas have been found to yield transfectants. The J558L mouse plasmacytoma cell line, which secretes λ_1 light chains but expresses no heavy chain, was shown by Oi *et al.* (1983) to give a significantly higher yield of stably transfected clones than other cell lines tested when using *gpt* as the selective marker. We have used J558L extensively in our work and have exploited the fact that it produces λ_1 light chains (Neuberger, 1983; Neuberger *et al.*, 1984, 1985). However, for many purposes it would be desirable to have a myeloma cell line that does not produce any immunoglobulin polypeptides of its own. Nonproducer cell lines that are ultimately derived from the mouse plasmacytoma MOPC 21 have indeed been used for this purpose (Boulianne *et al.*, 1984; Falkner and Zachau, 1984; Morrison *et al.*, 1984); however, the frequency of transfectants that is obtained differs from cell line to cell line.

B. Selectable Markers

Most work has been performed with one of two dominant selectable markers—*gpt* or *neo*. The *gpt* gene encodes a guanine/xanthine phosphoribosyltransferase activity and confers resistance to mycophenolic acid, whereas *neo* confers resistance to the antibiotic G418 (Mulligan and Berg, 1981; Southern and Berg, 1982). In the case of hgprt$^-$ cells (such as X63Ag8), the *gpt* gene can simply be used to revert the hgprt$^-$ phenotype. Recent experiments (Blochlinger and Diggelmann, 1984) have demonstrated that a gene conferring resistance to hygramycin can also be used as a selective marker in lymphocyte transfection.

C. Methods of Transfection

The methods that have been employed to introduce DNA into myelomas include calcium phosphate coprecipitation, DEAE–dextran-mediated DNA uptake, electroporation, and fusion with spheroplasts of the *Escherichia coli* cells that harbor the plasmid. DEAE–dextran-mediated DNA uptake works well for the introduction of DNA as measured by transient expression assays but gives only a poor yield of stable transfectants. In our work, we have mainly used spheroplast fusion as a means of DNA delivery and we describe the detailed procedure below.

D. Spheroplast Fusion of pSVgpt Plasmids into J558L

Spheroplast fusion was described by Schaffner in 1980 as a means of introducing viral DNA into mammalian cells. The procedure that we currently use for the stable transfection of J558L draws on our experience as well as on procedures described by Schaffner (1980), Sandri-Goldin *et al.* (1983), Gillies *et al.* (1983), Ochi *et al.* (1983b), and Oi *et al.* (1983).

1. Making Spheroplasts

A transformant of *E. coli* strain TG1 (Messing, 1979; Gibson, 1984) that harbors the appropriate pSV2gpt-derived plasmid is inoculated from a fresh plate into 100 ml of 2XTY medium containing 50 μg/ml ampicillin in a 250-ml conical flask and shaken at 37°C (2XTY is 16 g/liter tryptone, 10 g/liter yeast extract, 5 g/liter NaCl, pH 7.4). When the culture has reached late log phase (i.e., a 1 : 10 dilution of the culture gives an A_{650} of about 0.1 to 0.2), chloramphenicol (0.5 ml of a 34 mg/ml solution in ethanol) is added and the culture is left shaking overnight. The cells are pelleted in sterile 50-ml plastic tubes [2500 rpm (for an average rotor radius of 12 cm), 4°C, 15 min], and the supernatant discarded. The cells are resuspended in 2.5 ml of cold 20% (w/v) sucrose, 50 m*M* Tris-HCl, pH 8, and lysozyme (0.5 ml of a fresh 5 mg/ml solution in 250 m*M* Tris-HCl, pH 8) is added. The cells are swirled on ice for 5 min prior to the addition of 1 ml of cold 0.25 *M* NaEDTA, pH 8. After a further 5 min on ice, 1 ml of 50 m*M* Tris-HCl, pH 8, is added and the sample incubated with swirling at 37°C for up to 10 min. (Occasionally, particularly if the original bacterial culture was too thin, spheroplasts lyse at this stage. Sometimes, lysis does start within the first few minutes. If this does occur, the 10-min incubation is curtailed).

The spheroplasts are slowly diluted at room temperature with 20 ml of Dulbecco's modified Eagle's medium (DMEM) containing 10% (w/v) sucrose and 10 m*M* $MgCl_2$. Good conversion of the *E. coli* rods to spheroplasts can be confirmed by using phase contrast microscopy. A sample (5 ml) of the spheroplast suspension is now transferred to another 50-ml tube and the spheroplasts pelleted by centrifugation (2500 rpm, 10 min, room temperature).

2. Preparing Myeloma Cells

J558L myeloma cells growing in DMEM, 10% fetal calf serum (FCS) are harvested from 50 ml of culture (4–5 × 10^5 cells/ml) and resuspended in 10 ml of DMEM, 10% sucrose, 10m*M* $MgCl_2$. The J558L cells are carefully layered onto the spheroplast pellet so as to create minimum disturbance and the tube centrifuged at 1200 rpm for 5 min. The supernatant is carefully discarded.

3. Fusion

The pellet is broken by gently tapping the bottom of the tube and the tube is placed in a 200-ml beaker containing water at 37°C and kept there during the fusion. Polyethylene glycol 1500 (0.8 ml of a neutralized 50% solution in saline prewarmed to 37°C) is added to the pellet with a 1-ml pipette over a period of 30 sec, continually stirring the cells with the pipette tip. Stirring is continued for a further 30 sec, after which 1 ml of DMEM, taken from a tube containing 10 ml of DMEM kept at 37°C in a hot block, is added to the fusion mixture, continuously stirring as before. These 1-ml additions are repeated such that the 10 ml of DMEM is added over a total period of 30 sec prior to a final addition of 12 ml of prewarmed DMEM. The cells are spun down (1000 rpm, 8 min, room temperature) and then taken up in DMEM and plated in 24- or 96-well plates undiluted or at varying dilutions as required. After allowing the cells to settle, the supernatant (which contains debris from the fusion) is removed and the cells fed with DMEM, 10% FCS, gentamycin.

4. Selecting Transfectants

Selective medium (DMEM, 10% FCS, 5 μg/ml mycophenolic acid, 250 μg/ml xanthine, 15 μg/ml hypoxanthine, 2.5 m*M* HCl) is applied at 24 and 48 hr. (Stock solutions of xanthine can be prepared at 5 mg/ml in 0.05 *M* NaOH and autoclaved; mycophenolic acid is dissolved at 500 μg/ml in phosphate-buffered saline, neutralized, and sterile-filtered.) Multiple clones appear in each well after about 8–10 days. The transfection frequency varies from plasmid to plasmid and cell line to cell line; frequencies of around 10^{-3} can be obtained with J558L and *gpt*. The selective medium, in particular the concentration of mycophenolic acid, should be optimized for each cell line used. For some pre-B or B cell lines it may be necessary to include hypoxanthine–aminopterin–thymidine (HAT) or HAT with much (e.g., 100-fold) reduced aminopterin to achieve a sufficiently stringent selection.

5. Screening for Antibody Production

We usually screen for antibody production by radioimmunoassay of culture supernatants. Even if the original fusion is plated at 1/10 dilution, each well normally contains multiple clones. We therefore routinely clone cells from a well by limiting dilution (0.4 cells/well) without need for feeder cells. Before cloning, cells are grown for a few days in selective medium from which mycophenolic acid has been omitted. Cloning is then performed in DMEM, 10% FCS.

III. CHIMERIC ANTIBODIES AND IMMUNOGLOBULIN EXON SHUFFLING

The fact that the expression systems described in Section II utilize genomic immunoglobulin DNA (rather than cDNA) means that exon shuffling can easily be used for the construction of novel antibody molecules. This is of particular interest since the different domains of the antibody are encoded by separate exons. Thus chimeric antibodies have been among the first novel immunoglobulin molecules to be created by using DNA transfection (Boulianne *et al.,* 1984; Morrison *et al.,* 1984; Neuberger *et al.,* 1985). These chimeric antibodies, in which antigen-binding V regions of mouse origin are joined to human constant regions, are of significance given the difficulties experienced in making antigen-specific human monoclonal antibodies by more conventional procedures. Thus chimeric monoclonals should be of use in allowing controlled comparisons of the effector functions of the different human immunoglobulin classes and subclasses and, in the future, may also be employed in therapy.

IV. SUBSTITUTING NOVEL ACTIVITIES FOR THE ANTIBODY Fc PORTION

Many of the uses to which antibodies are put rely on the binding portion of the immunoglobulin to target a specific activity. Thus, in immunoassays the antibody is modified either by conjugating it with an enzyme or by radiolabeling to provide a way of quantitating the specific antigen. When used for affinity purification, either an antibody is linked to an insoluble support or use is made of the specific binding of the antibody to anti-antibodies or to protein A of *Staphylococcus aureus*. Similarly, in therapeutic applications, either use is made of the antibody's *in vivo* effector functions or novel moieties are conjugated to the immunoglobulin as in the case of immunotoxins (reviewed by Möller, 1982).

Immunoglobulin gene expression systems allow such complexes between antibodies and other moieties to be made by gene fusion rather than by chemical coupling of the proteins. Thus we have shown (Neuberger *et al.*, 1984) that the DNA encoding the antibody Fc portion can be excised and replaced by DNA encoding an enzymatic activity—the nuclease of *S. aureus*. This antibody/enzyme fusion, Fab–nuclease, was secreted from myeloma cells as an active enzyme-linked antibody conjugate. Presumably, a similar approach could be used to make a wide range of antibody/toxin conjugates.

V. SUMMARY AND PROSPECT

Efficient systems now exist to express the DNA that encodes antibodies. Use of this technology for the production of chimeric antibodies as well as recombinant antibodies that display novel effector functions has already been demonstrated in model systems. Molecules of this type may prove of use in both biology and medicine. We may expect that in the near future site-directed mutagenesis studies will yield important information about the contacts that antibodies make, both with their antigens and with the proteins that mediate their effector functions. In the longer term, it may well be that the main use of antibodies to the protein engineer will stem from the fact that their V regions will provide protein binding sites that can be used for the creation of entirely novel proteins.

REFERENCES

Blochlinger, K., and Diggelmann, H. (1984). *Mol. Cell. Biol.* **4,** 2929–2931.

Boss, M. A., Kenten, J. H., Wood, C. R., and Emtage, J. S. (1984). *Nucleic Acids Res.* **12,** 3791–3806.

Boulianne, G. L., Hozumi, N. and Shulman, M. J. (1984). *Nature (London)* **12,** 643–646.

Cabilly, S., Riggs, A. D., Pande, H., Shiveley, J. E., Holmes, W. E., Rey, M., Perry, L. J., Wetzel, R., and Heyneker, H. L. (1983). *Proc. Natl. Acad. Sci. U.S.A.* **81,** 3273–3277.

Deans, R. J., Denis, K. A., Taylor, A., and Wall, R. (1984). *Proc. Natl. Acad. Sci. U.S.A.* **81,** 1292–1296.

Falkner, F. G., and Zachau, H. G. (1982). *Nature (London)* **298,** 286–288.

Falkner, F. G., and Zachau, H. G. (1984). *Nature (London)* **310,** 71–74.

Gibson, T. J. (1984). Ph.D. Thesis, University of Cambridge.

Gillies, S. D., Morrison, S. L., Oi, V. T., and Tonegawa, S. (1983). *Cell* **33,** 717–728.

Kenten, J. H., Helm, B., Ishizaka, T., Cattini, P., and Gould, H. (1984). *Proc. Natl. Acad. Sci. U.S.A.* **81,** 2955–2959.

Messing, J. (1979). *NIH Recombinant DNA Tech. Bull.* **2,** 43–45.

Möller, G. (1982). *Immunol. Rev.,* Vol. 32.
Morrison, S. L., Johnson, M. J., Herzenberg, L. A., and Oi, V. T. (1984). *Proc. Natl. Acad. Sci. U.S.A.* 81, 6851–6855.
Mulligan, R. C., and Berg, P. (1981). *Proc. Natl. Acad. Sci. U.S.A.* **78,** 2072–2076.
Neuberger, M. S. (1983). *EMBO J.* **2,** 1373–1378.
Neuberger, M. S., Williams, G. T., and Fox, R. O. (1984). *Nature (London)* **312,** 604–608.
Neuberger, M. S., Williams, G. T., Jouhal, S. S., Mitchell, E. B., Flanagan, J. G. and Rabbitts, T. H. (1985). *Nature (London)* **314,** 268–270.
Ochi, A., Hawley, R. G., Hawley, T., Shulman, M. J., Traunecker, A., Köhler, G., and Hozumi, N. (1983a). *Proc. Natl. Acad. Sci. U.S.A.* **80,** 6351–6355.
Ochi, A., Hawley, R. G., Shulman, M. J., and Hozumi, N. (1983b). *Nature (London)* **302,** 340–342.
Oi, V. T., Morrison, S. L., Herzenberg, L. A., and Berg, P. (1983). *Proc. Natl. Acad. Sci. U.S.A.* **80,** 825–829.
Sandri-Goldin, R. M., Goldin, A. L., Levine, M., and Glorioso, J. C. (1981). *Mol. Cell. Biol.* **1,** 743–752.
Schaffner, W. (1980). *Proc. Natl Acad. Sci. U.S.A.* **77,** 2163–2167.
Southern, P. J., and Berg, P. (1982). *J. Mol. Appl. Genet.* **1,** 327–342.
Wood, C. R., Boss, M. A., Kenten, J. H., Calvert, J. E., Roberts, N. A., and Emtage, J. S. (1985). *Nature (London)* **314,** 446–449.

20

Proteolytic Processing of the Poliovirus Polyprotein by Two Virus-Encoded Proteinases

HARUKA TOYODA, MARTIN J. H. NICKLIN, MICHAEL G. MURRAY, AND ECKARD WIMMER

Department of Microbiology
School of Medicine
State University of New York at Stony Brook
Stony Brook, New York 11794

I. INTRODUCTION

Posttranslational modification of polypeptides through intra- or extracellular proteolytic cleavage is an important process by which functional proteins are formed *in vivo*. The proteinases involved in such posttranslational modification are not well characterized. Posttranslational, proteolytic cleavages also play a crucial role in the replicative cycle of many viruses. For example, paramyxo- and myxoviruses "mature" as a result of proteolysis after release from the cell. Only one of their two envelope glycoproteins is cleaved at a specific region of the polypeptide chain. Cleavage occurs extracellularly and is catalyzed by proteinases from the host (Scheid and Choppin, 1984). On the other hand, it is known that a large number of plant and animal viruses encode one or more intracellular proteinases that process virus-encoded polypeptides to functional structural and nonstructural proteins. In this case, virus-infected cells become a convenient source of highly specific proteolytic enzymes for the study of substrate recognition and cleavage. Moreover, genetic manipulations of the virus genome can lead to new phenotypes with respect to proteoly-

PROTEIN ENGINEERING
Applications in Science,
Medicine, and Industry

sis. Such studies have yielded important information concerning viral proliferation and host cell–virus interactions.

Picornaviruses, a large family of human and animal pathogenic agents, are unique viruses in that all of their known gene products originate from a single polypeptide, called the polyprotein (Jacobson and Baltimore, 1968) by proteolytic processing (Summers and Maizel, 1968; Holland and Kiehn, 1968; Kitamura *et al.,* 1981, and references therein). At least two virus-encoded proteinases are involved in the cleavages. The properties and function of these virus-specific proteinases will be discussed in this chapter.

Picornaviridae encompass four genera, the enteroviruses, rhinoviruses, cardioviruses, and aphthoviruses. Each genus is represented by well-known pathogens, such as poliovirus or enterovirus 72 (causing poliomyelitis or infectious hepatitis, respectively), human rhinoviruses (causing the common cold), encephalomyocarditis virus (a pathogen of rodents), and foot-and-mouth disease virus (causing foot-and-mouth disease in many domesticated animals). Altogether there are over 200 serologically distinct species. These viruses are small, icosahedral, nonenveloped particles that consist of 60 copies each of four capsid proteins (VP1, VP2, VP3 and VP4), and a single-stranded RNA genome approximately 7500 nucleotides in length. The RNA is covalently bound to a small protein at the 5′ end and polyadenylated at the 3′ end (Kitamura *et al.,* 1981, and references therein; see Fig. 1). The genome RNA is of plus-strand polarity; that is, it has the same polarity as mRNA. Indeed, virion RNA and viral mRNA are identical molecules except that the mRNA does not carry the 5′ terminal VPg (Nomoto *et al.,* 1977). In the case of poliovirus, the most thoroughly studied of all picornaviruses, translation of the mRNA begins at nucleotide 743 and terminates at nucleotide 7370 (Kitamura *et al.,* 1981; Dorner *et al.,* 1982, Emini *et al.,* 1982). The translation product is a polyprotein consisting of 2209 amino acids (molecular weight 246,468) that undergoes numerous cleavages to yield the capsid proteins VP1, VP2, VP3, and VP4 and nonstructural proteins of the P2 and P3 region (Kitamura *et al.,* 1981; Pallansch *et al.,* 1984, and references therein; see Fig. 1).

II. THE CLEAVAGE SITES OF THE POLIOVIRUS POLYPROTEIN

As revealed by sequence analyses of the genomic RNA (Kitamura *et al.,* 1981; Racaniello and Baltimore, 1981; Toyoda *et al.,* 1984) and of the poliovirus-specific proteins isolated from infected cells (Semler *et al.,*

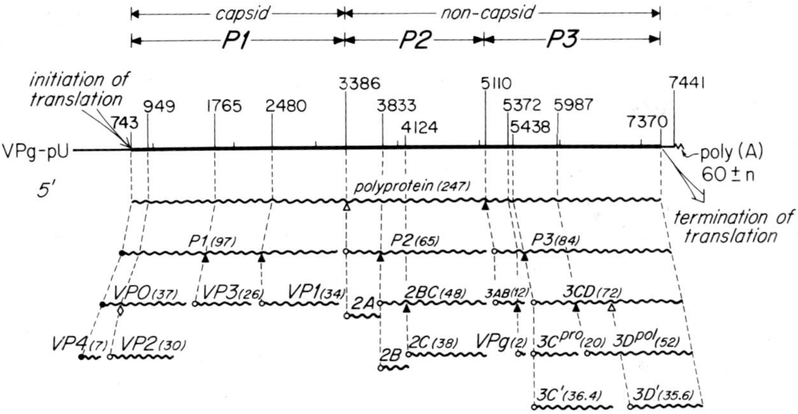

Fig. 1. Gene organization and polypeptide processing of poliovirus. Virion RNA, terminated at the 5′ end with the genome-linked protein VPg and at the 3′ end with poly(A), is shown as a solid line, the translated region being more pronounced than the noncoding regions. Arrows indicate the sites at which initiation (743) and termination (7370) of translation occur. The numbers above the virion RNA line refer to the first nucleotide of the codon specifying the amino-terminal amino acid of each virus-specific protein. The coding region has been divided into three regions (P1, P2, P3), corresponding to the primary cleavage products of polyprotein. Polypeptides are presented as wavy lines. Numbers in parentheses are molecular weights calculated from the amino acid sequences. Carboxy-terminal "trimming" does not occur. Open circles indicate glycine in all cases except for VP2, where it is serine. The carboxy-terminal amino acid of 3CD is phenylalanine. Filled circles indicate that the amino terminus is known to be blocked. Filled triangles, Gln–Gly; open triangles, Tyr–Gly; open diamond, Asn–Ser cleavage sites. Polypeptides 3C′ and 3D′ are products of an alternative cleavage mode of P3-2. (Modified after Kitamura *et al.*, 1981; Hanecak *et al.*, 1982; Semler *et al.*, 1981a,b, 1982, 1983).

1981a,b, 1982, 1983; Larsen *et al.*, 1982; Dorner *et al.*, 1982; Emini *et al.*, 1982; Adler *et al.*, 1983; Hanecak *et al.*, 1982; Pallansch *et al.*, 1984), the processing of poliovirus polyprotein adheres to the following principles. (1) Cleavages occur at three different amino acid pairs: Q-G, Y-G, and N-S (see Fig. 1). (2) The Q-G cleavages are the most common, although only 8–9 of 13 potential Q-G signals are recognized in proteolysis (see below). The polyprotein contains 10 Y-G pairs, but only two Y-G cleavages have been observed. Of these, only that between P1 and P2 may be biologically important; the other is an alternative cleavage of 3CD (see below). Of the four cleavage products of 3CD, only 3C and 3D have been shown to be essential viral proteins and no function has been assigned to 3C′ and 3D′. There are six N-S pairs in the polyprotein and several cluster in the N-terminal region of the polypeptide P1, the precursor of the capsid proteins. However, only one in VP0 is cleaved (see Fig. 1). (3) All proteolytic cleavages occur by hydrolysis of a single peptide bond. Neither proteolytic "trimming" of the major products nor removal of small oligopeptides has been observed in the case of poliovirus (Emini *et al.,* 1982). (4) Because not all potential amino acid pairs are cleaved, other determinants must be recognized by the proteinase. Alignment of the sequences surrounding the cleavage sites yields some evidence of additional primary specificity. In poliovirus, at least, both of the Y-G sites that are cleaved are preceded by a threonine residue. Position −4 in relation to the cleaved Q-G sites is alanine, proline, threonine, or valine in all cases and in all three serotypes of poliovirus. The residues occupying a similar position in relation to the uncleaved Q-G bonds are lysine, serine, and tyrosine (Fig. 2). In rhinovirus 14, position −4 of the cleaved "Q-G" sites (which also include a Q-A and an E-G site that can be aligned with Q-G sites of polio) features alanine, threonine, valine, and isoleucine. In two foot-and-mouth disease strains, O_1K and $A_{10}61$, the corresponding residues at position −4 of the E-G or E-S sites are proline, alanine, valine, or isoleucine. Thus position −4 appears to be occupied by alanine or a residue which has a deviation from a straight chain at the β position. It seems likely, but does not necessarily follow, that these viral proteinases have specificity toward residue −4 (as in the bacterial proteinase subtilisin) because these residues might impose constraints on the conformation of the substrate without interacting directly with the enzyme. Determination or accurate prediction of the conformation of the substrate or investigation of the specificity of the isolated enzyme toward peptide substrates may make it possible to distinguish between these two possibilities. Palmenberg *et al.* (1984) have already noticed that the Q-G and Q-S pairs of the polyprotein of encephalomyocarditis virus that probably correspond to the cleaved Q-G sites of poliovirus are always flanked on one side by a proline residue, and in this case there is clearly no specificity toward residue −4.

III. THE KINETICS OF INDIVIDUAL CLEAVAGES

The kinetics of the cleavages at the different sites have not been studied in detail for poliovirus. The half-life of individual precursor polypeptides, however, has been determined for encephalomyocarditis virus (EMC) (Rueckert, 1976, and references therein) and found to be between 1 and 12 min. This result indicates that some polypeptides are cleaved much more rapidly than others. A precise cascade of cleavages is not followed. That is, although precursor ABC is usually cleaved to A + BC, and BC to B + C, cleavage to AB + C, and AB to A + B may also occur.

Pulse-chase experiments have shown that the cleavage between P1 and P2 (of the Y-G site of poliovirus; Fig. 1) occurs before the synthesis of the

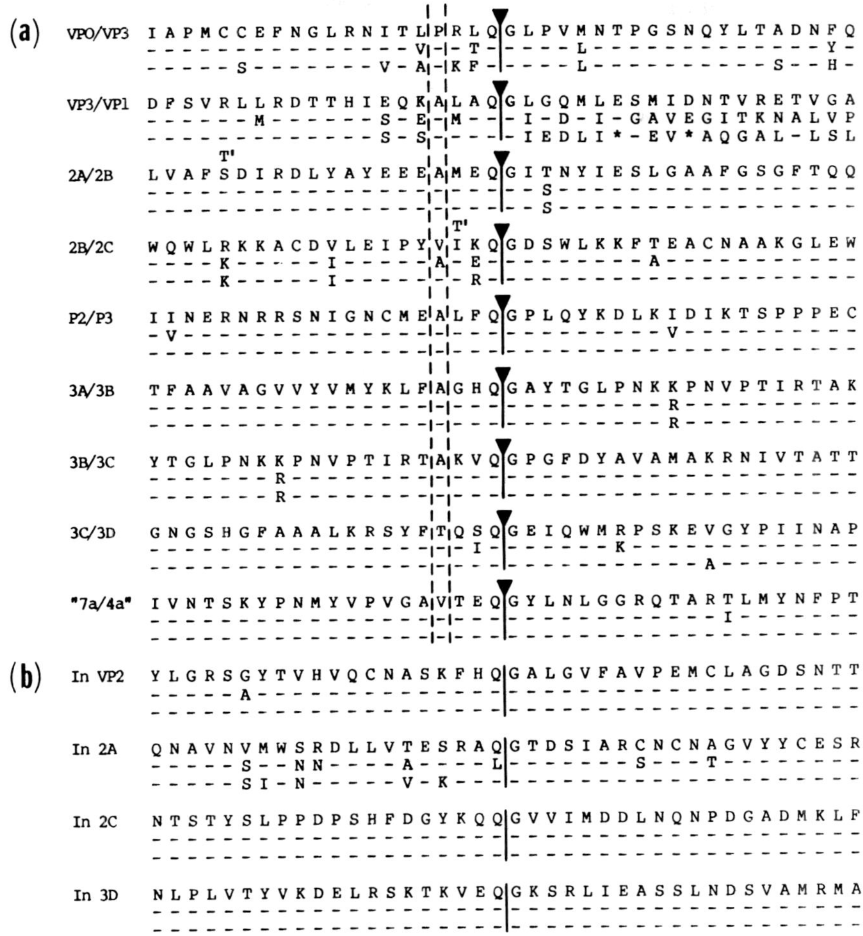

Fig. 2. Alignment of Q/G sites of the polyprotein of poliovirus. The sequence of the region surrounding each Q/G site in type 1 (Mahoney) is given in full. Primed characters (′) above the line refer to differences encountered in type 1 (Sabin). The second and third lines indicate only the differences between type 1 (Mahoney) and types 2 and 3 (Sabin), respectively. (a) Sites that are apparently substrates for proteinase 3C; (b) sites that are not cleaved.

polyprotein has been completed. Hence, the completed polyprotein can be observed in infected cells only if protein synthesis is carried out at supraoptimal temperature (43°C) or in the presence of several amino acid analogs (Baltimore, 1971, and references therein). As will be shown below, this cleavage is affected by a proteolytic active site which is located in 2A (Fig. 1). Cleavages at the Q-G sites between 2B and 2C and between 2C and P3 are also very rapid, an observation that led to the erroneous suggestion that polypeptide 2C might be extracistronic (Abraham and Cooper, 1975).

It is not generally possible to relate the kinetics of any particular cleavage to the physical properties of the precursor or the function of the product polypeptides. There are three exceptions: (1) 3AB, a membrane-bound polypeptide, may be cleaved to 3A + 3B (VPg) only after nucleotidylation of 3AB has occurred (Semler *et al.*, 1982; Takegami *et al.*, 1983). (2) Cleavage between P1 and P2 at the Y-G site occurs while the polyprotein is in *statu nascendi* and separates the precursors of the structural and the nonstructural proteins. (3) Cleavage at the N-S site in VP0 occurs only during the final step of maturation of the viral particle, when a procapsid (consisting of $[(VP0, VP3, VP1)_5]_{12}$) and virion RNA have been combined (by an unknown mechanism) to yield virions: $[(VP4, VP2, VP3, VP1)_5]_{12}$ + RNA (Putnak and Phillips, 1981).

IV. PROTEINASES INVOLVED IN PROTEOLYTIC PROCESSING

A. Overview of the Enzymes

Soon after the discovery of the mode of biosynthesis of picornavirus proteins by Summers, Maizel, Baltimore, Holland, and their colleagues (Summers and Maizel, 1968; Jacobson and Baltimore 1968; Holland and Kiehn, 1968), a number of papers were published that suggested that both cellular and virus-encoded proteinases participate in a variety of proteolytic processing steps. The evidence for cellular proteinase, however, was always weak, whereas that for the involvement of virus-encoded proteinases was more convincing (for a discussion, see Adler *et al.*, 1981). Pelham and Jackson and their colleagues, making use of their improved *in vitro* translation system (Pelham and Jackson, 1976), showed that certain EMC-specific proteins are processed in the cell-free system only if the EMC RNA was translated completely (Pelham, 1978). This observation was interpreted to mean that (1) a viral proteinase must be produced *in vitro* for proteolysis to be observed and (2) the viral proteinase maps in the 3′-terminal portion of the viral RNA. The proteinase was subsequently identified as protein 3C of EMC (Palmenberg *et al.*, 1979). Hanecak *et al.* (1982) then showed that the corresponding protein 3C of poliovirus is the proteinase responsible for cleavages at the Q-G amino acid pairs but not at the Y-G amino acid pairs.

The fastest cleavage of the polyprotein of poliovirus is that between P1 and P2 at a Y-G amino acid pair. As mentioned before, this cleavage occurs while the polyprotein is in *statu nascendi*. Cleavage between P1 and P2 is also rapid when picornavirus RNAs are translated *in vitro*. Thus, P1 is produced so rapidly *in vitro* that the polyprotein can usually not be detected (see, for example, Pelham, 1978). Based on these observations, it was originally suggested that a cellular proteinase is responsible for the P1–P2 cleavage. Evidence was obtained and interpreted as meaning that such cellular proteinase may be ribosome bound (for example, see Korant *et al.*, 1980a). In view of recent experiments, this is unlikely. As will be outlined below, the cleavage between P1 and P2, as well as the cleavage yielding 3C′ and 3D′ (both Y-G signals), is catalyzed by the poliovirus-specific polypeptide 2A, a protein that maps immediately adjacent to P1 (see Fig. 1). Since the cleavage between P1 and P2 is so fast even when very little protein has been translated, it is tempting to suggest that the reaction is intramolecular, as has been suggested for certain cleavages that are made by proteinase 3C (Palmenberg and Rueckert, 1982).

The nature of the proteinase cleaving VP0 at the N-S pair remains to be determined. It is conceivable that VP0 may also be capable of intramolecular cleavage provided proper conditions, e.g., association with genome RNA or specific components thereof or with cofactors, are met during maturation. Putnak and Phillips (1981) suggested that a virus-encoded protein, the "morphopoietic factor," is involved in the morphogenesis of procapsid, the morphogenetic precursor of virions consisting of VP0, VP3, and VP1. A protein that catalyzes morphogenesis might also participate in the maturation of the virion, that is, in the cleavage of VP0. Unfortunately, the nature of the "morphopoietic factor," particularly its position in the genetic map of poliovirus, is unknown.

B. The Q-G Proteinase 3C

Rothberg *et al.* (1978) exhaustively labeled the proteins that are present in purified, infectious poliovirus RNA to the extent that they were able to detect VPg (a peptide of only 22 residues) at a level of one copy per particle. No trace of 3C was found in these experiments, so one can conclude that proteinase is not imported by the infectious virion RNA and must therefore be synthesized *de novo*. In a cell-free translation system programmed with picornaviral RNA, it takes 15–30 min before the enzymatic activity of 3C can be observed. The only reliable assay for 3C is currently processing of P1 to VP0, VP3, and VP1. Although this is tedious, the efficiency of the cell-free systems (Pelham and Jacobson, 1976; Pelham, 1978) has allowed the identification of the virus-encoded polypeptide responsible for P1 processing, partial purification of the proteolytic activity, and determination of the properties of the protease with the use of specific inhibitors (Pelham, 1978; Gorbalenya and Svitkin, 1983; Palmenberg and Rueckert, 1982). Argos *et al.* (1984), in a detailed computer-aided study of polypeptides of picornaviruses and cowpea mosaic virus (a plant virus related in its gene organization and gene expression to picornaviruses), identified conserved sequences within all known 3C polypeptides. Examination of the conserved residues led to the conclusion that the proteinase is probably of the cysteine (thiol) class. A cysteine (147 in poliovirus) and a histidine (161) close to the carboxy terminus of the protein are the only conserved residues that could form the catalytic site of a proteinase of known catalytic class (Fig. 3). It had been shown previously that general-purpose thiol-specific reagents are capable of inhibiting the partially purified 3C from cells infected with encephalomyocarditis virus (Pelham, 1978; Gorbalenya and Svitkin, 1983). The 3C proteinases are not homologous to any of the plant and animal proteinases of this class.

Hanecak *et al.* (1982) studied the effect of monospecific antibodies against poliovirus 3C in a cell-free translation system. These antibodies inhibited Q-G cleavages but not Y-G cleavages (Fig. 4). The precise map position of 3C was obtained by protein and nucleic acid sequencing (Semler *et al.*, 1981a). This allowed the expression in *E. coli* of a cloned segment of poliovirus coding for a portion of 3AB, all of 3C, and a portion of 3D (Hanecak *et. al.*, 1984). Analysis of the expressed fusion polypeptide (Fig. 5) revealed rapid processing to 3C (Fig. 6). A mutation produced within the coding region of 3C by the insertion of an oligonucleotide "linker" abolished proteolysis (Hanecak *et al.*, 1984). These data provided strong evidence that enzymes of *E. coli* are not involved in the proteolysis. Instead, it was concluded that 3C can cleave itself out of precursor molecules. The 3C is itself flanked by Q-G sites and must therefore be active even in its precursors. Palmenberg and Rueckert (1982) investigated the processing of EMC precursors in a cell-free system and studied the effect of dilution of the reaction mixture on the rate of appearance of certain peptides. They showed that the progress of the reactions in which 3C and 3D are generated from P3 and 3CD is independent of the dilution of the reaction mixture. Based on this evidence, Palmenberg and Rueckert suggested that the release of 3C from its precursors involves intramolecular cleavage.

```
FMD   E/S G A P P T D L Q K M V M G N T K P V E L N L D G K T V A I - C C -
EMC   Q/G P N P V M D F E K Y V A K H V T A P I G F V Y P T G V S T Q - T C
Polio Q/G P - G - - - F D - Y A V A M A K - R - N I V T A T T S K G E F T -
CMPV                    Q/G S V A I M S K C R A N L V F G - G - - T N L Q I
                        +     + +                 + +

FMD   A T G V F G T A Y L V P R H L F - - A E - K Y D - K I M L D G R A M
EMC   L L V R - G R T L V V N R H M - - - A E S D W T - S I V V R G V T H
Polio M L G V - H D N - V A I L P - - T H A - S P G E - S I V I D G K E V
CPMV  V M V P - G R R F L A C K H F F T H I K T K L R V E I V M D G R R Y
      +                 + +               +               * + + . *     +

FMD   T D S D Y R V F E F E I K V - K G Q D - - M L S - D A A L M V - L H
EMC   A R S T V K I L A I - A K A G K E T D V S F I R L S S G P L F R D N
Polio E I L D A K A L - - E D Q A G T N L E I T I I T L K R N E K F R D I
CPMV  Y H - Q F D P A N I - Y D I P - D S E L V L Y S H P S L E D V S H S
              + .   +         . +           +   +                 +

FMD   R G N - C V - R D I T K H F R D T A R M K K G - T P V V G V V N N A
EMC   T S K - F V - - - - - K A G - D V L - - P T G A A P V T G I M N T -
Polio R P H - I - - - P - - - T Q - I T - E T N D G V L I V N T - - - S -
CPMV  C W D L F C W D P D K E L P - S V F - - - - G A D F L S C K Y N - -
              +                                   +       +

FMD   D V G R L I - F S G E A L T Y K D I V V C M D - G - D T M P G - - -
EMC   D I P M M Y - - T G T F L K A - G V S V P V E T G - Q T F N H - - -
Polio K Y P N M Y V P V G A V T E Q - G - Y L N L - G G R Q T A K T - - -
CPMV  K F G G F Y E A Q Y A D I K V R - T K K E C L T I - Q S G N Y V N K
      . + +   + +                               +         + +

FMD   - - - L F A Y K A A T R A G Y C G G A V L A K D G A D T F I V G T H
EMC   - - - C I H Y K A N T R K G W C G S A L L A D L G G S K K I L G I H
Polio - - - L M - Y N F P T R A G Q C G G V I T C T - G - - - K V I G M H
CPMV  V S R Y L E Y E A P T I P E D C G S L V I A H I G G K H K I V G V H
            + +   * . +   +         * *   + +   +     *         + + *   *
                                    *                                   *

FMD   S A G G N G - V G Y C S C V S R - S M L - Q K M - K A H V D P - E P
EMC   S A G - - - S M G I A A A - - - - S I V S Q E M I R A V V N A F E P
Polio V G G - N G S H G F A A A L K R - S Y F T Q S - - - - - - - - - - -
CPMV  V A G I Q G K I G C A S L L P P L E P I A Q A - - - - - - - - - - -
          *         + * + +   +             +   *

FMD   H H E/G
EMC   - - Q/G
Polio - - Q/G
CPMV  - - Q/G
          +
```

Fig. 3. Alignment of the amino acid sequence of the proteinases of EMC, polio, FMD, and CPMV. Residues 952–1155 of the CPMV sequence are given (Argos *et al.*, 1984). The sequences shown are immediately adjacent to the VPg sequence at the amino terminus and the polymerase at the carboxy terminus. (*) Residue is conserved in all four sequences. (+) Residues at this position in all four sequences are members of one of the following sets of residues: D, E, Q, N (acidic and polar); K, R (basic); T, S (polar); C, F, I, L, M, V, A, W, H, Y (hydrophobic); P, G, N, D (strong turn formers). (·) Residues are polar or charged (D, E, K, R, Q, N) at a given position. (/) Cleavage site of poliovirus and implied for the other viruses. (**) Residues are the conserved cysteine and histidine, which are probably involved in the catalytic mechanism of the enzymes.

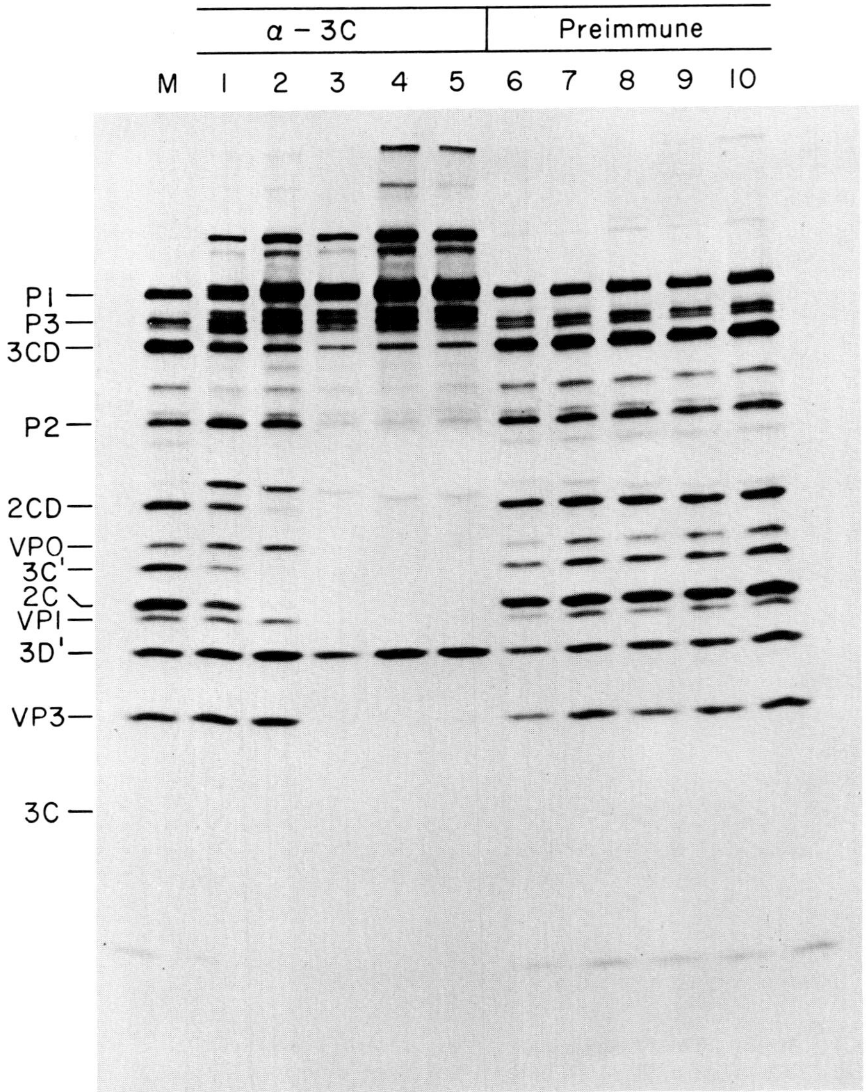

Fig. 4. Effect of anti-3C antibody on the processing of poliovirus proteins *in vitro*. Translation reaction mixtures were preincubated on ice for 1 hr in the presence of anti-3C or preimmune IgG. [^{35}S]Methionine was added (246 μCi/ml) and the samples were incubated for 17 hr at 30°C. Aliquots (10 μl) were withdrawn from each reaction and analyzed. Effect on adding increasing amounts of anti-3C IgG—lane 1, 15 μg; lane 2, 50 μg; lane 3, 150 μg; lane 4, 250 μg; lane 5, 300 μg—or preimmune IgG—lane 6, 15 μg; lane 7, 50 μg; lane 8, 150 μg; lane 9, 250 μg; lane 10, 300 μg. Poliovirus proteins synthesized in an extract of infected HeLa cells in the absence of antisera are displayed in lane M.

C. The Y-G Proteinase 2A

As pointed out above, the possibility of a second polypeptide with proteinase activity within the polyprotein was ignored for some time, although little, if any, direct evidence existed for a cellular enzyme that supposedly would cleave between P1 and P2. Toyoda *et al.* (1986) constructed plasmids for expression in *E. coli* that would yield a polypeptide comprising almost the entire capsid region plus varying lengths of the adjacent polypeptide of region P2 (Fig. 7). Expression of the proteins (in this case derived from poliovirus, type 3, strain Sabin) and immunoprecipitation of cleavage products with anti-poliovirus (type 3) antibodies showed that cleavage between P1 and P2 occurred, but only if the construct expresses all of 2A. Moreover, if the coding sequence of 2A was disturbed by deletion or insertion, in frame, at different sites, proteolysis was abolished (Fig. 7). Antibodies that were raised against purified 2A immunoprecipitated a polypeptide from lysates of *E. coli* corresponding to sequences containing 2A when proteolytic processing had occurred (Toyoda *et al.*, 1986). Also, in cell-free translation of poliovirus RNA, anti-2A antibodies inhibited the production of 3C′ and 3D′ (Toyoda *et al.*, 1986), which arises from 3CD by cleavage at a Y-G pair (Hanecak *et al.*, 1982; see Fig. 1). It should be pointed out that the inhibition of synthesis of 3D′ reaches no more than 65% compared to control translations, an observation suggesting weak interaction with or partial steric hindrance of the active site. Moreover, the Y-G cleavage between P1 and P2 is not inhibited at all by anti-2A, which might be explained if cleavage occurs intramolecularly in *statu nascendi* much more rapidly than the antibody associates to block the action of the enzyme.

Studies in which inhibitors are used to determine the catalytic class of proteinase 2A have not been done. Examination of the primary sequence of 2A, however, suggests that it may be a cysteine proteinase of the same group as 3C. Close to the carboxy terminus there are a cysteine and a histidine residue and the surrounding sequences, ···DCGGI··· and ···QHGVI···, are similar to those of 3C, ···QCGGV··· and ···MHVGG··· (D. Diamond, unpublished observations).

V. IS FOLDING THE MAJOR DETERMINANT OF PROCESSING?

Apart from the indication, discussed above, that both 3C and 2A might have extended recognition sites for their substrates, there are likely to be

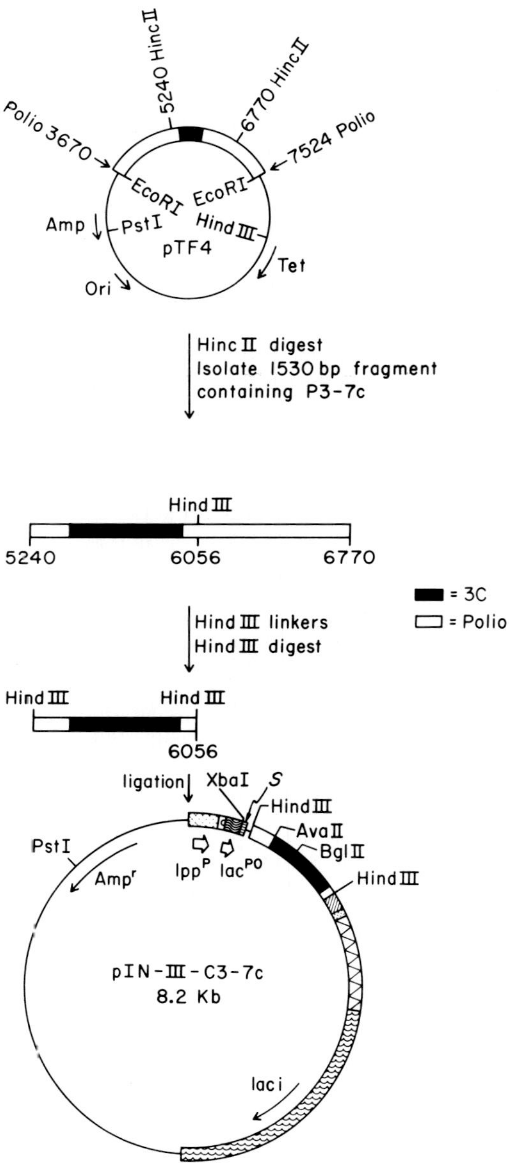

Fig. 5. Schematic representation of the construction of plasmid pIN-III-C3-7c. The approximate locations of several restriction endonuclease cleavage sites are shown for plasmid pTF4. The open bars represent poliovirus genomic sequences flanking the region of the viral genome encoding P3-7c (3C) designated by a solid bar. The positions of the origin of replication (ori), the β-lactamase gene of pBR322 (Amp), and the gene for tetracycline resistance are indicated by arrows. The construction of the pIN-III-C3-7c recombinant

other features of the substrates, apart from the primary structure, which identify the correct cleavage sites. Indeed, in some picornaviruses other than poliovirus, even the primary specificity is less precise. The Q-G sites are also found in the polyprotein of coxsackieviruses, rhinoviruses, and EMC, but in several cases the glycine is replaced by serine, threonine, or alanine (Palmenberg *et al.*, 1984; Callahan *et al.*, 1985; Skern *et al.*, 1985; Stanway *et al.*, 1984; Tracy *et al.*, 1985). Sometimes the glutamine residue is replaced by a glutamic acid residue. The E-G sites predominate in foot-and-mouth disease virus (FMDV; Robertson *et al.*, 1985, and references therein) but a single Q-G site may be cleaved by the 3C of FMDV. The second indication of more elaborate specificity is that few host proteins appear to be degraded as a result of poliovirus infection, as can be shown by two-dimensional gel electrophoresis (Korant *et al.*, 1980b; M. J. H. Nicklin, J. R. Putnak, and E. Wimmer, unpublished results), yet there must be many proteins with superficial Q-G or Y-G bonds. Third, neither the proteinase from EMC nor that from FMDV can process poliovirus proteins despite the similarity in the cleavage sites, at least between poliovirus and EMC.

It is likely that the accessibility of the substrate peptide bond at the surface of the protein is important in determining specificity, but the three observations mentioned above suggest that there must be further factors, such as the conformation of the substrate, or perhaps the location of the active enzyme and its substrates. The conformation of a given piece of primary structure, containing a cleavage site, might also change as a consequence of processing at another site. For instance, the polypeptide 3CD can undergo two alternative cleavages, each by the specific virus-encoded proteinase 3C and 2A (Rueckert, 1976; Hanecak *et al.*, 1982). As shown in Fig. 8, 3CD contains several potential Q-G and Y-G sites, but only one Y-G and usually only one Q-G site is used (filled triangle and filled circle). One Q-G site (half-filled triangle) is occasionally cleaved to yield polypeptide P3-4a (not shown in Fig. 8; see Semler *et al.*, 1983). Once cleavage has occurred at either of the major sites, to yield 3C and 3D or 3C′ and 3D′, the products appear to be quite stable. Only a small

---

vector is described in the Experimental Procedures of Hanecak *et al.*, 1984. The locations of the lipoprotein promoter sequence and the lactose promoter–operator region are indicated by blocked arrows. The box filled with diagonal lines represents the 3′ untranslated regions of bacterial lipoprotein and contains the stop codon for protein synthesis. A dotted box designates the location of transcription termination signals. The box filled with wavy lines represents the gene for the lactose repressor (*lac* I). The *S* denotes the location of the region encoding the signal peptide for lipoprotein.

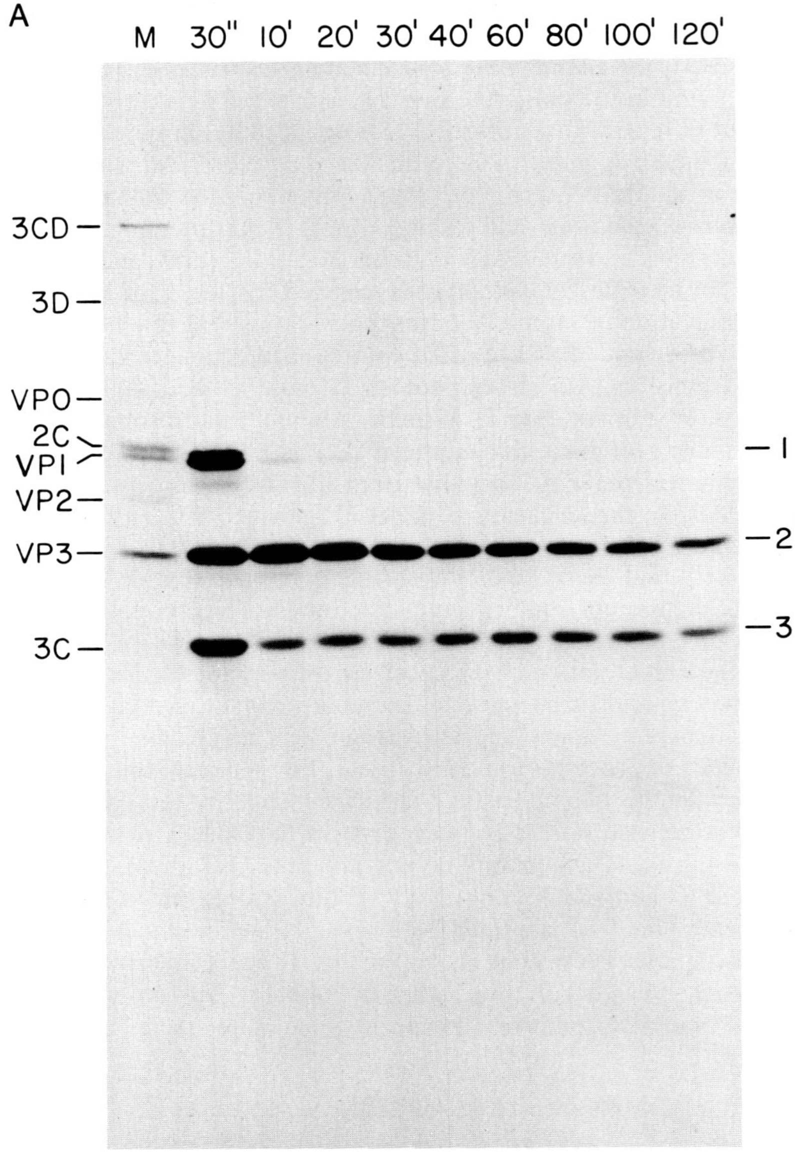

Fig. 6. (A) Pulse-chase analysis of polypeptides expressed by bacterial clone 1c transformed with pIN-III-C3-7c. The bacterial culture was pulsed for 30 sec and then chased with media containing unlabeled methionine for the indicated times. Aliquots of 5 ml were removed and placed in stop buffer. The preparation of bacterial extracts and immunoprecipitation of expressed proteins with anti-P3-7c (3C) serum are described in the Experimental Procedures of Hanecak *et al.*, 1984. Precipitated proteins were analyzed on a 12.5% SDS–

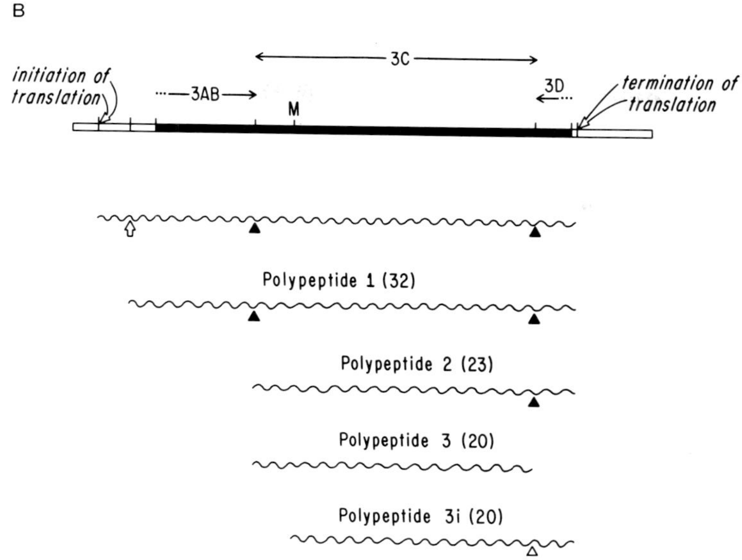

Fig. 6B. (Continued)

trace of the polypeptide that results from cleavage at both sites (P3-7d, not shown in Fig. 8; see Pallansch *et al.*, 1984) can be detected in infected cells, an observation suggesting that the cleavages are mutually inhibitory.

VI. CONCLUSION

Proteolytic cleavage of picornavirus polyproteins is a complex process that involves at least two virus-encoded proteinases. Although these proteinases have a preference for cleavage at specific amino acid pairs, the

---

polyacrylamide gel. Lane M represents an extract of [^{35}S]methionine-labeled poliovirus-infected HeLa cells not immunoprecipitated. (B) Processing scheme for poliovirus-specific polypeptides synthesized in bacteria. The pIN-III-C3-7c transcript is shown as a bar. The open bars represent pIN-III-C3-specific regions present within the transcript. The poliovirus genomic RNA sequence is shown as a filled bar. The methionine residue present within protein P3-7c (3C) at nucleotide 5516 is indicated by M. Polypeptide molecules are denoted by wavy lines. An arrow indicates the site of cleavage of the signal peptide. Cleavages between Gln–Gly pairs of amino acids are indicated by filled triangles. The cleavage between a Gln–Gly amino acid pair within polypeptide 3i that does not occur is shown as an open triangle. The numbers in parentheses are the molecular masses (in kilodaltons) calculated from the amino acid sequences, assuming an average molecular mass of 110 for each amino acid.

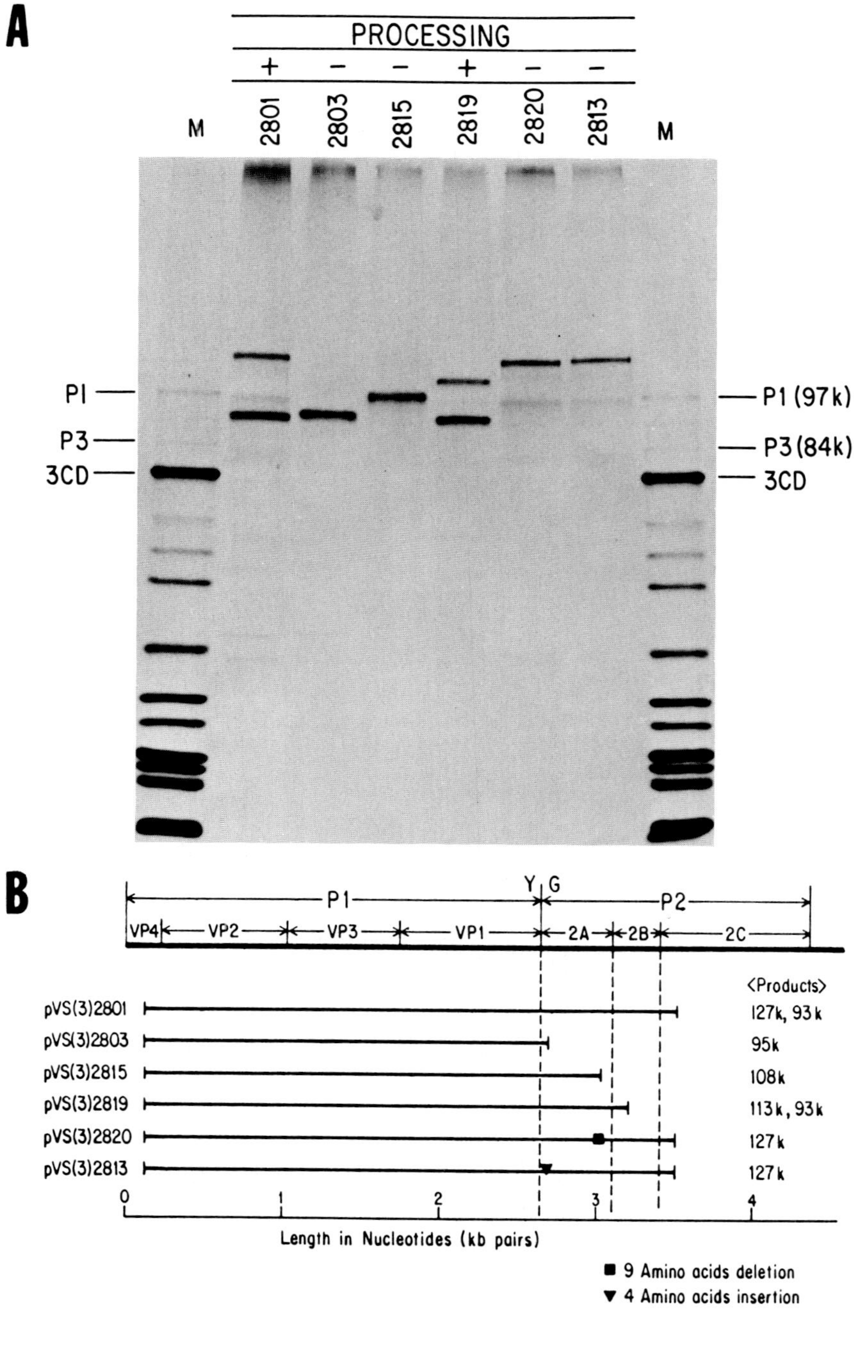

A
PROCESSING
+ − − + − −
M
2801
2803
2815
2819
2820
2813
M
P1
P3
3CD
P1 (97k)
P3 (84k)
3CD
B
P1
Y G
P2
VP4
VP2
VP3
VP1
2A
2B
2C
<Products>
pVS(3)2801
127k, 93k
pVS(3)2803
95k
pVS(3)2815
108k
pVS(3)2819
113k, 93k
pVS(3)2820
127k
pVS(3)2813
127k
0
1
2
3
4
Length in Nucleotides (kb pairs)
■ 9 Amino acids deletion
▼ 4 Amino acids insertion

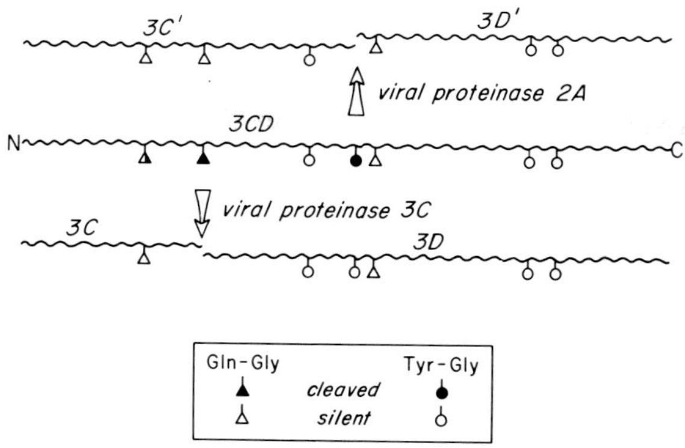

Fig. 8. Alternative cleavage pathways of poliovirus polypeptide 3CD. Gln–Gly pairs are cleaved by the viral proteinase 3C (Hanecak *et al.*, 1982). The Tyr–Gly pair is cleaved by the viral proteinase 2A (Toyoda *et al.*, 1986). Not all potential cleavage sites are recognized by the proteinases. Of the Gln–Gly pairs, one (filled triangle) is a preferred cleavage site and yields 3C and 3D; another (partially filled triangle) is used rarely and yields P3-4a (not shown here; Semler *et al.*, 1983). The P3-4a is processed further to 3D. The third Gln–Gly (at nucleotides 6486–6491) is apparently not used as a cleavage site. Tyr–Gly cleavage, on the other hand, occurs only at a single site (filled circle), and this yields the stable polypeptides 3C′ and 3D′ at the expense of the virus-specific RNA polymerase 3D and proteinase 3C. All four Tyr–Gly sites in the protein 3D are rendered inactive, presumably by protein folding, since the viral RNA polymerase is a stable end product of processing.

precise mechanism of substrate recognition is unknown. An understanding of the details of the specificity of these enzymes could form the basis of the development of new antiviral drugs.

ACKNOWLEDGMENTS

We appreciate the collaboration and stimulating discussions of many of our colleagues, particularly Bert L. Semler, Ronnie Hanecak, Naokazu Takeda, Richard J. Kuhn, David C. Diamond, Carl W. Anderson, John J. Dunn, and F. William Studier. We thank Akio Nomoto and Philip D. Minor for a sample of poliovirus type 3 specific antisera. Martin J. H. Nicklin

---

Fig. 7. Immunoprecipitation of polypeptides expressed in bacteria harboring derivatives of pVS(3)2801. Several derivatives of pVS(3)2801 were prepared and are shown in (B). Proteins expressed from each clone were examined by pulse-labeling with [^{35}S]methionine 1 hr after induction with phage λ (Toyoda *et al.*, 1986). Extracts were prepared and immunoprecipitated with monoclonal antibodies against VP1 of Sabin 3. Precipitated proteins were analyzed on an 8% SDS–polyacrylamide gel (A). Lane M represents an *in vivo* lysate of [^{35}S]methionine-labeled poliovirus-infected HeLa cells. In (A), processing is indicated by a (+) or (−).

is supported by a fellowship from the Damon Runyon–Walter Winchell Cancer Fund, DRG-848. This work was supported in part by National Institutes of Health grants AI15122 and CA28146.

REFERENCES

Abraham, G., and Cooper, P. D. (1975). *J. Gen. Virol.* **29,** 215–221.

Adler, C. J., Semler, B. L., Rothberg, P. G., Kitamura, N., and Wimmer, E. (1981). *In* "International Symposia on Metabolic Interconversion of Enzymes" (H. Holzer, ed.), pp. 356–371. Springer-Verlag, Berlin and New York.

Adler, C. J., Elzinga, M., and Wimmer, E. (1983). *J. Gen. Virol.* **64,** 349–355.

Argos, P., Kamer, G., Nicklin, M. J. H., and Wimmer, E. (1984). *Nucleic Acids Res.* **12,** 7251–7276.

Baltimore, D. (1971). *Perspect. Virol.* **7,** 1–14.

Callahan, P. L., Mizutani, S., and Colonno, R. J. (1985). *Proc. Natl. Acad. Sci. U.S.A.* **82,** 732–736.

Dorner, A. J., Dorner, L. F., Larsen, G. R., Wimmer, E., and Anderson, C. W. (1982). *J. Virol.* **42,** 1017–1028.

Emini, E. A., Elzinga, M., and Wimmer, E. (1982). *J. Virol.* **42,** 194–199.

Gorbalenya, A. E., and Svitkin, Y. V. (1983). *Biokhimiya Moscow* **48,** 442–453.

Hanecak, R., Semler, B. L., Anderson, C. W., and Wimmer, E. (1982). *Proc. Natl. Acad. Sci. U.S.A.* **79,** 3973–3977.

Hanecak, R., Semler, B. L., Ariga, H., Anderson, C. W., and Wimmer, E. (1984). *Cell* **37,** 1063–1073.

Holland, J. J., and Kiehn, E. D. (1968). *Proc. Natl. Acad. Sci. U.S.A.* **60,** 1015–1022.

Jacobson, M. F., and Baltimore, D. (1968). *Proc. Natl. Acad. Sci. U.S.A.* **61,** 77–84.

Kitamura, N., Semler, B. L., Rothberg, P. G., Larsen, G. R., Adler, C. J., Dorner, A. J., Emini, E. A., Hanecak, R., Lee, J. J., van der Werf, S., Anderson, C. W., and Wimmer, E. (1981). *Nature (London)* **291,** 547–553.

Korant, B. D., Chow, N. L., Lively, M. O., and Powers, J. C. (1980a). *Ann. N.Y. Acad. Sci.* **343,** 304–318.

Korant, B. D., Langner, J., and Powers, J. (1980b). *In* "Biosynthesis, Modification and Processing of Cellular and Viral Protein" (G. Koch and D. Richter, eds.), pp. 257–288. Academic Press, New York.

Larsen, G. R., Anderson, C. W., Dorner, A. J., Semler, B. L., and Wimmer, E. (1982). *J. Virol.* **41,** 340–344.

Nomoto, A., Kitamura, N., Golini, F., and Wimmer, E. (1977). *Proc. Natl. Acad. Sci. U.S.A.* **74,** 5345–5349.

Pallansch, M., Kew, O. M., Semler, B. L., Omilianowski, D. R., Anderson, C. W., Wimmer, E., and Rueckert, R. R. (1984). *J. Virol.* **49,** 873–880.

Palmenberg, A. C., and Rueckert, R. R. (1982). *J. Virol.* **41,** 244–249.

Palmenberg, A. C., Pallansch, M. A., and Rueckert, R. R. (1979). *J. Virol.* **32,** 770–778.

Palmenberg, A. C., Kirby, E. M., Janda, M. R., Drake, N. L., Duke, G. M., Potratz, K. F., and Collett, M. S. (1984). *Nucleic Acids Res.* **12,** 2969–2985.

Pelham, H. R. B. (1978). *Eur. J. Biochem.* **85,** 357–462.

Pelham, H. R. B., and Jackson, R. J. (1976). *Eur. J. Biochem.* **67,** 247–256.

Putnak, J. R., and Phillips, B. A. (1981). *Microbiol. Rev.* **45,** 287–315.

Racaniello, V. R., and Baltimore, D. (1981). *Science* **214,** 916–919.

Robertson, B. H., Grubman, M. J., Weddell, G. N., Moore, D. M., Welsh, J. D., Fischer,

T., Dowbenko, D. J., Yansura, D. G., Small B., and Kleid, D. G. (1985). *J. Virol.* **54,** 651–660.

Rothberg, P. G., Harris, T. J. R., Nomoto, A., and Wimmer, E. (1978). *Proc. Natl. Acad. Sci. U.S.A.* **75,** 4868–4872.

Rueckert, R. R. (1976). *In* "Comprehensive Virology" (H. Fraenkel-Conrat and R. R. Wagner, eds.), Vol. 6, pp. 131–213. Plenum, New York.

Scheid, A., and Choppin, P. W. (1984). *In* "Concepts in Viral Pathogenesis" (A. L. Notkins and M. B. A. Oldstone, eds.), Vol. 1, pp. 26–31. Springer-Verlag, Berlin and New York.

Semler, B. L., Anderson, C. W., Kitamura, N., Rothberg, P. G., Wishart, W. L., and Wimmer, E. (1981a). *Proc. Natl. Acad. Sci. U.S.A.* **78,** 3464–3468.

Semler, B. L., Hanecak, R., Anderson, C. W., and Wimmer, E. (1981b). *Virology* **114,** 589–594.

Semler, B. L., Anderson, C. W., Hanecak, R., Dorner, L. F., and Wimmer, E. (1982). *Cell* **28,** 405–412.

Semler, B. L., Hanecak, R., Dorner, L. F., Anderson, C. W., and Wimmer, E. (1983). *Virology* **126,** 624–635.

Skern, T., Sommergruber, W., Blaas, D., Gruender, P., Fraundorfer, F., Pieler, C., Fogy, I., and Kuechler, E. (1985). *Nucleic Acids Res.* **13,** 2111–2126.

Stanway, G., Hughes, P. J., Mountford, R. C., Minor, P. D., and Almond, J. W. (1984). *Nucleic Acids Res.* **12,** 7859–7875.

Summers, D. F., and Maizel, J. V. (1968). *Proc. Natl. Acad. Sci. U.S.A.* **59,** 966–971.

Takegami, T., Kuhn, R. J., Anderson, C. W., and Wimmer, E. (1983). *Proc. Natl. Acad. Sci. U.S.A.* **80,** 7447–7451.

Toyoda, H., Kohara, M., Kataoka, Y., Suganuma, T., Omata, T., Imura, N., and Nomoto, A. (1984). *J. Mol. Biol.* **174,** 561–585.

Toyoda, H., Nicklin, M. J. H., Murray, M. G., Dunn, J. J., Studier, F. W., and Wimmer, E. (1986). *Cell* **45,** 761–770.

Tracy, S., Liu, H.-L., and Chapman, N. M. (1985). *Virus Res.* **3,** 263–270.

IV

Applications

21

Enzymatic Reactions in Organic Media

ALEXANDER M. KLIBANOV

Department of Applied Biological Sciences
Massachusetts Institute of Technology
Cambridge, Massachusetts 02139

I. INTRODUCTION

In nature, enzymes function in aqueous solutions. Therefore, it is not surprising that virtually all studies in enzymology thus far have employed water as the reaction medium. However, from the biotechnological standpoint there are numerous advantages of conducting enzymatic conversions in organic solvents as opposed to water; these include the (i) high solubility of most organic compounds in nonaqueous media, (ii) ability to carry out new reactions impossible in water due to kinetic or thermodynamic restrictions, (iii) greater stability of enzymes, (iv) relative ease of product recovery from organic solvents compared to water, and (v) insolubility of enzymes in organic media, which permits their easy recovery and reuse and thus eliminates the need for immobilization.

Conventional wisdom dictates that water is required for enzyme action. This conclusion stems from the fact that water participates (directly or indirectly) in all noncovalent interactions maintaining the native, catalytically active enzyme conformation (Kauzmann, 1959; Tanford, 1961; Schulz and Schirmer, 1979; Cantor and Schimmel, 1980); hence removal of water should drastically distort that conformation and inactivate the enzyme. Although this reasoning is undoubtedly correct, it misses the point somewhat because the real question should be not *whether* water is indeed required but *how much* water is needed. It is hard to imagine that an enzyme molecule can "see" more than just a monolayer or so of water

PROTEIN ENGINEERING
Applications in Science,
Medicine, and Industry

around it. As long as this water is localized about the enzyme molecules, the rest (i.e., the bulk) of the water can probably be replaced by an organic solvent without adversely affecting the enzyme. Since the absolute amount of water contained in that monolayer is very small, this situation is equivalent to an enzyme functioning in a nearly anhydrous organic medium.

Over the past 2 years we have experimentally examined the above rationale and proved it valid. We have found that a number of enzymes can vigorously function as catalysts in organic solvents (and we have yet to encounter one which cannot). Using enzymes in nonaqueous media, one can carry out processes (e.g., in the area of preparative organic synthesis or chemical analyses) that are very difficult in water. In this chapter we briefly describe some key catalytic properties of enzymes in nearly anhydrous organic solvents and outline some possible biotechnological applications.

II. LIPASES IN ORGANIC SOLVENTS

An ideal initial model for the investigation of enzymatic reactions in organic solvents should satisfy the following criteria: (i) the enzyme should be readily available and inexpensive; (ii) the enzyme should be cofactor-independent, as common cofactors are insoluble in organic solvents; (iii) the substrates should be soluble in organic media; and (iv) water should be a nonreactant in the enzymatic process. Lipases (EC 3.1.1.3) meet all these requirements. Although the natural lipolytic reaction is hydrolysis of carboxylic esters, lipases can also catalyze transesterifications (which do not involve water as a substrate). Therefore, we investigated transesterifications (and other reactions not requiring water) in organic solvents catalyzed by three different lipases—porcine pancreatic, yeast (*Candida cylindracea*), and mold (*Mucor* sp.).

A. Activity versus Water Content

Tributyrin (glyceryl tributyrate) and heptanol were dissolved in 2-pentanone, followed by dehydration of the mixture and addition of a known amount of water. Then a lipase powder was added and the suspension formed [proteins are insoluble in this, as well as in almost all other organic solvents (Singer, 1962)] was shaken at 20°C; periodically aliquots were withdrawn and assayed by gas chromatography. The most striking finding was (Zaks and Klibanov, 1985) that all three lipases were capable of catalyzing the transesterification reaction in 2-pentanone containing less than 1% water. The enzymatic activities of porcine pancreatic and mold lipases were only slightly dependent on the concentration of water in the range 0.02–1%, while that of yeast lipase gradually decreased as the water content was lowered in that range. It should be stressed, however, that despite the quantitative differences, all three lipases exhibited robust catalytic activity even *at a concentration of water as low as 0.02%,* i.e., in 99.98% organic medium. This enzymatic activity was sustained over extended periods of time and was sufficient to nearly complete the reaction. Lipase-catalyzed transesterification in 2-pentanone was not merely an artifact due to the presence of protein, because when the lipases were inactivated by the active center-specific reagent diethyl *p*-nitrophenyl phosphate (Maylie *et al.,* 1969), all three exhibited no catalytic activity in the organic medium.

B. pH Dependence

Since enzymatic reactions in aqueous solutions are strongly pH-dependent, one of the most intriguing aspects of the described phenomenon is that of the "pH" in a nearly anhydrous organic milieu. To elucidate it, the following experiment was designed. Porcine pancreatic lipase was dissolved in an aqueous buffer and the pH was adjusted to a given value. Then the enzyme was precipitated with cold acetone, recovered, and dried under vacuum. The enzymatic activity of the resulting lipase preparation was assayed both in water (at pH 8.4) and in a mixture of heptanol and tributyrin (containing only 0.02% water) as a function of the pH of the aqueous solution from which the enzyme was precipitated. It was found that the hydrolytic activity of lipase in *water* was independent of the "pH history" of the enzyme sample. This was not surprising because the assay solution was always maintained at pH 8.4 (the optimum for the hydrolytic activity). At the same time, the catalytic activity of lipase in the *organic medium* very much depended on the pH from which the enzyme was precipitated, with the maximum approximately coinciding with the pH optimum of the enzymatic activity in water. Thus, the enzyme "remembers" the pH of the last aqueous solution it has been exposed to (i.e., the enzyme's ionogenic groups acquire the corresponding ionization states, which then remain both in the solid state and in organic solvents). When it was assayed in the organic medium, the specific activity of the commercial sample of porcine pancreatic lipase was one-third of that of the enzyme precipitated from pH 8.4; in all subsequent experiments the porcine pancreatic lipase precipitated from that pH was used.

C. Activity in Different Solvents

We have found (Zaks and Klibanov, 1985) that all three lipases are catalytically active in a wide variety (about two dozen) of different organic solvents. The only solvents in which all the enzymes were completely inactive were dimethyl sulfoxide and formamide. Since these are the only ones that can *dissolve* proteins (in all others the lipases existed exclusively as suspensions), one can hypothesize that enzymes in these solvents change their conformation and consequently inactivate.

The lipases are diverse in their sensitivity to the nature of the organic solvent. For instance, while the catalytic activity of porcine pancreatic lipase is comparable in different solvents, that of the yeast lipase varies by almost three orders of magnitude, being highest in water-immiscible solvents (paraffins, butyl ether, toluene, and carbon tetrachloride) and lowest in water-miscible ones (acetonitrile, tetrahydrofuran, dioxane, pyridine, and acetone). These data can be explained in terms of the model described in Section I: a certain amount of water molecules is required for the enzymatic activity. In the case of porcine pancreatic lipase this water is very tightly bound to the enzyme, and even hydrophilic solvents do not strip it. On the other hand, this essential water is apparently bound much more loosely to the yeast lipase and therefore it is partitioned into water-miscible (but not water-immiscible) solvents. Hence, the main factor in the effect of organic solvents seems to be their interaction not with the enzyme molecule itself, but with the enzyme-bound water.

D. How Active Are Lipases in Organic Media?

We have established (Zaks and Klibanov, 1985) that lipase-catalyzed transesterifications obey Michaelis–Menten kinetics. It was of vital interest to know the absolute enzymatic activity in nearly anhydrous organic solvents. To that end, we have determined the acceleration effect [defined as the ratio of the rate of the lipase-catalyzed transesterification (at a given enzyme concentration) to that of the reaction in the absence of the enzyme] of the lipases in the transesterification reaction in hexadecane. For comparison, acceleration effects of the lipases in the hydrolysis of tributyrin in water were also determined. It turned out that the catalytic power of each lipase in the transesterification reaction in hexadecane and in the hydrolysis reaction in water were *of the same order of magnitude* (Zaks and Klibanov, 1985). That is, in organic solvents, the enzymes seem to display their full inherent catalytic power, not just a small fraction of it. This strongly suggests that the enzyme conformations in organic solvents and in water are nearly identical.

E. Diversity of Lipase-Catalyzed Reactions in Organic Solvents

We found that in addition to transesterifications, several other processes can be carried out enzymatically in nonaqueous media (Zaks and Klibanov, 1985). For example, porcine pancreatic lipase can catalyze reactions of esterification, aminolysis, acyl exchange, thiotransesterification, and oximolysis. It should be stressed that some of these processes can be conducted only in nonaqueous solvents; for instance, the esterification reaction between butyric acid (0.3 *M*) and heptanol (0.3 *M*) catalyzed by 10 mg/ml porcine pancreatic lipase in water (0.05 *M* glycine buffer, pH 8.4) stops at a degree of conversion less than 0.1%, whereas the degree of conversion of the same reaction in hexane after 2 hr exceeds 90%.

F. Remarkable Thermostability of Lipase in Organic Media

We recently succeeded in answering the question, "Why do enzymes irreversibly thermoinactivate at high temperatures?" (Ahern and Klibanov, 1985). Upon heating, enzyme molecules first partially unfold (this process is reversible) and then can undergo one of the following irreversible reactions: aggregation, formation of incorrect (scrambled) structures, destruction of disulfide bonds, hydrolysis of the polypeptide chain at aspartic acid residues, and deamidation of asparagine residues. All of these processes require water. Hence one should expect that enzymes will be very thermostable in water-free environments, e.g., in organic solvents, where the above processes will not take place.

We experimentally verified the rationale described above (Zaks and Klibanov, 1984) by experiments with dry porcine pancreatic lipase in various environments. In water or 0.1 *M* phosphate buffer (pH 8.0) at 100°C, the enzyme lost its activity almost instantaneously. When the lipase powder was placed in a tributyrin–heptanol mixture containing only 0.8% water, its thermoresistance increased greatly. Moreover, the dry enzyme in an even drier organic medium (0.015% water) was remarkably stable; its half-life at 100°C was more than 12 hr. Similar results were obtained with another enzyme, lipase from *Candida cylindracea:* whereas the half-life of the enzyme powder (3% water content) in the wet tributyrin–heptanol mixture at 100°C was less than 2 min, it was almost 1.5 hr in the dry reaction mixture.

Thermal stability of both wet and dry porcine pancreatic lipase decreased as the water content of the tributyrin–heptanol mixture increased. However, at water concentrations below approximately 0.3%, the stability of the dry lipase at 100°C remained constant on further dehydration of the system, apparently indicating that the rate-limiting step in the irreversible thermoinactivation of the enzyme no longer involved water. The thermostability in nearly anhydrous organic media depended on the nature of the solvent: the half-lives of lipase at 100°C in heptanol, decanol, toluene, hexadecane, and tributyrin were 5, 12, 6, 10, and 26 hr, respectively.

The dry lipase, when placed in the dry organic environment, withstood heating at 100°C for many hours (as determined by assaying at 20°C). To determine whether the enzyme could also act as a catalyst under such harsh conditions, we measured the lipase-catalyzed transesterification between tributyrin and heptanol at 100°C (no reaction took place in the absence of the enzyme). We found that not only was the dry lipase catalytically active in the reaction mixture containing 0.015% water at 100°C, but its activity exceeded that at 20°C in the same system by a factor of 5. This proved to be a general phenomenon: heptanol was replaced by a number of other primary and secondary alcohols, and in all cases the lipase-catalyzed transesterification in this substrate mixture was faster at 100 than at 20°C. These data indicate that lipase maintains its catalytically active conformation at 100°C.

Thus, upon dehydration lipase becomes extremely thermostable. In addition, the enzyme becomes more selective (Zaks and Klibanov, 1984). A search for other changes in catalytic properties of enzymes brought about by dehydration is currently under way (e.g., see Zaks and Klibanov, 1986).

III. OXIDOREDUCTASES IN ORGANIC SOLVENTS

The ability to work as a catalyst in organic media is by no means limited to lipases. For instance, Price and associates have shown that chymotrypsin (Dastoli *et al.*, 1966) and xanthine oxidase (Dastoli and Price, 1967a,b) are catalytically active when suspended in organic solvents. We have found that several oxidoreductases, including mushroom polyphenol oxidase, horseradish peroxidase, *Pseudomonas* cholesterol oxidase, and horse liver alcohol dehydrogenase, function vigorously in organic media.

The basic rules of the behavior of enzymes in organic solvents are similar to those described above for lipases. It is always beneficial to precipitate an enzyme from an aqueous solution having the right (i.e., optimal for the enzymatic activity) pH prior to placing it in organic media. As one would expect from the general model described in Section I, the "best" solvents for enzymatic activity are water-immiscible, hydrophobic ones, e.g., paraffins, toluene, and chloroform. These solvents have no affinity for water and therefore do not strip the essential water molecules from the enzyme globule. When more hydrophilic solvents are used, there is significant partitioning of the essential water into them, which results in enzyme inactivation. This partitioning can be minimized if, in the case of water-immiscible solvents, the medium is presaturated with an appropriate aqueous buffer before the enzyme is added.

In many instances, in particular if the specific activity of the enzyme to be used is high, the apparent enzymatic activity in organic solvents is drastically reduced due to intraparticle diffusional limitations. This problem can be alleviated, as is well known in the area of heterogeneous catalysis, if finer particles of the catalyst are used. We have determined that enzymes deposited as a thin layer on glass powder are much more effective as catalysts in organic solvents than the same amounts of the suspended free enzymes.

We have not yet found an enzyme which cannot function in organic solvents. Therefore, we venture to predict that perhaps all enzymes can work in organic solvents if used correctly. In the next section we outline some biotechnological applications of enzymes in nonaqueous media.

IV. BIOTECHNOLOGICAL APPLICATIONS

A. Lipases

As described in Section II, lipases can catalyze reactions in organic solvents that are not feasible in water (because they are suppressed by hydrolysis), e.g., transesterifications and esterifications. We have successfully used this phenomenon for preparative purposes. For example, if a glycol is employed as a nucleophile in a porcine pancreatic lipase-catalyzed transesterification in organic media, the enzyme displays a remarkable regioselectivity, acylating only the primary hydroxyls; as a result, a number of primary monoesters of glycols have readily been produced on a preparative scale (Cesti *et al.*, 1985).

A great challenge to organic chemists is the synthesis of optically active compounds. We have found that porcine pancreatic and yeast lipase catalyze the reactions of transesterification and esterification, respectively, in organic solvents in a highly stereoselective manner. When a racemic alcohol or racemic acid was used as the substrate, lipase would convert only one optical isomer to the corresponding ester, thereby affording facile preparative resolution of racemates (Kirchner *et al.*, 1985).

B. Polyphenol Oxidase

The enzyme polyphenol oxidase catalyzes hydroxylation of phenols with O_2 to catechols and subsequent dehydrogenation to *o*-quinones. Since regioselective oxidation of aromatic compounds is an often desirable but difficult problem in organic chemistry, the above enzymatic reactions appear to be of great preparative interest. However, attempts to use polyphenol oxidase as a practical catalyst have been unsuccessful because the instability of *o*-quinones in water causes rapid polymerization to polyaromatic pigments, also inactivating the enzyme. Inspection of the mechanism of *o*-quinone polymerization reveals that water is an essential component of that reaction. This seems to rule out polyphenol oxidase as a useful catalyst because enzymatic processes take place in aqueous solutions. However, we have discovered that mushroom polyphenol oxidase can vigorously function in chloroform, quantitatively converting a number of phenols to stable *o*-quinones, which can then be nonenzymatically reduced to catechols, thus affording an efficient and regioselective net hydroxylation of phenols (Kazandjian and Klibanov, 1985).

C. Enzymatic Assays in Organic Media

Because of their keen substrate specificity, enzymes are uniquely suited for analytical applications and have therefore been used for quantitative determinations of numerous compounds of clinical, biochemical, and industrial interest. Thus far, all such enzymatic analyses have been performed in aqueous solutions because of the commonly assumed requirement for water as the reaction medium in enzyme-catalyzed reactions. This limitation severely restricts analytical applications of enzymes, for in many instances one needs to determine compounds in organic media, e.g., in industrial chemical reactors and waste streams or when analytes are insoluble in water.

We have found (Kazandjian *et al.,* 1986) that horseradish peroxidase can act vigorously as a catalyst in a number of water-immiscible organic solvents. The rates of peroxidase-catalyzed oxidation of *p*-anisidine with H_2O_2 in toluene, benzene, ethyl and butyl acetates, and ether are 10–25% of that in water (pH 7.0) at the same reactant concentrations. Peroxidase was coupled with cholesterol oxidase (which was also found to be catalytically active in organic media), and the bienzymatic system was successfully used for the accurate, reliable, and reproducible determination of cholesterol in toluene.

The same approach can be applied to the determination of other compounds in organic solvents by coupling the corresponding oxidase to horseradish peroxidase.

D. Concluding Remarks

The applications described above illustrate the point that many enzymatic processes, while very difficult or impossible in water, can readily be carried out in nonaqueous media. Realization of the fact that enzymes can work in organic solvents should therefore create new opportunities in the area of applied enzymology and result in many novel enzyme-based processes.

REFERENCES

Ahern, T. J., and Klibanov, A. M. (1985). *Science* **228,** 1280–1284.

Cantor, C. R., and Schimmel, P. R. (1980). "Biophysical Chemistry," Part I. Freeman, San Francisco.

Cesti, P., Zaks, A., and Klibanov, A. M. (1985). *Appl. Biochem. Biotechnol.* **11,** 401–407.

Dastoli, F. R., and Price, S. (1967a). *Arch. Biochem. Biophys.* **118,** 163–165.

Dastoli, F. R., and Price, S. (1967b). *Arch. Biochem. Biophys.* **122,** 289–291.

Dastoli, F. R., Musto, N. A., and Price, S. (1966). *Arch. Biochem. Biophys.* **115,** 44–47.

Kauzmann, W. (1959). *Adv. Protein Chem.* **14,** 1–63.

Kazandjian, R., and Klibanov, A. M. (1985). *J. Am. Chem. Soc.* **107,** 5448–5450.

Kazandjian, R., Dordick, J. S., and Klibanov, A. M. (1986). *Biotechnol. Bioeng.* **28,** 417–421.

Kirchner, G., Scollar, M. P., and Klibanov, A. M. (1985). *J. Am. Chem. Soc.* **107,** 7072–7076.

Maylie, M. F., Charles, M., Sarda, L., and Desnuelle, P. (1969). *Biochim. Biophys. Acta* **178,** 196–198.

Schulz, G. E., and Schirmer, R. H. (1979). "Principles of Protein Structure." Springer-Verlag, Berlin and New York.

Singer, S. J. (1962). *Adv. Protein Chem.* **17,** 1–68.

Tanford, C. (1961). "Physical Chemistry of Macromolecules." Wiley, New York.

Zaks, A., and Klibanov, A. M. (1984). *Science* **224,** 1249–1251.

Zaks, A., and Klibanov, A. M. (1986). *J. Am. Chem. Soc.* **108,** 2767–2768.

22

Antibody Targeting of Toxin Polypeptides

A. E. FRANKEL,[1] M. J. BJORN, AND J. L. WINKELHAKE

Departments of Protein Chemistry and Pharmacology
Cetus Corporation
Emeryville, California 94608

I. INTRODUCTION

Immunotoxins are hybrid molecules consisting of an antibody covalently linked to polypeptide toxins. The antibody portion guides the molecule to the surface of target cells, and on entry into the cell the toxin moiety interferes with protein synthesis. Although a number of investigators have linked anti-tumor antibodies to intact toxins, subunits, or toxinlike peptides and observed selective tumor cell killing *in vitro*, there have been few reports of the *in vivo* efficacy of such molecules (1–21). Since most clinical applications of immunotoxins depend on *in vivo* activity, we have focused on understanding the parameters that affect biological activity of toxin–antibody conjugates in an animal model.

The *in vivo* variables that could affect immunotoxin treatment include host factors related to the animal species and host immune competence, as well as factors related to the tumor site, size, heterogeneity, and growth characteristics. In addition, the antibody, linker, and toxin chosen contribute to the ultimate efficacy. Such molecule-specific variables include the route and mechanism of conjugate clearance from the bloodstream, the effect of the immunotoxin on normal tissue cells

[1] Present address: Department of Medicine, Hematology-Oncology Division, Duke University Medical Center, Durham, North Carolina 27710.

PROTEIN ENGINEERING
Applications in Science, Medicine, and Industry

TABLE I

Physiochemical Characteristics of Immunotoxins

Antibody	Component linker	Toxin
Class	Reducibility	Net hydrophilicity
Affinity	Site specificity	p*I*
Antigen copy number per cell		Presence of hydrophobic domains
Antigen location and mobility		Size
		Glycosylation

involved in clearance, the ability of the conjugate to enter the interstitial space of tumor nodules, and its cytotoxic potency once bound to tumor cells.

Protein engineering may be used to optimize the antibody, linker, and toxin to achieve success *in vivo* (see Table I).

With regard to the antibody, the availability of hybridoma-derived monoclonal antibodies (MAb's) permits screening of large numbers of antibodies with defined specificities. Monoclonal antibodies have been identified which bind a majority of tumors of a given type but bind very few normal tissue substructures (22–32). Such antibodies are anticipated to localize to tumors in patients, and a number of recent reports confirm such activity (33–35).

The MAb's not only must bind a cell surface antigen but also should cause the antibody–toxin conjugate to enter a cellular compartment from which the toxin moiety can cross the membrane and enter the cytosol. Not all antibody–toxin conjugates have this property (36–39).

Two groups of peptide toxins are being used in the preparation of immunotoxins. One group is derived from bacteria and includes diphtheria toxin and *Pseudomonas* exotoxin A (40). The other group consists of plant-derived proteins called phytotoxins (41). Some phytotoxins, like ricin, have two polypeptide chains, A and B. The A chain has the enzymatic activity which inactivates eukaryotic 60 S ribosomal subunits. The B chain has a binding site for galactose residues present on glycoproteins on cells. The B chains appear to have an additional function besides binding, which is to facilitate A chain translocation across the cell membranes. Other phytotoxins include hemitoxins such as pokeweed antiviral protein, gelonin, and saporin. These toxins consist only of A-chain-like polypeptides and thus cannot bind normal cells per se (42,43).

These toxins have been linked to antibody in three ways. In one case, the bacterial toxins and two-chain phytotoxins are linked to antibody by derivatizing both the antibody and toxin. The normal tissue binding site on the bacterial toxin or phytotoxin has been blocked either by the derivatizing reagent (44), the antibody (39), or the inclusion of exogenous lactose in the case of ricin and abrin conjugates (45). In the second case, two-chain phytotoxins and diphtheria toxin are separated into the A chain and B chain, and isolated A chain is linked to antibody (2,46–59). In the third case, plant hemitoxins and antibody are derivatized and conjugated (10,11,60–63). Currently, toxins have been selected for immunotoxin preparation based on two properties. On the one hand, there has been an emphasis on choosing toxins lacking a normal tissue binding site, such as isolated A chains or hemitoxins. On the other hand, investigators have sought to use extremely potent toxins which have good translocational properties in addition to enzymatic activity. This was achieved by using the B chain along with the A chain in conjugates or by delivering B chain via a secondary antibody (64).

However, the *in vivo* efficacy of immunotoxins may depend on properties of the toxin other than binding and internalization. These properties include the toxin isoelectric point, hydrophobicity, glycosylation, and site and nature of the cross-linker, which may determine the stability and eventual tumor localization *in vivo* of the antibody conjugate. To address this issue, several investigators have linked a variety of peptide toxins to a single antibody and compared their *in vivo* efficacy (1,2,10,11,21).

The linker is usually a heterobifunctional reagent. An example is the reagent *N*-succinimidyl 3-(2-pyridyldithio) propionate (SPDP) used to establish linkages containing a disulfide bond. Most linkers are not site-specific on the antibody molecule and create linkages with unknown stability *in vivo*. To avoid random modification of the antibody, one group has made linkers specific for the antibody oligosaccharides located in the Fc region (65). Conjugates with such a linker should restrict the attachment site of toxin on the antibody and would not block antigen binding sites. The disulfide-based conjugates may dissociate in the circulation *in vivo* by a thiol exchange process. To avoid this problem, attempts have been made to utilize alternative disulfide linkers with varying chemical structure (66). An alternative approach has been to choose toxins which are biologically active with other modes of conjugation such as thioether linkages (20).

Based on these considerations, there are multiple ways to engineer *in vivo* active immunotoxins. We started by first systematically examining MAb's to identify ones with optimal binding to breast cancer cells. We selected a single toxin, ricin toxin A chain (RTA), and a single class of linkers, disulfide, for this study.

II. MONOCLONAL ANTIBODIES FOR BREAST CANCER IMMUNOTOXINS

Two-month-old female Balb/c mice were immunized with fresh breast carcinoma membrane extracts or live breast tumor cell lines. After booster injections, splenocytes were removed and fused with SP2/0 mouse myeloma cells under established conditions (67). Hybridoma supernatants were screened for binding on breast tumor and normal tissue membranes, breast tumor and fibroblast cell lines, and breast carcinoma frozen sections by indirect immunoassays (68).

A total of 122 hybridomas were cloned, and the purified antibody from each was evaluated sequentially for (a) normal tissue binding, (b) breast tumor binding, (c) cytotoxicity after coupling to RTA, and (d) *in vivo* efficacy of the RTA conjugates.

Of the 122 MAb's, 39 bound 8 or fewer of 128 normal tissue substructures. The normal tissue reactivity pattern of antibodies which bound 3 or fewer of 128 normal tissues is shown in Table II. The most common normal tissue cross-reactivity was observed with kidney, skin, pancreas, esophagus, and stomach. Further evaluations revealed that in each case the MAb's bound predominantly epithelial substructures.

These 39 antibodies were tested on breast tumor sections from 21 patients and on breast tumor cell lines from 14 patients. Twenty-three of 39 bound 30% or more of breast tumor sections and 33 of 39 bound 30% or more of breast tumor cell lines. Thirty-eight of 39 bound either 30% of cell lines or tumor sections. These investigations identified a number of MAb's which were reactive with few normal tissue substructures but reactive with a large number of breast carcinomas. However, these antibodies were infrequent (38 of 22,000 hybridoma antibodies screened and evaluated). Interestingly, a number of these MAb's recognized a few surface and intracellular antigens (69).

The 38 MAb's meeting our criteria of tumor range and selectivity were derivatized with SPDP and conjugated to RTA (46). Sixteen of the 38 conjugates were toxic to at least one of four breast tumor cell lines *in vitro* at a concentration of 10 nM or less. Nine of 17 conjugates were very active *in vitro,* with an LD_{50} (inhibitory dose 50%) less than or equal to 1 nM. For 3 of the 17 antibodies which made active RTA conjugates *in vitro* there was a discrepancy between antibody cell line binding determined by indirect immunofluorescence and that determined by RTA conjugate cytotoxicity (see Table III). Antigen appeared to be present at approximately the same level on several cell lines, and yet antibody–RTA conjugates were much more toxic to one of the cell lines. Since other antibody–RTA conjugates could efficiently kill the resistant cell lines, this

TABLE II

Binding of the 14 Most Selective Antibodies to 21 Normal Tissues[a]

Tissue	Monoclonal antibodies								
	451B7 452F2 736G9 741F8 769E11	520C9 758G5	759E3	761B10	369F10	788G6	41B4	452E12	493D1
Pancreas	0	0	0	0	0	0	0	0	0
Esophagus	0	0	0	0	0	0	0	0	0
Lung	0	0	0	0	0	0	0	0	0
Colon	0	0	0	0	0	0	0	0	0
Brain	0	0	0	0	0	0	0	0	0
Heart	0	0	0	0	0	0	0	0	0
Ovary	0	0	0	0	0	0	0	0	0
Bone marrow	0	0	0	0	0	0	0	0	0
Bladder	0	0	0	0	0	0	0	0	0
Stomach	0	0	0	0	1G	0	0	0	0
Liver	0	0	0	0	0	0	0	1	0
Uterus	0	0	0	0	0	0	0	0	1L
Normal breast	0	0	0	0	0	0	1E	2	0
Tonsil	0	0	1E	0	0	1FE	1E	0	0
Skin	0	0	0	0	1S	0	1W	2S	2S
Kidney	1T	1T	0	1T	0	2T	0	0	2T
Platelets	0	0	0	0	0	0	0	0	0
Red blood cells	0	0	0	0	0	0	0	0	0
Monocytes	0	0	0	0	0	0	0	0	0
Lymphocytes	0	0	0	0	0	0	0	0	0
Granulocytes	0	0	0	0	0	0	0	0	0

[a] E, Epithelium; FE, focal epithelium; 1, weak binding; 2, strong binding; 0, no binding; G, glands; L, lumen; S, sebaceous glands; T, tubules; W, sweat glands.

could not be due to ribosomal resistance to RTA. Based on measurements of cytotoxicity on a larger group of 111 MAb's, several antibody properties were predictive of active RTA conjugates (see Fig. 1) (70). Antibodies with higher affinity and antigen copy number, antibodies of the immunoglobulin G (IgG) class, and antibodies binding cell surface antigens made more toxic RTA conjugates. Nevertheless, one-third of IgG antibodies which recognized surface antigens and bound them with high affinity still made inactive conjugates.

For studies with an animal model, we had to scale up the production of RTA conjugates at least 10-fold. The antibody was derivatized by conven-

TABLE III

***In Vitro* Cytotoxicity and Binding of Monoclonal Antibody–RTA Conjugates on Four Breast Cancer Cell Lines**[a]

	$TCID_{50}$ (n*M*)/binding[b]				
Conjugate	MCF7	CAMAI	SKBR3	BT20	CC95
113F1–RTA	0.5/++	0.6/+	15/+++	6/+++	>50/0
398D6–RTA	>20/+	1/+	>20/+	>20/0	>50/0
317G5–RTA	0.4/+	10/+++	20/++	2/++	>50/0
760C5–RTA	>50/++	>50/++	1/+++	>50/ND	>50/0

[a] Breast cancer-selective monoclonal antibodies were purified and conjugated to RTA and assayed as the breast cancer cell lines (MCF7, CAMA1, SKBR3, BT20) and the fibroblast cell line (CC95) as described (46). Binding was measured with an indirect live cell immunofluorescence assay with purified antibody as described (6f).

[b] The $TCID_{50}$ is the concentration of immunotoxin (in nanomoles per liter) causing 50% inhibition of protein synthesis. Binding ranged from 0 to +++; ND, not determined.

tional procedures with iminothiolane to minimize the loss of positive charge and was conjugated to reduced RTA (71). After affinity purification and sizing chromatography, we had sufficient material of defined purity (less than 10% free antibody or free RTA) for *in vivo* studies. These iminothiolane–RTA conjugates were indistinguishable from RTA conjugates prepared by SPDP in *in vitro* cytotoxicity assays.

A number of the 17 antibody–RTA conjugates that were active *in vitro* and several additional antibody–RTA conjugates are being prepared for *in vivo* studies. Toxicity measurements in nude mice demonstrated an LD_{50} for a single i.v. dose of 400 μg for RTA and 200 μg for RTA conjugates. MAb alone were not toxic at doses up to 10 mg. Using an every-other-day dosing schedule with six i.v. injections, the cumulative LD_{10}–LD_{50} values were 100 and 200 μg, respectively, in nude mice.

These selected antibodies which made active RTA conjugates *in vitro* are being tested in nude mouse tumor models.

Control experiments with the animal model demonstrated that phosphate-buffered saline (PBS), RTA, and antibodies alone had no effect on MX-1 tumor growth. The tumors grew fivefold over a 14-day period with a coefficient of variation of less than 50%. Animals showed weight loss of no more than 20% over the test period. Cyclophosphamide and diphtheria toxin, serving as positive controls, caused complete tumor regression when given at the cumulative LD_{10} dose.

After considering of all four criteria—selectivity, broad range, *in vitro*

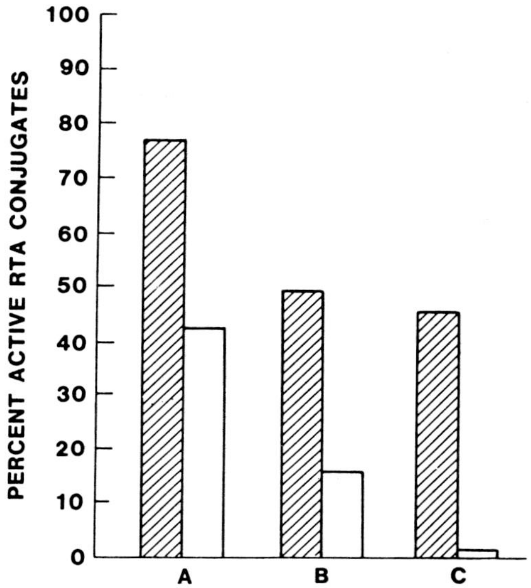

Fig. 1. Effect of physicochemical characteristics on *in vitro* activity of immunotoxins. Purified antibodies were conjugated to native ricin toxin A chain as described (46). RTA conjugates were assayed on four breast cancer cell lines and two fibroblast cell lines. Active conjugates inhibited protein synthesis by 50% after a 20-hr incubation of cells with ≤ 10 n*M* immunotoxin in one to four breast cancer cell lines. Antibody affinity (K_a) and antigen copy number per cell (n) were determined as described (70). Antibody class and surface or intracellular localization were determined as described (22). (A) Hatched bar represents antibodies with $nK_a \geq 10^{12}$ and light bar represents antibodies with $nK_a < 10^{12}$. (B) Hatched bar represents IgG antibodies and light bar IgM. (C) Hatched bar represents antibodies binding cell surface antigens and light bar represents antibodies binding intracellular antigens.

potency, and *in vivo* efficacy—we hope to identify MAb's as clear candidates for clinical development of immunotoxins.

III. TOXIN AND LINKER FOR BREAST CANCER IMMUNOTOXINS

We are currently conjugating a number of alternative bacterial toxins and phytotoxins to our most promising MAb's and testing these immunotoxins *in vitro* and *in vivo* on breast tumors. The toxins vary in p*I*, hydrophobicity, glycosylation, and site of cross-linkage. We hope that such a systematic evaluation coupled with studies in pharmacokinetics will elucidate the factors contributing to the optimal efficacy of immunotoxins *in vivo*.

IV. DISCUSSION

The identification of MAb's with little normal tissue reactivity and binding to a broad range of tumor types was considered essential for the eventual clinical development of immunotoxins, because it was thought that such highly potent molecules would have significant toxicity if they bound to vital normal tissues. This concern was based on two assumptions: (a) the immunotoxins would penetrate normal tissue capillaries as well as tumor capillaries and (b) normal tissue cells and tumor cells would internalize the immunotoxin with equal efficiency. Only clinical experience with a number of antibody conjugates with varying normal tissue reactivity will determine whether these assumptions are correct. In the interim, the safest approach is to choose antibodies with maximal tumor selectivity *in vitro*.

We found a number of MAb's reactive to breast cancer that satisfied the requirement of tumor selectivity. However, it is probable that none of the antibodies is absolutely breast tumor-selective as only a subset of human normal tissues was examined. Further, investigations of human fetal tissue reactivity have not been carried out to determine whether these MAb's recognize oncofetal antigens. Our breast cancer-selective MAb's paralleled those described by Colcher, Edgington, and others (72–86). However, the epitopes and antigens we identified appear different from those previously described. If our specificity criteria are broadened to accept antibodies which bind 10 or fewer of 128 normal tissue substructures and more than 30% of breast tumors, then we can include as candidates a number of antibodies identified by us and by others that react with epitopes on a high-molecular-weight mucin (76–78).

The finding that most of the normal tissue cross-reactivity of the selective antibodies is with epithelial substructures coincides with the epithelial sweat gland origin of the normal breast ductal system. These breast ductal cells are the target of transformation in breast carcinoma (87). The screening method we employed was useful in rapidly evaluating 20,000 hybridomas to choose those with tumor selectivity.

The antibodies were derivatized with SPDP and conjugated to RTA. The RTA conjugates were tested for cytotoxicity in four breast tumor cell lines and in two fibroblast cell lines as controls. Four tumor cell lines were chosen so that conjugates that were selectively active on a subset of breast tumors would not be overlooked. This appeared to be a reasonable concern as 10 of 14 antibody–RTA conjugates were toxic to only one or two of four cell lines (the other three antibody–RTA conjugates were tested on only one breast tumor cell line). We observed antibody–RTA conjugates that bound approximately equally well to two cell lines and yet were toxic to only one of them. Since both cell lines were sensitive to RTA (based on their sensitivity to other RTA conjugates), we presume that the antigen or details of its internalization after immunotoxin binding differ between the two cell lines. This indicates a new relationship between surface antigens and the host cell. The development of such variant tumor cells needs careful investigation since such variants would escape immunotoxin treatment.

Only 40% of the RTA conjugates were active *in vitro*. Some antibodies had a high enough affinity to be measured. Most of these (7 of 11) made RTA conjugates cytotoxic *in vitro*. In addition, antibodies of the IgG class that bound cell surface antigens tended to make RTA conjugates that were active *in vitro*. A number of antibodies made inactive conjugates. We believe that some of these immunotoxins may fail to enter a cellular compartment and allow toxin transmembrane passage. By screening our antibodies only as RTA conjugates, we may have failed to include antibodies which may not work with RTA but may work with another toxin. There are two published examples, anti-T cell antibodies W 3/25 and TA-1, where inactive RTA conjugates could be converted to active immunotoxins by changing the toxin moiety to ricin (38,39). Other investigators have used lysosomotropic agents such as chloroquine, ammonium chloride, or carboxylic acid ionophores to enhance *in vitro* cytotoxicity of RTA conjugates (44,59,62,87–89). Either by changing the toxin to whole ricin or by adding enhancers, the transmembrane penetration of the toxin A chain may be facilitated. We had 17 MAb's which met our criteria for selectivity, tumor range, and RTA conjugate *in vitro* toxicity. Nevertheless, future investigations with alternative toxins or enhancers may reveal additional MAb's with optimal binding properties that yield active toxin conjugates *in vitro*.

Conjugates based on intact ricin have limited applicability *in vivo*, since the ricin B chain binds directly to normal tissues (7). Protein modification by photoaffinity labeling (90) or iodination (91) has been used with limited success in an attempt to reduce such normal-tissue binding. Another logical approach is the use of recombinant DNA technology to produce ricin molecules that are devoid of a binding site but retain domains responsible for internalization and ribosomal inhibition.

The RTA conjugates *in vivo* were more toxic on a molar basis than RTA alone. The 10-fold difference may be partly ascribed to rapid renal excretion of the small (30-kilodalton) RTA molecule through the kidneys. A number of RTA conjugates made from antibodies to different antigens showed approximately equal toxicity *in vivo*, suggesting that the determining factor in their toxicity must have been damage to some vital normal tissue not related to antigenic specificity. The failure to increase the

LD_{50} dosage by administering the conjugate in divided doses over 10 days suggests either that tissue recovery from a single RTA conjugate injection is delayed (more than 48 hr) or that there is minimal recovery. Further study of alternative schedules suggests that dosing every third or fourth day allows the delivery of a maximal cumulative dosage which is approximately twice the single-dose LD_{50} (unpublished observations).

Current studies are focusing on the *in vivo* antitumor efficacy of anti-breast cancer monoclonal antibody toxin conjugates.

It is too early to predict what the principal factors are that lead to *in vivo* efficacy of particular antibodies in RTA conjugates. Some of these factors may be related to the antigen and its access in particular tumor models. Model dependence of immunotoxins has also been described by Moolten (5) and Krolick (6). Hamsters were injected subcutaneously with either SV40-transformed fibrosarcoma cells or SV40-induced lymphoma cells (5). When tumors were palpable, hamsters were treated with rabbit antibody to SV40 coupled to diphtheria toxin. After three weekly injections subcutaneously in the flank opposite to tumor implantation, the hamster with the lymphomas showed 56% complete regressions compared to none for the fibrosarcomas. The vasculature and antigen accessibility or density of the lymphoma may have been more important for immunotoxin therapy than those of the fibrosarcoma. In another system mice were injected with mouse lymphoma cells (6). Six to eight weeks later, one group of mice was treated with total lymphoid irradiation and splenectomized. Both untreated and treated mice were then given goat antibody to the lymphoma cell idiotype linked to RTA, i.v. in two doses. The chemoirradiated mice were cured by immunotoxin. The untreated lymphomatous mice died within weeks. This model suggests that a lower *in vivo* tumor burden improves immunotoxin efficacy.

Factors other than antibody and tumor biology may influence the *in vivo* efficacy of immunotoxins. We have not yet extended our evaluation *in vivo* to alternative toxin and linkers with several MAb's. However, several investigators, including Hwang and Thorpe, have looked at several different toxins linked to the same antibody in a single *in vivo* system. Hwang and Bernhard injected guinea pigs intradermally with hepatocarcinoma cells. Seven days later the animals had palpable tumors and were treated intravenously with immunotoxins composed of D3, an MAb to the hepatocarcinoma cells, linked to DTA or abrin A chain. The researchers noted significant tumor growth inhibition of similar extent for each construction (1,2). Furthermore, Thorpe *et al.* have investigated the therapeutic value of immunotoxin treatment in a leukemia model system, using immunocompromised mice with thymectomy, irradiation, and syngeneic bone marrow transplantation (10,11,21). Among three constructs with varying toxin moieties (gelonin, abrin, and saporin), only the saporin–antibody conjugate prolonged survival dramatically—up to 5 weeks.

One of the potential problems in *in vivo* treatment with immunotoxins is the production of antibodies against toxin polypeptides (8) and murine IgG. Any prolonged treatment plan (more than 14 days) in animals or humans will have to consider this potential complication. An immune response against a toxin polypeptide may be circumvented by switching between immunologically non-cross-reactive toxin peptides (92). Alternatively, it may be possible to induce tolerance to the immunotoxin by using immunosuppressive drugs and radiation (93). An immune response to the murine Ig is also likely to develop. Under those circumstances, it may be possible to use alternative sources of antibody (human or rat hybridomas). Recently, investigators have engineered an antibody molecule having a mouse Fab fragment (antigen binding site) attached to human Fc framents (94). Such chimeric antibodies are expected to produce a different and perhaps reduced immune response.

Our approach has allowed us to find first-generation immunotoxins with biological activity in the absence of very clear rules for optimal immunotoxin engineering. Further success with both animal models and patients will depend on identifying some of these rules and applying them to improve immunotoxin behavior. When this is accomplished, this new class of anticancer therapeutics should hold promise for patients with cancer and other diseases caused by defined cell populations.

ACKNOWLEDGMENT

The authors thank S. Ramakrishnan and D. Ring for advice and criticism on this manuscript and Ms. Patty Glennon for preparation of the manuscript. We thank W. Laird, W. Bloch, R. Ferris, and T. Batter for the ricin A chain, antibody, and antibody conjugates used in the work referred to here.

REFERENCES

1. Bernhard, M. *et al.* (1983). *Cancer Res.* **43,** 4420–4428.
2. Hwang, K. *et al.* (1984). *Cancer Res.* **44,** 4578–4586.
3. Seto, M. *et al.* (1982). *Cancer Res.* **42,** 52909–5215.
4. Moolten, F. *et al.* (1972). *J. Natl. Cancer Inst.* (*U.S.*) **49,** 1057–1062.
5. Moolten, F. *et al.* (1975). *J. Natl. Cancer Inst.* (*U.S.*) **55,** 473–477.
6. Krolick, K. *et al.* (1982). *J. Exp. Med.* **155,** 1797–1809.
7. Weil-Hillman, G. *et al.* (1985). *Cancer Res.* **45,** 1328–1336.
8. Ramakrishnan, S. *et al.* (1984). *Cancer Res.* **44,** 1398–1404.
9. Kishida, K. *et al.* (1983). *Cancer Immunol. Immunother.* **16,** 93–97.
10. Thorpe, P. *et al.* (1981). *Eur. J. Biochem.* **116,** 447–454.

11. Thorpe, P. *et al.* (1985). *J. Natl. Cancer Inst.* (*U.S.*) **75,** 151–159.
12. Oladapo, J. *et al.* (1984). *Gut* **25,** 619–623.
13. Blythman, H. *et al.* (1981). *Nature* (*London*) **290,** 145–146.
14. Trowbridge, I. *et al.* (1981). *Nature* (*London*) **294,** 171–173.
15. Bumol, T. *et al.* (1983). *Proc. Natl. Acad. Sci. U. S. A.* **80,** 529–533.
16. Jansen, F. *et al.* (1980). *Immunol. Let.* **2,** 97–102.
17. Imai, L. *et al.* (1983). *Protides Biol. Fluids* **30,** 365–368.
18. Blythman, H. *et al.* (1983). pp. 99–106. Academic Press, Orlando.
19. Jansen, F. *et al.* (1982). *Immunol. Rev.* **62,** 185–216.
20. Edwards, D. *et al.* (1982). *Biochim. Biophys. Acta* **717,** 272–277.
21. Thorpe, P. *et al.* (1983). *Haematol. Blood Transfus.* **28,** 107–111.
22. Frankel, A. *et al.* (1985). *Biol. Respir. Mod.* (in press).
23. Vorki, N. *et al.* (1984). *Cancer Res.* **44,** 681–687.
24. Cuttitta, F. *et al.* (1981). *Proc. Natl. Acad. Sci. U. S. A.* **78,** 4591–4595.
25. Morgan, A. C. *et al.* (1981). *Hybridoma* **1,** 27–36.
26. Bast, R. C. *et al.* (1983). *N. Engl. J. Med.* **309,** 883–887.
27. Herlyn, M. *et al.* (1979). *Proc. Natl. Acad. Sci. U. S. A.* **76,** 1438–1442.
28. Frankel, A. *et al.* (1982). *Proc. Natl. Acad. Sci. U. S. A.* **79,** 903–907.
29. Metzgar, R. *et al.* (1982). *Cancer Res.* **42,** 601–608.
30. Koho, H. *et al.* (1984). *Cancer Immunol. Immunother.* **17,** 165–172.
31. Veda, R. *et al.* (1981). *Proc. Natl. Acad. Sci. U. S. A.* **78,** 5122–5126.
32. Sekine, T. *et al.* (1984). *Cancer* **75,** 106–108.
33. Chatal, J.-F. *et al.* (1984). *J. Nucl. Med.* **25,** 307–314.
34. Mach, J. P. *et al.* (1981). *Immunol. Today* **2,** 239–249.
35. Moldofsky, P. *et al.* (1984). *N. Engl. J. Med.* **311,** 106–108.
36. Cawley, D. *et al.* (1980). *Cell* **22,** 563–570.
37. Uchida, T. *et al.* (1980). *J. Biol. Chem.* **255,** 6687–6698.
38. Vallera, D. *et al.* (1984). *Transplantation* **37,** 387–392.
39. Thorpe, P. *et al.* (1984). *Eur. J. Biochem.* **140,** 63–71.
40. Pappenheimer, A. M. (1977). *Annu. Rev. Biochem.* **46,** 69–94.
41. Olsnes, S. *et al.* (1974). *Nature* (*London*) **249,** 627–831.
42. Irvin, J. (1983). *Pharmacol. Ther.* **21,** 371–387.
43. Stirpe, F. *et al.* (1980). *J. Biol. Chem.* **255,** 6947–6953.
44. Fitzgerald, D. *et al.* (1983). *Proc. Natl. Acad. Sci. U. S. A.* **80,** 4134–4138.
45. Youle, R. *et al.* (1980). *Proc. Natl. Acad. Sci. U. S. A.* **77,** 5483–5486.
46. Bjorn, M. *et al.* (1985). *Cancer Res.* **45,** 1214–1221.
47. Krolick, K. *et al.* (1981). *Cancer Immunol. Immunother.* **12,** 39–41.
48. Weils, J. *et al.* (1984). *Cancer Res.* **44,** 129–133.
49. Masuho, Y. *et al.* (1982). *J. Biochem.* (*Tokyo*) **91,** 1583–1591.
50. Krolick, K. *et al.* (1980). *Proc. Natl. Acad. Sci. U. S. A.* **77,** 5419–5423.
51. Gilliland, D. *et al.* (1980). *Proc. Natl. Acad. Sci. U. S. A.* **77,** 4539–4543.
52. Casellas, P. *et al.* (1982). *Int. J. Cancer* **30,** 437–443.
53. Tsukasaki, K. *et al.* (1985). *Cancer Res.* **45,** 1834–1838.
54. Raso, V. *et al.* (1982). *Cancer Res.* **42,** 457–464.
55. Griffin, T. *et al.* (1982). *J. Natl. Cancer Inst.* (*U.S.*) **69,** 799–805.
56. Seon, B. (1984). *Cancer Res.* **44,** 259–264.
57. Villemez, C. *et al.* (1984). *Biochem. Biophys. Res. Commun.* **125,** 25–29.
58. Esworthy, R. *et al.* (1984). *J. Biol. Chem.* **259,** 11496–11504.
59. Casellas, P. *et al.* (1985). *Blood* **65,** 289–297.
60. Masuho, Y. *et al.* (1982). *Biochem. Biophys. Res. Commun.* **105,** 462–469.

61. Ramakrishnan, S. *et al.* (1984). *Cancer Res.* **44,** 201–208.
62. Ramakrishnan, S. *et al.* (1984). *Science* **223,** 58–61.
63. Uckun, F. *et al.* (1985). *Cancer Res.* **45,** 69–75.
64. Vitteta, E. *et al.* (1984). *J. Exp. Med.* **160,** 341–346.
65. Rodwell, J. *et al.* (1984). *Fed. Proc. Fed Am. Soc. Exp. Biol.* **43,** 1773.
66. Thorpe, P. (1985). Personal communication.
67. Buck, D. *et al.* (1982). *In Vitro* **18,** 377–381.
68. Frankel, A. *et al.* (1985). *J. Biol. Respir. Mod.* **4,** 437–446.
69. Ring, D. *et al.* (1985). In preparation.
70. Frankel, A. *et al.* (1985). pp. 269–292. Martinus Nijhoff, Dordrecht, Netherlands (in press).
71. Bloch, W. *et al.* (1985). In preparation.
72. Schlom, J. *et al.* (1984). *Cancer* **54,** 2777–2794.
73. Yuan, D. *et al.* (1982). *J. Natl. Cancer Inst.* (*U.S.*) **68,** 719–728.
74. Foster, C. *et al.* (1982). *Virchows Arch. A: Pathol. Anat. Histol.* **394,** 279–293.
75. Imam, A. *et al.* (1985). *Cancer Res.* **45,** 263–271.
76. Taylor-Papiadimitriou, J. *et al.* (1981). *Int. J. Cancer* **28,** 17–21.
77. Papsidero, L. *et al.* (1983). *Cancer Res.* **43,** 1741–1747.
78. Ceriani, R. *et al.* (1983).*Somatic Cell Genet.* **9,** 415–427.
79. Menard, S. *et al.* (1983). *Cancer Res.* **43,** 1295–1300.
80. Soule, H. *et al.* (1983). *Proc. Natl. Acad. Sci. U. S. A.* **80,** 1332–1336.
81. Thompson, C. *et al.* (1983). *J. Natl. Cancer Inst.* (*U.S.*) **70,** 409–419.
82. Ciocca, D. *et al.* (1982). *Cancer Res.* **42,** 4256–4258.
83. Feller, W. *et al.* (1985). *Proc. Am. Assoc. Cancer Res.* **26,** 149.
84. Dairkee, S. *et al.* (1985). *Proc. Am. Assoc. Cancer Res.* **26,** 292.
85. White, C. *et al.* (1985). *Cancer Res.* **45,** 1337–1343.
86. Rasmussen, B. *et al.* (1985). *Cancer Res.* **45,** 1424–1427.
87. Casellas, P. *et al.* (1984). *J. Biol. Chem.* **259,** 9359–9364.
88. Raso, V. *et al.* (1984). *J. Exp. Med.* **160,** 1234–1240.
89. Akiyama, S. *et al.* (1985). *Cancer Res.* **45,** 1005–1007.
90. Houston, L. L. (1983). *J. Biol. Chem.* **258,** 7208–7212.
91. Uhr, J. (1984). *J. Immunol.* **133,** i–x.
92. Houston, L. *et al.* (1983). *J. Biol. Chem.* **258,** 9601–9604.
93. Moolten, F. *et al.* (1975). *J. Natl. Cancer Inst.* (*U.S.*) **55,** 709–712.
94. Morrison, S. *et al.* (1984). *Proc. Natl. Acad. Sci. U. S. A.* **81,** 6851–6855.

23

Production of Novel Antibiotics by Gene Cloning and Protein Engineering

RICHARD H. BALTZ, JEFFREY T. FAYERMAN, THOMAS D. INGOLIA, AND R. NAGARAJA RAO

Lilly Research Laboratories
Eli Lilly and Company
Indianapolis, Indiana 46285

I. INTRODUCTION

The production of commercially important antibiotics has been carried out traditionally by microbial fermentation and purification. In some cases, further chemical, enzymatic, or microbiological modifications of the antibiotic are required to obtain the desired product. Antibiotic-producing microorganisms include many actinomycetes (e.g., *Streptomyces, Nocardia, Actinoplanes, Actinomadura,* and *Micromonospora*), lower fungi (e.g., *Penicillium* and *Cephalosporium*), and bacilli. Gene cloning and protein engineering may be useful as a means to efficiently produce novel or hybrid antibiotics and to increase production yields of existing antibiotics. This chapter considers the use of protein engineering to develop novel antibiotics for the pharmaceutical and animal health industries.

Although protein engineering is not yet applied in industrial actinomycetes and fungi, the current status of gene cloning methodology and biochemical information indicate that protein engineering may well be applied in the very near future. Important advances have been made in the development of efficient transformation and cloning procedures for *Streptomyces* (Thompson *et al.*, 1980; Chater *et al.*, 1981; Kieser *et al.*, 1982;

PROTEIN ENGINEERING
Applications in Science,
Medicine, and Industry

Baltz and Matsushima, 1983; Matsushima and Baltz, 1985; Fayerman *et al.,* 1985), and genetic transformation of *Cephalosporium acremonium* has been achieved (Queener *et al.,* 1985). Also, many vectors have been developed for cloning in *Streptomyces* (Hopwood *et al.,* 1979; Hershberger, 1982; Kieser *et al.,* 1982; Richardson *et al.,* 1982; Katz *et al.,* 1983; Larson and Hershberger, 1984; Jones *et al.,* 1984; Jones and Fayerman, 1984; Larson and Hershberger, 1986). Although the gene cloning procedures may vary for different species of *Streptomyces* and fungi, the recent developments mentioned above have established some principles which facilitate the development of new recipient strains in suitable cloning hosts.

Since a major goal of gene cloning in antibiotic-producing microorganisms is to construct recombinants that can produce novel compounds, *in vitro* mutagenesis and protein engineering might be used to modify key enzymes to function more efficiently by improving or eliminating inhibitor binding sites, by improving either the K_m or K_{cat}, or by changing sustrate specificity. In this chapter, we discuss progress toward the production of potentially useful novel antibiotics of the clinically important macrolide, glycopeptide, and β-lactam classes. The approaches and cloning procedures described for these classes should have broader applications to other important classes of antibiotics and indeed to other nonantibiotic secondary metabolites produced by actinomycetes and fungi.

II. MACROLIDE ANTIBIOTICS

A. Cloning Hosts and Transformation

Genetic transformation of *Streptomyces* protoplasts was first reported by Bibb *et al.* (1978). This procedure was further modified by Thompson *et. al.* (1982b) and is very efficient for cloning genes in *S. lividans*. Indeed, several antibiotic biosynthesis and resistance genes have been cloned in *S. lividans* by this procedure (Hopwood *et al.,* 1985, 1986; Vara *et al.,* 1985).

Two other streptomycetes, *S. ambofaciens* and *S. fradiae,* are potentially useful industrial cloning hosts. *Streptomyces ambofaciens* produces an important macrolide antibiotic, spiramycin (Ōmura and Nakagawa, 1975). It is relatively nonrestricting, unlike most *Streptomyces* species (Cox and Baltz, 1984), and it can be transformed by plasmid DNA (Richardson *et al.,* 1982; Baltz and Matsushima 1983; Matsushima and Baltz, 1985).

Streptomyces fradiae produces tylosin, another important macrolide antibiotic (Baltz, 1982; Baltz *et al.*, 1983; Seno and Hutchinson, 1986). It is also readily converted to protoplasts that regenerate viable cells under restricted conditions (Baltz, 1978; Baltz and Matsushima, 1981). Many mutants blocked in different steps in the biosynthesis of tylosin, and suitable for complementation, have been isolated and characterized (Baltz and Seno, 1981).

Our initial goal was to develop and apply gene cloning methodology to these macrolide antibiotic-producing *Streptomyces* to improve antibiotic yields and to produce hybrid antibiotics (Baltz, 1982). Initial attempts to transform *S. fradiae* by the procedure of Bibb *et al.* (1978) were very discouraging. Only about one transformant was obtained per microgram of plasmid DNA (Baltz and Matsushima, 1983; Matsushima and Baltz, 1985). The problem of inefficient transformation was due in part to a multiplicity of restriction systems in *S. fradiae* and in part to certain aspects of the transformation procedure (Matsushima and Baltz, 1985, 1986). To improve the efficiency of plasmid transformation in *S. fradiae,* the protoplast transformation procedure has been modified substantially, and a highly efficient procedure has been developed (Matsushima and Baltz, 1985). Similar modifications have been made to improve the efficiency of transformation of *S. ambofaciens* (Matsushima and Baltz, 1985). The transformation procedures developed for *S. ambofaciens* and *S. fradiae* now yield efficiencies of 10^6 and 10^7 transformants per microgram of modified plasmid DNA. However, if vector DNA is unmodified for the *S. fradiae* restriction systems, then transformation frequencies are reduced by at least 10^3-fold (Matsushima and Baltz, 1985). Because of this problem, mutants defective in restriction were sought. It appears that *S. fradiae* produces at least six restriction systems, and mutants defective in all of these systems have been identified after sequential mutation and selection (Matsushima and Baltz, 1986). Nonrestricting mutants of *S. fradiae* containing different mutations in the tylosin biosynthetic genes can now be used as cloning hosts.

B. Hybrid Macrolides

Macrolide antibiotics are produced by a number of different actinomycetes, including *Streptomyces, Micromonospora,* and *Nocardia*. They are composed primarily of 14- or 16-membered lactones to which are attached one to three sugars. Within the macrolide family of antibiotics there is a wealth of diversity in lactone structure, sugar structure and attachment, and other modifications (Ōmura and Nakagawa, 1975). This biosynthetic, and therefore genetic, diversity thus provides a rich source of genes for recombinant DNA studies.

There are two general well-established methods of generating new macrolide structures. These methods, chemical and biological (Ōmura and Nakagawa, 1975; Ōmura and Tanaka, 1983, Sakakibara and Ōmura, 1984), can be interactive. With regard to the biological methods, some of the enzymes involved in the synthesis of macrolides have loose substrate specificity, and this feature has been exploited in the generation of new macrolides (Sadakane *et al.,* 1982). In this method, potential substrates are added to fermentation broths containing strains impaired in macrolide biosynthesis, either physiologically by the use of cerulenin or genetically by the use of blocked mutants (Ōmura *et al.,* 1980). Blocked mutants not only serve as recipient strains for the bioconversion of substrates, but also serve as a source of potential substrates for bioconversion. The use of bioconversions to generate useful structures suffers, however, from the inherent inefficiencies associated with the need to carry out two separate fermentations. With the advent of gene cloning technology it might be possible to carry out a single fermentation instead of two successive fermentations. A few examples of potentially interesting hybrid macrolide structures that have been produced by bioconversions are discussed below.

One part of the macrolide structure which is an attractive target for modification is the 4″ acyl group (see Fig. 1). The importance of the 4″ acyl group in increasing the efficacy of macrolide antibiotics was recognized as early as 1967 (Nakazawa *et al.,* 1967). Several chemical modifications have been made at this position. It has been shown, for example, that 3-*O*-acetyl-4″-*O*-isovaleryl tylosin binds to macrolide-resistant ribosomes (Tsuchiya *et al.,* 1982a,b; Okamoto *et al.,* 1979). Strains that can carry out some of these acylations have been identified and the properties of acylated derivatives of tylosin have been characterized (Okamoto *et al.,* 1980a,b).

Incorporation of unusual sugars into macrolide structures is another target for cloning and protein engineering, especially since it is difficult to exchange sugar groups via organic synthesis. For example, exchanging sugar groups on tylosin and spiramycin could yield interesting hybrid structures. Spiramycin and tylosin are both 16-membered macrolides. Although their lactone rings are different, they have two sugars, mycaminose and mycarose, attached in a similar manner to the lactone ring. In addition, spiramycin has forosamine attached to carbon 9 (see Fig. 1) while tylosin has mycinose attached at carbon 23. Ōmura *et al.* (1983) were able to generate a novel macrolide, chimeramycin, by feeding tylactone, an intermediate in the tylosin pathway (Baltz *et al.,* 1983), to the spiramycin-producing *S. ambofaciens* in the presence of cerulenin, an inhibitor of fatty acid biosynthesis. Chimeramycin differs from tylosin in

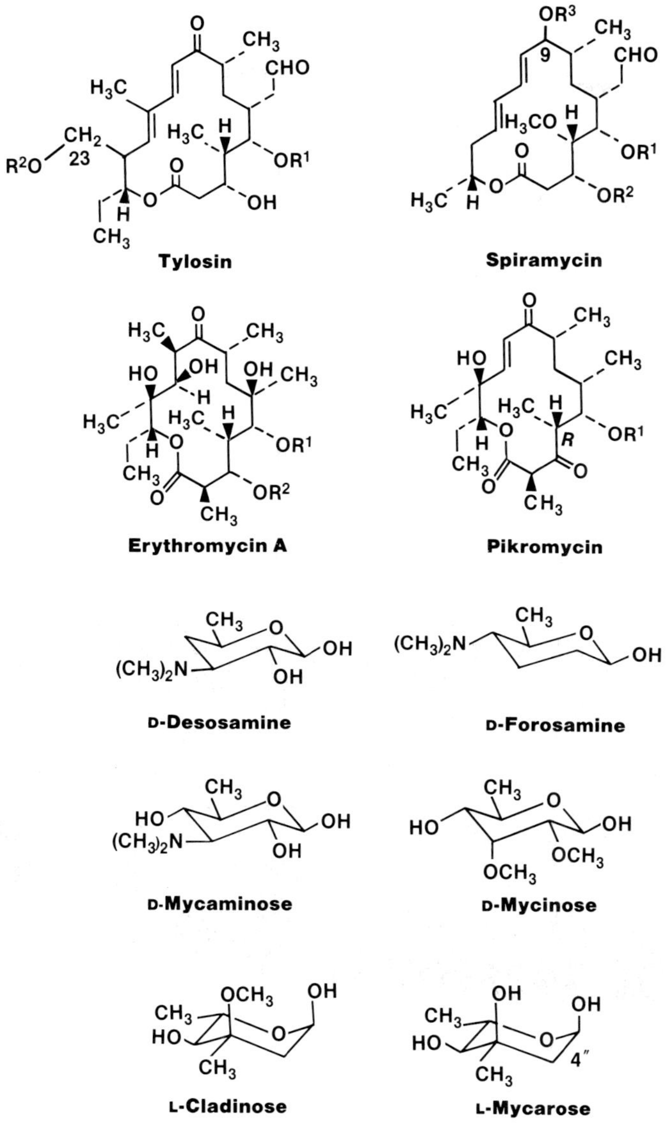

Fig. 1. Structures of some 16- and 14-membered macrolide antibiotics. Tylosin has the disaccharide mycarosyl-mycaminose at R^1 and mycinose at R^2. Spiramycin has mycarosyl-mycaminose at R^1; H, acetyl, or propionyl at R^2 (for spiramycins I, II, and III, respectively); and forosamine at R^3. Erythromycin A has desosamine at R^1 and cladinose at R^2. Pikromycin has desosamine at R^1. The stereochemistry is from Jones *et al.* (1982) and Ōmura (1984).

that it lacks mycinose but contains forosamine. It also contains a methyl group at carbon 23 (see Fig. 1). Chimeramycin could be produced only when *S. ambofaciens* ATCC15154 was used as a host for bioconversion experiments and not when *S. ambofaciens* KA-1028 was used as a host (Sadakane *et al.*, 1982). This example shows that the two strains of *S. ambofaciens* may produce biosynthetic enzymes of varying levels or varying substrate specificity. Protein engineering might provide an opportunity to change the substrate specificity so that diverse substrates might be accepted by these enzymes.

Another example of a potentially interesting sugar exchange among macrolide structures is incorporation of desosamine into tylactone. Pikromycin is a 14-membered macrolide produced by *S. felleus* (Ōmura and Tanaka, 1983) that contains desosamine as its sugar moiety. By feeding tylactone to a pikromycin producer in the presence of cerulenin to inhibit the synthesis of pikromycin lactone ring, a new structure, β-D-desosaminyl tylactone, was obtained (Ōmura *et al.*, 1980). This compounds has, however, already been identified as a natural product from *Micromonospora* (Kinumaki *et al.*, 1977).

The importance of sugar residues in determining the biological activity of macrolides has been well recognized (Sakakibara and Ōmura, 1984). Some of the enzymes involved in the transfer of sugar residues clearly have somewhat relaxed substrate specificity. The three different examples of bioconversions generating new macrolide structures discussed here provide potential targets both for gene cloning and gene transfer experiments and for site-directed mutagenesis. The next section describes the progress that has been made in the cloning of specific genes for use in generating novel macrolides.

C. Cloning of Erythromycin Biosynthetic Genes

Streptomyces erythraeus produces the macrolide erythromycin (Seno and Hutchinson, 1986). Erythromycin is composed of a 14-membered lactone, erythrolonolide, and two sugars, cladinose and desosamine (Fig. 1). These sugars differ from those observed in tylosin (mycarose, mycaminose, and mycinose) and spiramycin (mycarose, mycaminose, and forosamine).

The erythromycin biosynthetic genes have been cloned on a cosmid cloning vector which is bifunctional for *Escherichia coli* and *Streptomyces* (Stanzak *et al.* 1986). A cosmid library of *S. erythraeus,* the producer of erythromycin, was prepared in *E. coli* after *in vitro* packaging in bacteriophage λ particles. Clones of *E. coli* were identified which contained DNA sequences that hybridized to a sequence containing an erythromycin resistance gene initially cloned by Thompson *et al.* (1982a), and the DNA was isolated. Cosmid DNA (pKC488) isolated from one of the *E. coli* clones, when transformed into *S. lividans,* caused it to produce erythromycin. This cosmid is unstable in *S. lividans* and apparently undergoes rearrangements or deletions that can eliminate some of the biosynthetic steps, thereby causing the genetically altered derivatives of *S. lividans* to produce biologically active intermediates or side products of the erythromycin pathway. The cosmid DNA prepared from *E. coli* was used to transform two different blocked mutants (Seno and Hutchinson, 1986) of *S. erythraeus.* The transformants produced a biologically active erythromycin compound or compounds rather than the products of the blocked mutants. The data indicate that the cosmid pKC488 contains structural genes for erythromycin biosynthesis, and not simply a regulatory gene or genes which activate silent genes encoding an erythromycin biosynthetic pathway in *S. lividans.*

It is interesting to note that the erythromycin resistance gene is very closely linked to the biosynthetic genes and that these genes are contained on a 35-kilobase (kb) sequence of DNA. It has been estimated that biosynthesis of erythromycin requires about 30 enzyme steps (Stanzak *et al.*, 1986). Since the average gene requires about 1 kb of DNA, the 35-kb insert in the cosmid should be large enough to encode the entire pathway (Stanzak *et al.* 1986).

If genetic linkage of antibiotic resistance genes and biosynthesis genes is common in *Streptomyces,* cosmid cloning and identification of clones by hybridization to independently cloned resistance genes should be a very efficient procedure for isolating antibiotic biosynthetic genes (Chater, 1984). Once the biosynthetic genes have been isolated, one can recombine genes from different pathways and modify specific genes to alter their substrate specificities. Since the kinetic parameters of these potential heterologous enzyme reactions may be relatively inefficient, they suggest interesting practical applications for site-directed mutagenesis and protein engineering.

III. GLYCOPEPTIDE ANTIBIOTICS

Glycopeptide antibiotics of the vancomycin/actaplanin class are characterized by a seven-amino-acid core (Fig. 2). Either all seven or five of the seven amino acids are aromatic. The core, or aglycone, is modified at various positions by sugar or chloride residues. Cultures frequently make several different glycopeptide antibiotics that have identical cores and a variety of similar modifications to the cores. Often, only one of the glyco-

A

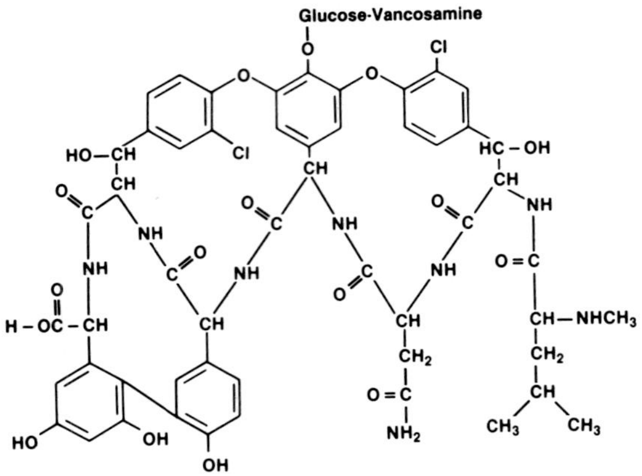

B

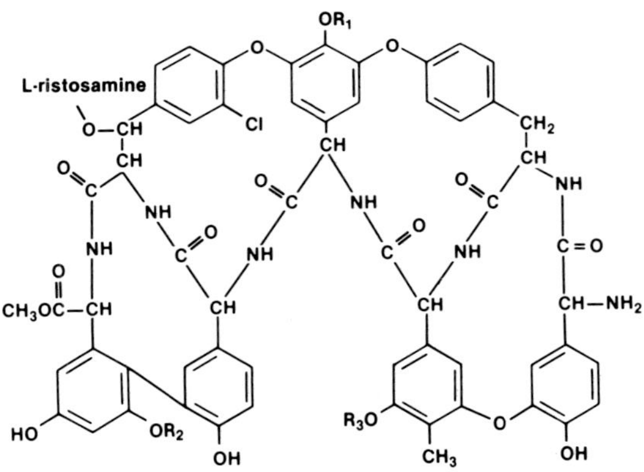

	R_1	R_2	R_3
A	mannosyl-glucose	mannose	mannose
B_1	rhamnosyl-glucose	mannose	mannose
B_2	glucose	mannose	mannose
B_3	mannosyl-glucose	mannose	H
C_1	rhamnosyl-glucose	mannose	H
G	glucose	mannose	H

Fig. 2. Chemical structures of four known glycopeptide antibiotics. (A) Vancomycin (Williamson and Williams, 1981); (B) some of the known actaplanins (Hunt *et al.*, 1984a); (C) the major teichoplanin glycopeptide (Hunt *et al.*, 1984b); (D) the aglycone produced by *S. toyocaensis* (Hunt *et al.*, 1983).

C

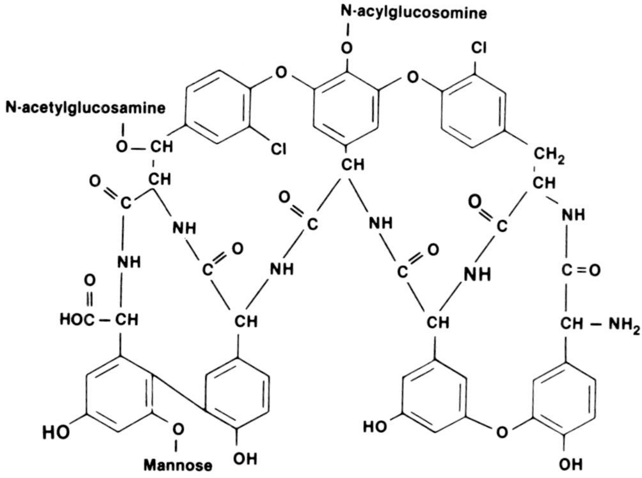

D

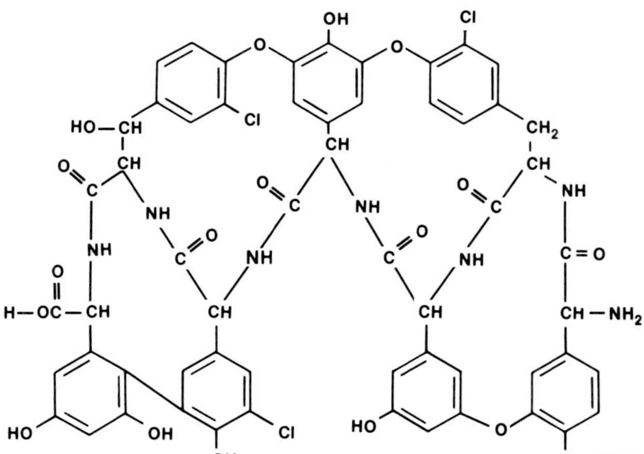

Fig. 2. Continued

peptides is the desired product. Chemical structures of a few of the known glycopeptide antibiotics appear in Fig. 2. Potential applications of genetic engineering are in developing cultures that produce only one desired glycopeptide antibiotic or glycopeptides of varying structures.

One of the requirements for successful genetic engineering is a suitable host strain. Unfortunately, nearly all glycopeptide antibiotics are produced by *Nocardia* and *Actinoplanes* species. Genetic engineering technology is not yet sufficiently developed for these species. However, two *Streptomyces* species, *S. virginiae* and *S. toyocaensis,* produce glycopeptide antibiotics. *Streptomyces virginiae* is difficult to protoplast and the protoplasts formed are poorly regenerated. *Streptomyces toyocaensis* can, however, be readily protoplasted and transformed (Fayerman *et al.*, 1985). In addition, *S. toyocaensis* appears to restrict heterologous DNA to a lesser extent (B. Schoner, personal communication) than other *Streptomyces* species (Cox and Baltz, 1984). Thus *S. toyocaensis* might serve as a good host for gene cloning. It was noted that, unlike some other *Streptomyces* species, DNA transformation frequencies were linear with respect to concentration of DNA up to at least 10 μg per assay. The linearity and frequency of transformation (up to $4 \times 10^5/\mu g$) indicate that *S. toyocaensis* might be used as a primary cloning host to prepare gene libraries from other glycopeptide-producing microorganisms.

IV. β-LACTAM ANTIBIOTICS

A. Structure and Biosynthesis

The four-membered β-lactam ring (Fig. 3) is the central feature of a large number of useful compounds known as β-lactam antibiotics, including penicillins and cephalosporins. Because of the important clinical roles for β-lactam antibiotics, extensive microbiological and chemical modifications have been carried out on these compounds. The goal of much of this research has been to isolate new compounds with improved clinical utility. Organic chemists have carried out many modifications of naturally occurring β-lactam compounds and have synthesized some *de novo.* While many advances have been made in the development of β-lactam chemistry, there are still technical and practical limitations to the purely chemical approach to certain β-lactam structures. Microbiological screens have turned up a wide variety of organisms that synthesize β-lactam compounds, including the fungi *C. acremonium* and *Penicillium chrysogenum* and the actinomycetes *Nocardia lactamdurans, Streptomyces clavuligerus,* and *Streptomyces lipmanii.* The utility of microbiological screens is limited, of course, by the number of different structures actually present in nature.

Some of the limitations of chemical and microbiological methods might be overcome by protein engineering, in that new compounds could be made or old compounds produced more efficiently. β-Lactam biosynthetic enzymes can be incubated *in vitro* with unusual substrates in an

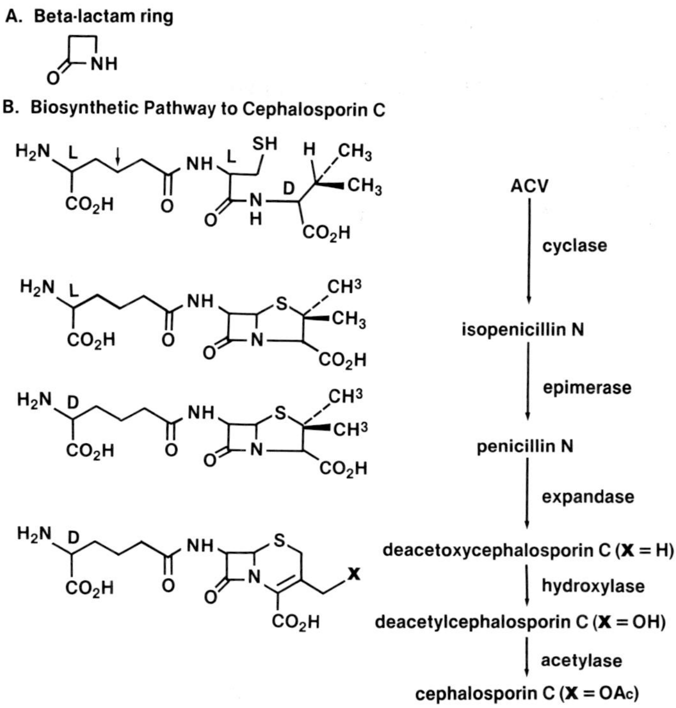

Fig. 3. (A) Structure of the β-lactam ring. (B) Biosynthetic pathway to cephalosporin C. The abbreviations used are described in the text. The arrow refers to the methylene group replaced by a sulfur atom in the experiments of Wolfe *et al.* (1984a).

attempt to produce novel antibiotics. If the native enzymes will not efficiently utilize a compound as a substrate, it may be possible to obtain a mutant that will.

The biosynthetic pathway of cephalosporin C is shown in Fig. 3. The tripeptide delta-(L-α-aminoadipyl)-L-cysteinyl-D-valine (ACV) is condensed to isopenicillin N by the isopenicillin N synthetase (cyclase). Isopenicillin N is epimerized to penicillin N, and the five-membered sulfur-containing (thiazolidine) ring of penicillin N is then expanded to a six-membered (dihydrothiazine) ring to form deacetoxycephalosporin C. Deacetoxycephalosporin C is finally hydroxylated and acetylated to form cephalosporin C. *In vitro* enzymatic synthesis of unusual β-lactams has already been accomplished by Wolfe *et al.* (1984a), using extracts from the eukaryotic *C. acremonium* and the prokaryotic *S. clavuligerus*. Both of these organisms synthesize, among other antibiotics, cephalosporin C.

B. Cloning and Use of Isopenicillin N Synthetase

The cyclase enzyme (Hollander *et al.*, 1984; Pang *et al.*, 1984) which catalyzes the condensation of ACV to isopenicillin N has been purified to homogeneity from *C. acremonium*. A partially purified preparation of cyclase has been used to condense unusual substrates to novel β-lactams by Wolfe *et al.* 1984b). They used a substrate with a carboxymethylcysteinyl moiety substituted for the α-aminoadipyl group of ACV (the normal substrate). The difference between the carboxymethylcysteinyl-containing substrate (called CMC-CV) and ACV is a sulfur in CMC-CV in place of a CH_2 group in ACV (at the position marked by the arrow in the drawing of ACV in Fig. 3). The *Cephalosporium* cyclase enzyme was able to efficiently convert the CMC-CV to the corresponding sulfur derivative of isopenicillin N.

The *Cephalosporium* cyclase gene has recently been cloned and expressed in *E. coli* (Samson *et al.*, 1985). About 20% of the protein in the engineered *E. coli* cells is active cyclase enzyme. Availability of such large amounts of cyclase enzyme will allow more extensive tests of the substrate specificity. In addition, mutagenesis of the cloned gene could allow selection of altered forms of the enzyme with altered substrate specificity. Screening for such mutants will be aided by the fact that peptide substrates of the cyclase enzyme have little or no antimicrobial activity, whereas the cyclized products of the reaction are likely to have antimicrobial activity. An example of this type of screen is shown in Fig. 4. Crude extracts from *E. coli* cells containing expression vectors with (plate A) or without (plate C) the cyclase gene were incubated with ACV. Half of the reaction was then incubated with penicillinase (β-lactamase). The reaction mixtures were then placed into wells in agar plates. The agar plates contained *Micrococcus luteus*, an organism sensitive to isopenicillin N. The penicillinase-treated fractions were put in wells at the 3 and 9 o'clock positions. Plate B is a control in which the ACV was left out of the assay, indicating that the antibiotic activity was not found in the recombinant *E. coli* cells themselves, but required addition of the ACV substrate. This type of biological assay could be useful for the identification of desired altered forms of cyclase enzyme. Furthermore, the availability of the cloned gene and high expression levels in *E. coli* should facilitate protein engineering with the cyclase enzyme.

Coupling of enzymatic reactions *in vitro* to produce cephalosporins from ACV has also been accomplished. Wolfe *et al.* (1984a) immobilized proteins from crude extracts of *S. clavuligerus*. The ACV was converted to deacetylcephalosporin C by these immobilized enzymes, indicating that the cyclization, epimerization, expansion, and hydroxylation en-

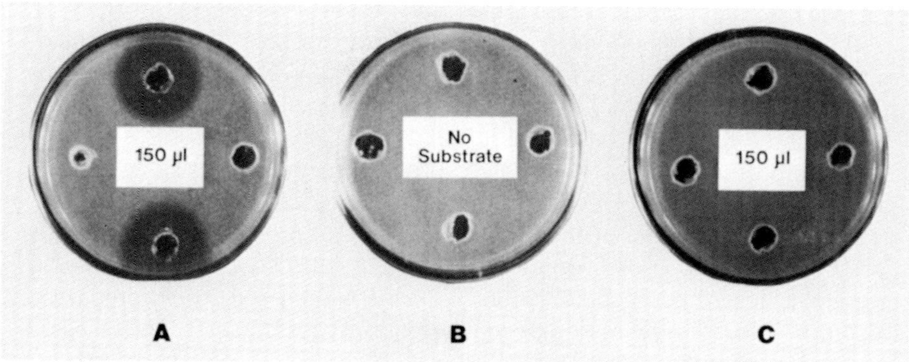

Fig. 4. Demonstration of active isopenicillin N synthetase activity produced in *E. coli*. Reaction mixtures were placed in wells in an agar plate impregnated with *M. luteus*. Production of compounds with antimicrobial activity, in this case isopenicillin N, is demonstrated by a clear zone around the well. The reaction mixtures placed in the wells of plate A and plate B contained extracts of *E. coli* cells containing plasmid pIT337, an expression vector designed to express the cloned *C. acremonium* isopenicillin N synthetase gene. The substrate ACV (see Fig. 3) was left out of the reaction mixtures on plate B. The reaction mixtures placed in the wells of plate C contained extracts of *E. coli* cells containing plasmid pCZ106, a plasmid identical to pIT337 except that the bovine growth hormone gene replaces the isopenicillin N synthetase gene. In all cases, penicillinase was added to the portion of the reaction mixture placed in the wells at 3 and 9 o'clock to show that the antimicrobial activity produced was sensitive to inactivation by penicillinase cleavage.

zymes were active in the immobilized enzyme system. It is noteworthy that immoblized enzymes from *C. acremonium* did not convert ACV to deacetylcephalosporin C under conditions where the *Streptomyces* enzymes were active. Wolfe *et al.* (1984a) attributed the difference to instability in the *Cephalosporium* enzymes, especially the epimerase. Successful protein engineering may therefore depend on choosing sagaciously the best enzyme to be altered when more than one is available.

Besides the enzymes involved in the biosynthesis of penicillins and cephalosporins, there are other proteins involved with β-lactam compounds that are potential targets for protein engineering. Clavulanic acid is a β-lactam produced by *S. clavuligerus* that is a potent inhibitor of β-lactamases. Clavulanic acid improves the clinical efficiency of antimicrobial β-lactams because of its inhibition of β-lactamase. Protein engineering of the genes involved in clavulanic acid biosynthesis might lead to other more potent β-lactamase inhibitors or to more efficient production of clavulanic acid itself. At least one of the genes involved in clavulanic acid biosynthesis has been cloned (Bailey *et al.*, 1984), so protein engineering on this pathway could be feasible in the near future. Other targets for protein engineering are enzymes that would facilitate the production of the β-lactam compounds in current use. For example, removal of the α-aminoadipyl side chain is important in industrial processing of cephalosporins. The D-amino-acid oxidase activity described by Szwajcer and Mosbach (1985) might be useful in this type of processing.

An important part of a complete protein engineering strategy is the ability to transform DNA back into the producing organisms. Transformation systems for some of the important β-lactam producers have already been worked out. Transformation of actinomycetes such as *Streptomyces* has been described in this chapter. Transformation of *C. acremonium* is also possible with the aminoglycoside hygromycin B as a selective agent and a hygromycin B resistance gene as a dominant selectable marker (Queener *et al.,* 1985). Transformation of *Penicillium* has not yet been reported, but methods effective for *C. acremonium* might also be applicable in *Penicillium*.

The key ingredients for protein engineering of enzymes active with β-lactam compounds are now in place. *In vitro* biosynthesis of β-lactam compounds, even in immobilized systems, has been accomplished. Several β-lactam biosynthetic enzymes have been purified, thereby facilitating cloning of the genes encoding the enzymes. One of the enzymes in β-lactam biosynthesis, the isopenicillin N synthetase (cyclase) gene, has been cloned and expressed in *E. coli* (Samson *et al.,* 1985). Transformation systems are available to put DNA back into the producing organisms. These methods can now be combined in attempting to engineer proteins to aid in the production of β-lactam compounds.

V. PROSPECTS FOR THE FUTURE

In the past few years, much progress has been made in the development of gene cloning methods in the antibiotic-producing *Streptomyces*. Many cloning vectors have been constructed, efficient transformation systems have been developed for a (so far) limited number of species, and several antibiotic genes have been cloned. Further, a rudimentary transformation procedure has been developed for *C. acremonium,* and at least one biosynethetic gene for cephalosporin biosynthesis has been cloned and expressed in *E. coli*.

A major problem that remains in *Streptomyces* gene cloning is the widespread production of restriction endonucleases. In the case of *S. fradiae,* an exhaustive mutation and selection program was required to eliminate the many restriction systems which functioned to restrict incoming transforming DNA. Development of more general methods to overcome restriction should facilitate the expansion of recombinant DNA technologies to a much larger number of species in this important genus.

Another potential problem associated with the application of recombinant DNA technology is the expression of endogenous resistance to the novel antibiotic produced by a hybrid strain. In some cases, where the resistance mechanisms are well understood, appropriate cloning vectors might be constructed to contain multiple antibiotic resistance genes in order to protect the cells from the novel antibiotics being produced. However, in most cases the fundamental mechanisms of antibiotic resistance are not understood well enough to predict the combinations of genes that would be most effective in precluding self-destruction. Also, many of the important resistance genes have not yet been isolated. Therefore, much work remains to be done to isolate antibiotic resistance genes and to understand fundamental mechanisms of resistance to antibiotics in producing microorganisms.

Another problem to be addressed is the stability of cloned genes on *Streptomyces* plasmids. In some cases instability has hindered the analysis of cloned sequences in recombinant strains and has precluded the development of stable production strains. Therefore, additional work is needed to develop stable plasmid vectors, preferably ones that can be integrated into the chromosome when desired, to facilitate the analysis and practical scaleup of recombinant strains.

Once these technical problems are overcome, the cloning and expression of antibiotic genes may become relatively routine. Movement of key genes or gene clusters into a variety of genetic backgrounds to attempt to produce novel antibiotic structures will then be a major goal. Thus, broad-host-range conjugal plasmid vectors and transducing phage vectors will become very important. Once this additional technology is available, applications of the technology to produce hybrid antibiotics and other secondary metabolites such as antitumor agents, immune modulators, anthelmintic agents, and other biologically active compounds may become commonplace. Since many unnatural enzymatic bioconversions might be very inefficient and not amenable to production scaleup, the use of site-directed mutagenesis and protein engineering may also become an important adjunct of gene cloning to improve the efficiency of unnatural enzyme reactions and to direct the evolution of novel biosynthetic pathways.

ACKNOWLEDGMENT

The authors thank Mrs. Cheryl Alexander for typing the manuscript.

REFERENCES

Bailey, C. R., Butler, M. J., Normansell, I. D., Rowlands, R. T., and Winstanley, D. J. (1984). *Biotechnology* **2,** 808–811.

Baltz, R. H. (1978). *J. Gen. Microbiol.* **107,** 93–102.

Baltz, R. H. (1982). *In* "Genetic Engineering of Microorganisms for Chemicals" (A. Hollaender, R. D. DeMoss, S. Kaplan, J. Konisky, D. Savage, and R. S. Wolfe, eds.), pp. 431–444. Plenum, New York.

Baltz, R. H., and Matsushima, P. (1981). *J. Gen. Microbiol.* **127,** 137–146.

Baltz, R. H., and Matsushima, P. (1983). *Experientia, Suppl.* **46,** 143–148.

Baltz, R. H., and Seno, E. T. (1981). *Antimicrob. Agents Chemother.* **20,** 214–225.

Baltz, R. H., Seno, E. T., Stonesifer, J., and Wild, G. M. (1983). *J. Antibiot.* **36,** 131–141.

Bibb, M. J., Ward, J. M., and Hopwood, D. A. (1978). *Nature (London)* **274,** 398–400.

Chater, K. F. (1984). *In* "Microbial Development" (R. Losick and L. Shapiro, eds.), pp. 89–115. Cold Spring Harbor Lab., Cold Spring Harbor, New York.

Chater, K. F., Hopwood, D. A., Kieser, T., and Thompson, C. J. (1981). *Curr. Top. Microbiol. Immunol.* **N96,** 69–95.

Cox, K. L., and Baltz, R. H. (1984). *J. Bacteriol.* **159,** 499–504.

Fayerman, J. T., Jones, M. D., and Richardson, M. A. (1985). *In* "Microbiology—1985" (L. Lieve, ed.), pp. 414–420. Am. Soc. Microbiol., Washington, D.C.

Hershberger, C. L. (1982). *Annu. Rep. Ferment. Processes* **5,** 101–126.

Hollander, I. J., Shen, Y.-Q., Heim, J., Demain, A. L., and Wolfe, S. (1984). *Science* **224,** 610–612.

Hopwood, D. A., Bibb, M. J., Ward, J. M., and Westpheling, J. (1979). *In* "Plasmids of Medical, Environmental and Commercial Importance" (K. N. Timmis and K. Puhler, eds.), pp. 245–258. Elsevier/North-Holland Biomedical Press, Amsterdam.

Hopwood, D. A., Malpartida, F., Kieser, H. M., Ikeda, H., Duncan, J., Fujii, I., Rudd, B. A. M., Floss, A. G., and Ōmura, S. (1985). *Nature (London)* **314,** 642–644.

Hopwood, D. A., Kieser, T., Lydiate, D. J., and Bibb, M. J. (1986). *In* "The Bacteria: A Treatise on Structure and Function" (L. E. Day and S. W. Queener, eds.), Vol. 9, pp. 159–229. Academic Press, New York.

Hunt, A. H., Occolowitz, J. L., Debano, M., Molloy, R. M., and Maciak, G. M. (1983). *Abstr. Intersci. Conf. Antimicrob. Agents Chemother.,* p. 441.

Hunt, A. H., Elzey, T. K., Merkel, K. E., and Debono, M. (1984a). *J. Org. Chem.* **49,** 641–645.

Hunt, A. H., Molloy, R. M., Occolowitz, J. L., Marconi, G. G., and Debono, M. (1984b). *J. Am. Chem. Soc.* **106,** 4891–4895.

Jones, M. D., and Fayerman, J. T. (1984). *J. Antibiot.* **37,** 1727–1728.

Jones, M. D., Mabe, J. A., Nakatsukasa, W. M., and Fayerman, J. T. (1984). *Plasmid* **11,** 92–95.

Jones, N. D., Chaney, M. D., Kirst, H. A., Wild, G. M., Baltz, R. H., Hamill, R. L., and Paschal, J. W. (1982). *J. Antibiot.* **35,** 420–425.

Katz, E., Thompson, C. J., and Hopwood, D. A. (1983). *J. Gen. Microbiol.* **129,** 2703–2714.

Kieser, T., Hopwood, D. A., Wright, H. M., and Thompson, C. J. (1982). *Mol. Gen. Genet.* **185,** 223–238.

Kinumaki, A., Harada, K., Suzuki, T., Suzuki, M., and Okuda, T. (1977). *J. Antibiot.* **30,** 450–454.

Larson, J. L., and Hershberger, C. L. (1984). *J. Bacteriol.* **157,** 314–317.

Larson, J. L., and Hershberger, C. L. (1986). In preparation.

Matsushima, P., and Baltz, R. H. (1985). *J. Bacteriol.* **163,** 180–185.
Matsushima, P., and Baltz, R. H. (1986). In preparation.
Nakazawa, S., Shimoyawa, M., Maeda, A., and Matsubayashi, M. (1967). *J. Antibiot.* **20,** 259–261.
Okamoto, R., Nomura, H., Tsuchiya, M., Tsunekawa, H., Fakumoto, T., Mui, T., Sawa, T., Takeuchi, T., and Umezawa, H. (1979). *J. Antibiot.* **32,** 542–544.
Okamoto, R., Fukumoto, T., Nomura, H., Kiyoshima, K., Nakamaro, K., Takamatso, A., Naganawa, H., Takeuchi, T., and Umezawa, H. (1980a). *J. Antibiot.* **33,** 1300–1308.
Okamoto, R., Tsuchiya, M., Nomura, H., Iguchi, H., Kiyoshima, K., Hori, S., Inui, T., Sawa, T., Takenchi, T., and Umezawa, H. (1980b). *J. Antibiot.* **33,** 1309–1315.
Ōmura, S., ed. (1984). "Macrolide Antibiotics: Chemistry, Biology and Practice." Academic Press, New York.
Ōmura, S., and Nakagawa, A. (1975). *J. Antibiot.* **28,** 401–433.
Ōmura, S., and Tanaka, Y. (1983). *In* "Macrolides" (L. C. Vining, ed.), pp. 179–206. Addison-Wesley, Reading, Massachusetts.
Ōmura, S., Ikeda, H., Matsubara, H., and Sadakane, N. (1980). *J. Antibiot.* **33,** 1570–1572.
Ōmura, S., Sadakane, N., Tanaka, Y., and Matsubara, H. (1983). *J. Antibiot.* **36,** 927–930.
Pang, C.-P, Chakravarti, B., Adlington, R. M., Ting, H.-H., White, R. L., Layatilake, G. S., Baldwin, J. E., and Abraham, E. P. (1984). *Biochem. J.* **222,** 789–795.
Queener, S. W., Ingolia, T., Kaster, K., Skatrud, P., and Chapman, J. (1985). *In* "Microbiology—1985" (L. Lieve, ed.), pp. 468–472. Am. Soc. Microbiol., Washington, D.C.
Richardson, M. A., Mabe, J. A., Beerman, N. E., Nakatsukasa, W. M., and Fayerman, J. T. (1982). *Gene* **20,** 451–457.
Sadakane, N., Tanaka, Y., and Ōmura, S. (1982). *J. Antibiot.* **35,** 680–687.
Sakakibara, H., and Ōmura, S. (1984). *In* "Macrolide Antibiotics: Chemistry, Biology and Practice" (S. Ōmura, ed.), pp. 85–125. Academic Press, New York.
Samson, S. M., Belagaje, R., Blankenship, D. T., Chapman, J. L., Perry, D., Skatrud, P. L., Van Frank, R. M., Abraham, E. P., Baldwin, J. E., Queener, S. W., and Ingolia, T. D. (1985). *Nature (London)* **318,** 191–194.
Seno, E. T., and Hutchinson, C. R. (1986). *In* "The Bacteria: A Treatise on Structure and Function" (L. E. Day and S. W. Queener, eds.), Vol. 9, pp. 231–279. Academic Press, New York.
Stanzak, R., Matsushima, P., Baltz, R. H., and Rao, R. N. (1986). *Biotechnology* **4,** 229–232.
Szwajcer, E., and Mosbach, K. (1985). *Biotechnol. Lett.* **7,** 1–7.
Thompson, C. J., Ward, J. M., and Hopwood, D. A. (1980). *Nature (London)* **286,** 525–527.
Thompson, C. J., Kieser, T., Ward, J. W., and Hopwood, D. A. (1982a). *Gene* **20,** 51–62.
Thompson, C. J., Ward, J. M., and Hopwood, D. A. (1982b). *J. Bacteriol.* **151,** 668–677.
Tsuchiya, M., Hawada, M., Takeuchi, T., Umezawa, H., Yamamoto, K., Tanaka, H., Kiyoshiwa, K., Mori, S., and Okamota, R. (1982a). *J. Antibiot.* **35,** 661–672.
Tsuchiya, M., Sawa, T., Takeuchi, T., Umezawa, H., and Okamoto, R. (1982b). *J. Antibiot.* **35,** 673–679.
Vara, J., Malpartida, F., Hopwood, D. A., and Jimenez, A. (1985). *Gene* **33,** 197–206.
Williamson, M. P., and Williams, D. H. (1981). *J. Am. Chem. Soc.* **103,** 6580–6585.
Wolfe, S., Demain, A. L., Jensen, S. E., and Westlake, D. W. S. (1984a). *Science* **224,** 1386–1392.
Wolfe, S., Hollander, I. J., and Demain, A. L. (1984b). *Biotechnology,* **2,** 635–636.

24

Genetic Transformation of Plants

ERNEST G. JAWORSKI,* ROBERT T. FRALEY,* STEPHEN G. ROGERS,* ROBERT B. HORSCH,* ROGER N. BEACHY,† AND NAM-HAI CHUA‡

**Monsanto Company*
St. Louis, Missouri 63198
and
†Washington University
St. Louis, Missouri 63130
and
‡The Rockefeller University
New York, New York 10021-6399

I. INTRODUCTION

Advances in molecular biology and microbial genetics over the past decade have provided biologists with powerful tools for the genetic modification of plants and for the study of gene regulation and development. Plant cell transformation by infection with virulent strains of *Agrobacterium tumefaciens* has been studied extensively in a number of laboratories (Chilton *et al.*, 1977, 1980; Van Larabeke *et al.*, 1974; Kerr *et al.*, 1977; Braun, 1956; Yadav *et al.*, 1980; Willmitzer *et al.*, 1980). *Agrobacterium tumefaciens* is a naturally occurring soilborne bacterium that infects plant cells, producing a tumor (crown gall) at the infection site (De Cleene and DeLey, 1976). Virulent strains of the bacterium contain a large plasmid (150–250 kilobases) called the Ti or tumor-inducing plasmid (Zaenen *et al.*, 1974; Watson *et al.*, 1975). The Ti plasmid contains a segment of DNA, termed the T-DNA, which is transferred (Chilton *et al.*, 1977) and stably incorporated into the nuclear DNA (Chilton *et al.*, 1980; Willmitzer *et al.*, 1980) of transformed plant cells. The T-DNA is transcribed within the cells (Gurley *et al.*, 1979; Gelvin *et al.*, 1981; Drummond *et al.*, 1977;

PROTEIN ENGINEERING
Applications in Science,
Medicine, and Industry

Willmitzer *et al.*, 1982, 1983; Bevan and Chilton, 1982) and expresses a number of genes that are responsible for phytohormone synthesis (Barry *et al.*, 1984; Lichenstein *et al.*, 1984; Klee *et al.*, 1984) and the synthesis of novel metabolites called opines (Bevan *et al.*, 1983a; Depicker *et al.*, 1982; DeGreve *et al.*, 1983). The transfer and insertion process for T-DNA transformation in plants involves repeated nucleotide sequences ("T-DNA borders") which define the ends of the T-DNA segment (Zambryski *et al.*, 1982; Yadav *et al.*, 1982) as well as other genes of unknown function located in a specific region (*vir* region) outside the T-DNA (Hille *et al.*, 1982; Klee *et al.*, 1982).

The properties of the *Agrobacterium*/ Ti plasmid system have now been used to design a large number of new transforming vectors. Intermediate vectors containing selectable antibiotic resistance markers for introduction of foreign genes into the Ti plasmid have been constructed. The tumorigenic properties of *Agrobacterium* have been eliminated by "disarming" the Ti vector, and new efficient *in vitro* plant cell transforming techniques have been perfected.

II. VECTOR CONSTRUCTION

Chimeric genes were constructed (Fraley *et al.*, 1983; Bevan *et al.*, 1983b; Herrera-Estrella *et al.*, 1983) that contain the neomycin phosphotransferase NPT (ATP : Kanamycin 3′-*O*-phosphotransferase, EC 2.7.1.95) coding sequences from the bacterial transposon Tn5 (type II) joined to the 5′ and 3′ regulatory sequences of the nopaline synthase (nopaline dehydrogenase) gene derived from the Ti plasmid. It was known that expression of NPT in yeast (Jiminez and Davis, 1980) and mammalian cells (Colbere-Garapin *et al.*, 1981; Southern and Berg, 1982) conferred resistance to the aminoglycoside antibiotics and that plant cells were sensitive to such antibiotics (e.g., kanamycin and G418). The 5′ and 3′ regulatory regions of the nopaline synthase gene were well characterized (Bevan and Chilton, 1982; Depicker *et al.*, 1982), and it was known that the gene was expressed constitutively in plant tissues that had been transformed with the Ti plasmid (Tempé and Goldmann, 1982).

Direct cloning approaches using the Ti plasmid were impractical because of its large size, and for that reason it was necessary to design shuttle or intermediate vectors. Vectors were created either to integrate with a resident Ti plasmid via recombination (Fraley *et al.*, 1983; Bevan *et al.*, 1983b; Herrera-Estrella *et al.*, 1983) or to replicate independently of the Ti plasmid (deFramond *et al.*, 1983; Hoekema *et al.*, 1983; Bevan, 1984) as *trans* vectors. The integrative vector designated pMON200 (Fig. 1) has been used for the transfer of several foreign genes into *Agrobacte-*

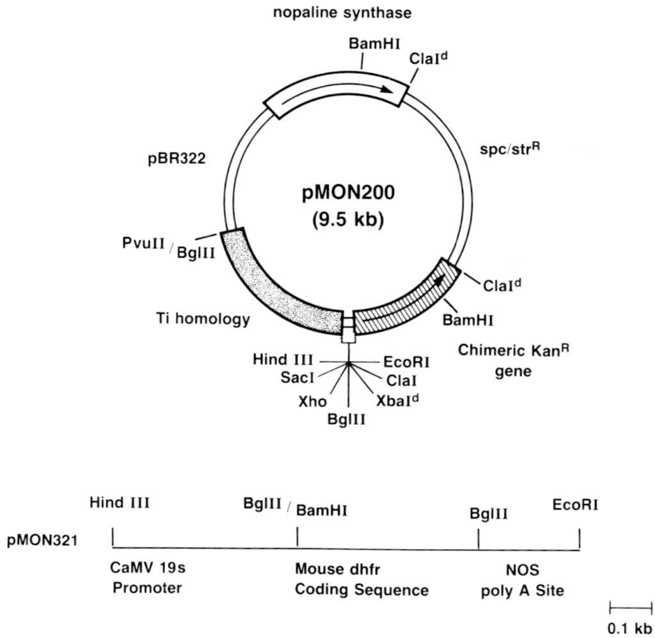

Fig. 1. Plasmid pMON200. The plasmid is a derivative of pMON120 (18) containing a modified chimeric Km^r gene which lacks an upstream ATG codon present in the bacterial leader sequence (23) and a synthetic multilinker with unique *Hin*dIII, *Xho*I, *Bgl*II, *Xba*I, *Cla*I, and *Eco*RI restriction sites.

rium cells. The essential characteristics of this plasmid include a segment of pBR322 DNA for replication in *Escherichia coli,* a portion of a Ti plasmid (pTiT37) containing the functional nopaline synthase gene for ease in scoring transformed plant cells, a spectinomycin/streptomycin-resistance determinant from Tn7 for selection in *A. tumefaciens,* a portion of DNA from another Ti plasmid (pTiA6) to provide homology for recombination with a resident octopine-type plasmid in *A. tumefaciens,* a synthetic multilinker containing unique restriction sites for gene insertion, and the chimeric kanamycin resistance gene (NOS/NPTII/NOS).

The pMON200 plasmid and its derivatives were introduced into *A. tumefaciens* by conjugation procedures, and homologous recombination between pMON200 and the wild-type octopine Ti plasmid produced the cointegrate shown in Fig. 2. The long transfer occurs on selection for phytohormone autotrophy (Fig. 2D) and facilitates the evaluation of the kanamycin marker. The short transfer (Fig. 2E) results from selection for kanamycin resistance and provides a system that allows transformed plant cells to be readily regenerated to intact plants (Horsch *et al.,* 1984).

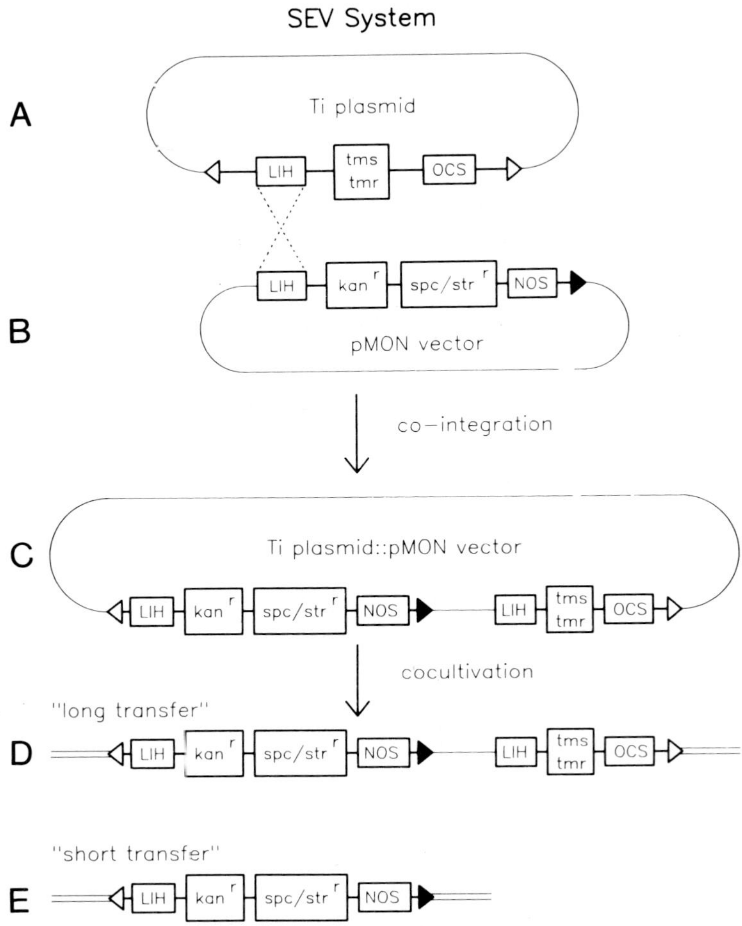

Fig. 2. Reciprocal recombination of (A) wild-type resident Ti plasmid (pTiB6S3) and (B) pMON200 or derivative (pMON128), yielding (C) the cointegrate pTiB6S3 : : pMON128. After cocultivation and selection for kanamycin-resistant plant cells, either (D) the entire hybrid T-DNA or (E) a truncated T-DNA without tumor genes is transferred into the plant genome.

III. DISARMED VECTOR

The system described above was useful for the study of gene expression and inheritance of traits but was not highly efficient for routine production of transformed plants since only 10% of the transformed callus tissue contained the short-transfer avirulent DNA (Horsch *et al.*, 1984).

A derivative of a virulent Ti plasmid was therefore constructed to form a cointegrate with the pMON200 intermediate vector or derivatives as shown in Fig. 3A. By double-crossover events between the pRK290 derivative and pTiB6S3 (Fig. 3A), the phytohormone biosynthetic (tumor) genes, the T_L-DNA right border, and all of the T_R-DNA were deleted and replaced by the Tn*903* kanamycin resistance marker (Km^r). Subsequent cointegrate formation between pMON200 and pTiB6S3SE (Fig. 3B) resulted in the generation of an avirulent selectable T-DNA system called the SEV or split end vector system.

IV. PLANT TRANSFORMATION

An *in vitro* transformation procedure was developed by using plant protoplasts incubated directly with *A. tumefaciens* bacterial cells (Fraley *et al.,* 1983, 1984; Marton *et al.,* 1979; Wullems *et al.,* 1981). The protoplasts were prepared from plant leaf tissue by conventional enzyme digestion. The bacteria attach to the protoplasts during cell wall regeneration and in some way, as yet unknown, transfer the T-DNA into the plant cells. Such cells are easily identified by selection for kanamycin resistance and are found at transformation frequencies as high as 50% within 3 weeks after protoplast isolation.

An improved alternative procedure was subsequently developed (Horsch *et al.,* 1985a) to obviate some of the problems of protoplast isolation and regeneration. In this procedure, whole leaf punches were infected with *Agrobacterium* and the subsequent callus formation around the circumference of the disk produced regenerated plantlets. This simple reproducible system resulted in transformed plantlets with 3–4 weeks.

Using both the protoplast cocultivation system (Horsch *et al.,* 1984) and the leaf disk method (Horsch *et al.,* 1985b), it was possible to demonstrate the stable maintenance and expression of foreign genes (Km^r). Furthermore, progeny seed derived from selfed plants were shown to inherit the kanamycin resistance in a simple Mendelian pattern (Horsch *et al.,* 1984).

V. GENE EXPRESSION

With the development of a transformation system which allowed the stable heritable introduction and expression of a foreign gene, a tool was available for the analysis of gene expression in general.

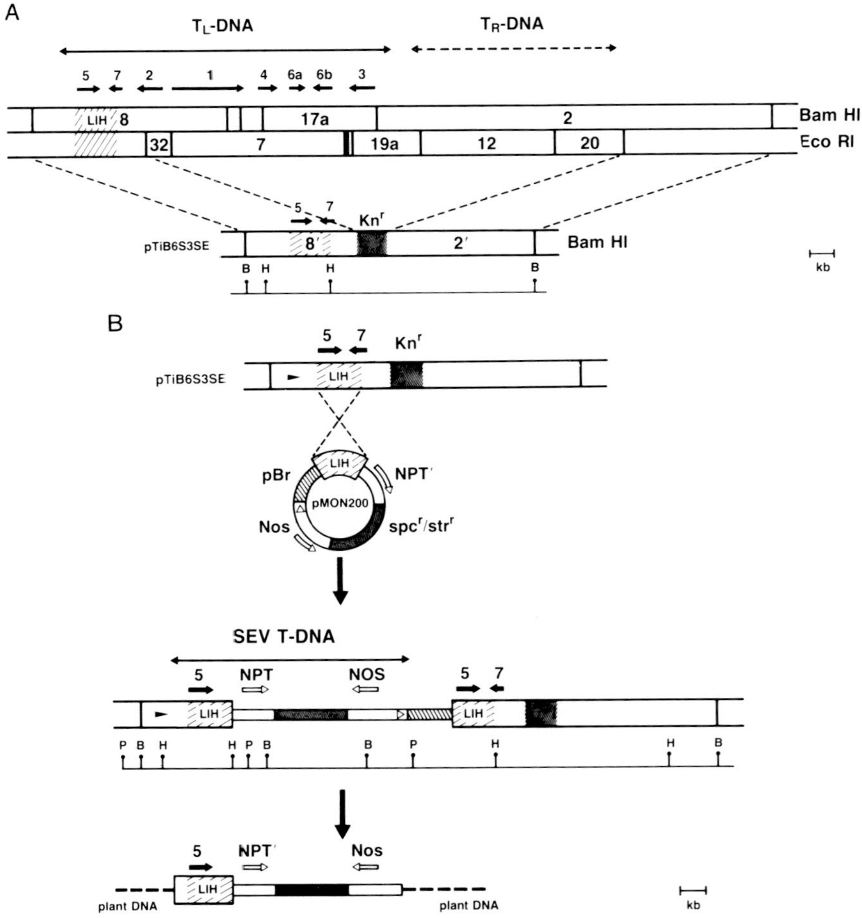

Fig. 3. (A) Construction of avirulent pTiB6S3SE acceptor plasmid. A 4.5-kb *Bam*HI–*Eco*RI fragment 8′: (derived from *Bam*HI fragment 8) and a 6.0-kb *Eco*RI–*Bam*HI fragment 2′: (derived from *Bam*HI fragment 2) were used as homologous DNA segments for double recombination. A 1.2-kb DNA fragment carrying the Km^r determinant from Tn*903* was introduced between the two homologous DNA fragments to provide a marker for subsequent genetic manipulations. The resulting 11.5-kb *Bam*HI fragment was inserted into pRK290 and introduced in *A. tumefaciens* by conjugation. The *A. tumefaciens* cells in which the double-crossover event had occurred, resulting in the replacement of *Eco*RI fragments 32, 7, 19a, 12, and 20 with the pMON195 T-DNA, were identified following introduction of the incompatible plasmid pPH1JI. The resulting derivative, pTiB6S3SE, was used as a Ti plasmid acceptor in subsequent experiments with pMON200. (B) Cointegration between pMON200 and pTiB6S3SE at their homologous LIH regions results in the formation of a short, avirulent SEV T-DNA containing chimeric Km^r and nopaline synthase genes for monitoring foreign DNA transfer and presence in plant cells. Transfer of the SEV T-DNA into plant cells utilizes the nopaline synthase right border sequence contained in the pMON vector.

A. Photosynthetic Genes

The first study involved a tissue-specific, light-regulated gene coding for the small subunit of ribulose-1,5-bisphosphate carboxylase (*rbcS*) from peas. The genomic clone encoding the pea *rbcS* gene was inserted into a pMON200 derivative vector and, after transfer into *A. tumefaciens,* was introduced into petunia cells by cocultivation (Broglie *et al.,* 1984). Northern and S_1 analyses of transformed petunia cells revealed that the pea *rbcS* gene was expressed in petunia under the control of its own promoter. Furthermore, gene expression was regulated by light in a manner identical to that seen for the endogenous *rbcS* gene in pea. Finally, the pea *rbcS* gene retained its tissue-specific expression pattern in leaves derived from regenerated transformed petunia plants (Nagy *et al.,* 1986).

Further analysis of deletion mutations in the 5′ regulatory region of the pea *rbcS* gene revealed that the promoter region contained two functional domains (Morrelli *et al.,* 1986). One region 450 bp 5′-upstream from the transcription initiation site was needed for high-level expression and a second region 10–30 bp from the transcription start site was necessary for light regulation.

In subsequent studies using *in vivo* radiolabeling followed by immunoprecipitation of ribulose-1,5-biphosphate carboxylase, it was demonstrated on two-dimensional gels that the heterologous *rbcS* protein of pea could be separated from the endogenous petunia *rbcS* protein. This indicated that the pea *rbcS* protein was correctly processed *in vivo* by petunia chloroplasts. Pea *rbcS* labeled *in vivo* was also recovered from holoenzymes that were immunoselected with the petunia anti-large subunit antibody, indicating that the pea small subunit could form a hybrid holoenzyme assembled with petunia large subunits.

B. Storage Protein Genes

The transfer of legume storage protein genes into solanaceous plants has been shown to result in tissue-specific accumulation of storage proteins in the seeds of transformed plants. Sengupta-Gopalan *et al.,* 1985) reported seed-specific accumulation of phaseolin protein in tobacco seeds, and, more recently, the regulated expression of soybean α'-conglycinin protein in petunia seeds has also been demonstrated (R. Beachy, unpublished results). Sedimentation analysis of extracted proteins from petunia seeds indicated that the α'-conglycinin protein migrates in a position expected for a properly assembled three-subunit protein complex.

These model systems for storage protein expression will permit determination of the structural sequences required for protein translocation, glycosylation, and assembly as well as the regulatory sequences essential

SPECIFIC APPLICATIONS

Improved protein quality of seed grain	Salt tolerance
Enhanced protein levels in forage crops	Metal tolerance (Al^{+3}, Pb^{+2})
Enhanced production of natural products: Amino acids, allelopathic chemicals, latex, enzymes, oils, starch	Herbicide and pesticide tolerance
Plant disease resistance	Enhanced photosynthetic carbon fixation
Insect tolerance	Improved response to fertilizers
Drought and flooding tolerance	Heat and cold tolerance

Fig. 4. Genetic engineering applications in plant agriculture.

for seed-specific expression. Such systems should prove useful in analyzing the processing, assembly, stability, and functioning of storage proteins which have been modified through mutagenesis to alter their amino acid composition and nutritional value.

C. Mammalian Genes

Most recently (S. Rogers and D. Eichholtz, unpublished studies), the expression of two mammalian genes in plant cells has also been demonstrated. A cDNA clone encoding α human chorionic gonadotropin (α-hCG) has been engineered as a chimeric gene under control of a viral promoter and shown to produce detectable levels [by radioimmunoassay (RIA)] of α-hCG protein in transformed plants. In addition, a mouse cDNA clone encoding a methotrexate-insensitive dihydrofolate reductase (dhfr) gene has been similarly engineered under control of a viral promoter and has been demonstrated to function efficiently as a dominant selectable marker in plant transformation experiments (see Fig. 1). These results with α-hCG and mouse dhfr suggest the broad utility of the *A. tumefaciens* transformation system for the study of gene regulation and expression in plants.

VI. APPLICATIONS AND NEEDS

The next decade promises to bring with it significant advances in the genetic modification of plants, including some with direct value to plant breeders. These will initially focus on such single-gene traits as herbicide tolerance and insect tolerance but will broaden to many other characteristics cited in Fig. 4. Protein engineering may well see its first application in simple amino acid codon shifts in the storage proteins of seed grains to enhance their content of lysine (corn) or methionine (soybeans). Gene amplification techniques may yield higher protein levels in forage crops.

What is clear at present is the striking lack of basic biochemical knowledge about plant systems that is so essential to our understanding of pathways involved in conferring some of the desirable traits listed in Table I. Efforts to improve this situation are being made, and modest expansions of competitive grants by the National Science Foundation and the U.S. Department of Agriculture are encouraging. But the full social and economic impacts of modern molecular biology will be felt only when more resources are directed toward improvements in plant tissue culture and regeneration; novel transformation systems (transposable elements, *Gemini* virus vectors); understanding of gene structure, organization, and function; selection, isolation, and characterization of agronomically important genes; and development of unique plant breeding techniques.

It is important, however, to close with the observation that the developments in plant molecular biology just in the past 2 years have been extremely rapid and illuminating. It is clear that key regulatory sequences related to tissue and developmentally specific gene regulation are closer at hand than was thought not long ago. Transformation is an accomplished fact. This has provided us with the tools to further develop our understanding of eukaryotic regulation of transcription and translation and to solve other important technical problems such as defining the role of "context DNA" in gene expression or developing a means for the precise direction of genes into specific target sites.

REFERENCES

Barry, G., Rogers, S., Fraley, R., and Brand, L. (1984). *Proc. Natl. Acad. Sci. U.S.A.* **81,** 4776–4788.

Bevan, M. (1984). *Nucleic Acids Res.* **12,** 8711–8721.

Bevan, M., and Chilton, M.-D. (1982). *J. Mol. Appl. Genet.* **1,** 539–546.

Bevan M., Barnes, W., and Chilton, M.-D. (1983a). *Nucleic Acids Res.* **11,** 369.

Bevan, M., Flavell, R., and Chilton, M.-D. (1983b). *Nature (London)* **304,** 184–187.

Braun, A. (1956). *Cancer Res.* **16,** 53–56.

Broglie, R., Coruzzi, G., Fraley, R., Rogers, S., Horsch, R., Niedermeyer, J., Fink, C., Flick, J., and Chua, N.-H. (1984). *Science* **224,** 838–843.

Chilton, M.-D., Drummond, M. H., Merlo, D. J., Sciaky, D., Montoya, A. L., Gordon, M., and Nester, E. (1977). *Cell* **11,** 263–271.

Chilton, M.-D., Saiki, R., Yadav, N., Gordon, M., and Quétier, F. (1980). *Proc. Natl. Acad. Sci. U.S.A.* **77,** 4060–4064.

Colbere-Garapin, G., Horodniceanu, F., Kourilsky, P., and Garapin, A.-C. (1981). *J. Mol. Biol.* **150,** 1–14.

De Cleene, M., and DeLey, J. (1976). *Bot. Rev.* **42,** 389.

deFramond, A., Barton, K., and Chilton, M.-D. (1983). *Bio/Technology* **1,** 262–270.

DeGreve, H., Dhaese, P., Seurinch, J., Lemmers, M., Van Montagu, M., and Schell, J. (1983). *J. Mol. Appl. Genet.* **1,** 499, 1983.

Depicker, A., Stachel, S., Dhaese, P., Zambryski, P., and Goodman, H. (1982). *J. Mol. Appl. Genet.* **1,** 561–574.

Drummond, M. H., Gordon, M. P., Nester, E. W., and Chilton, M.-D. (1977). *Nature (London)* **269,** 535.

Fraley, R., Rogers, S., Horsch, R., Sanders, P., Flick, J., Adams, S., Bittner, M., Brand, L., Fry, J., Galluppi, G., Goldberg, S., Hoffmann, N., and Woo, S. (1983). *Proc. Natl. Acad. Sci. U.S.A.* **80,** 4803–4807.

Fraley, R., Horsch, R., Matzke, A., Chilton, M.-D., Chilton, S., and Sanders, P. (1984). *Plant Mol. Biol.* **3,** 371–378.

Gelvin, S. B., Gordon, M. P., Nester, E. W., and Aronson, A. A. (1981). *Plasmid* **6,** 17.

Gurley, W. B., Kemp, J. D., Albert, M. J., Sutton, D. W., and Callis, J. (1979). *Proc. Natl. Acad. Sci. U.S.A.* **76,** 2828.

Herrera-Estrella, L., DeBlock, M., Messens, E., Hernalsteen, J. P., Van Montagu, M., and Shell, J. (1983). *EMBO J.* **2,** 987–995.

Hille, J., Klasen, I., and Shilperoort, R. (1982). *Plasmid* **7,** 107–118.

Hoekema, A., Hirsch, D., Hooykaas, J., and Schilperoort, R. (1983). *Nature (London)* **303,** 179–182.

Horsch, R., Fraley, R., Rogers, S., Sanders, P., Lloyd, A., and Hoffmann, N. (1984). *Science* **223,** 496.

Horsch, R., Fry, J., Hoffmann, N., Eichholtz, D., Rogers, S., and Fraley, R. (1985a). *Science* **227,** 1229.

Horsch, R., Fry, J., Hoffmann, N., Wallroth, M., Eichholtz, D., Rogers, S., and Fraley, R. (1985b). *Science* **227,** 1229.

Jiminez, A., and Davis, J. (1980). *Nature (London)* **287,** 860–871.

Kerr, A., Manigault, P., and Tempe, J. (1977). *Nature (London)* **265,** 560–561.

Klee, H., Gordon, M., and Nester, E. (1982). *J. Bacteriol.* **150,** 327–331.

Klee, H., Montaya, A., Hordyski, F., Lichenstein, C., Garfinkel, D., Fuller, S., Flores, C., Peschon, J., Nester, E., and Gorden, M. (1984). *Proc. Natl. Acad. Sci. U.S.A.* **81,** 1728.

Lichenstein, C., Klee, H., Montoya, A., Garfinkel, D., Fuller, S., Flores, C., Nester, E., and Gorden, M. (1984) *J. Mol. Appl. Genet.* **2,** 354.

Marton, L., Wullems, G., Molendijk, L., and Schilperoort, R. (1979). *Nature (London)* **227,** 129–131.

Morelli, G., Nagy, F., Fraley, R., Rogers, S., and Chua, N.-H. (1985). *Nature (London)* **315,** 200.

Nagy, F., Morelli, G., Fraley, R., Rogers, S., and Chua, N.-H. (1986). *EMBO J.* **4,** 3063.

Sengupta-Gopalan, C., Renhert, W., Barker, R., Hall, T., and Kemp, J. (1985). *Proc. Natl. Acad. Sci. U.S.A.* (in press).

Southern, P., and Berg, P. (1982) *J. Mol. Appl. Genet.* **1,** 327–341.

Tempé, J., and Goldmann, A. (1982). *In* "Molecular Biology of Plant Tumors" (G. Kahl, and J. S. Schell, eds.), p. 427. Academic Press, New York.

Van Larebeke, N., Engler, G., Holsters, M., Van der Elsacker, S., Zaenen, I., Schilperoort, R., and Schell, J. (1974). *Nature (London)* **252,** 169–170.

Watson, B., Currier, T. C., Gordon, M. P., Chilton, M.-D., and Nester, E. W. (1975). *J. Bacteriol.* **123,** 225.

Willmitzer, L., DeBeuckeleer, M., Lemmers, M., Van Montagu, M., and Schell, J. (1980). *Nature (London)* **287,** 359–361.

Willmitzer, L., Simon, G., and Schell, J. (1982). *EMBO J.* **1,** 139.

Willmitzer, L., Dhaese, P., Schreier, P., Schmalenbach, W., Van Montagu, M., and Shell, J. (1983). *Cell* **32,** 1045.

Wullems, G., Molendiji, L., Ooms, G., and Schilperoort, R. (1981). *Proc. Natl. Acad. Sci. U.S.A.* **78,** 4344–4348.

Yadav, N., Postle, K., Saiki, R., Thomashow, M., and Chilton, M.-D. (1980). *Nature (London)* **287,** 458–461.

Yadav, N., Vanderleyden, J., Bennet, D., Barnes, W., and Chilton, M.-D. (1982). *Proc. Natl. Acad. Sci. U.S.A.* **79,** 6322–6326.

Zaenen, I., Van Larebeke, N., Teuchy, H., Van Montagu, M., and Schell, J. (1974). *J. Mol. Biol.* **86,** 109.

Zambryski, P., Depicker, A., Kruger, K., and Goodman, H. (1982) *J. Mol. Appl. Genet.* **1,** 361–370.

25

Genetic Engineering of Bioinsecticides

DAVID R. WILCOX,*
ANNAPUR G. SHIVAKUMAR,* BRIAN E. MELIN,†
MAHLON F. MILLER,‡ TERRY A. BENSON,*
CYNTHIA W. SCHOPP,* DIANNE CASUTO,*
GERARD J. GUNDLING,* TIMOTHY J. BOLLING,*
BRIAN B. SPEAR,§ AND J. LAWRENCE FOX*

** Department of Molecular Biology*
† Department of Entomology Research
‡ Department of Electron Microscopy
§ Agricultural Research and Development
Abbott Laboratories
Abbott Park, Illinois 60064

I. INTRODUCTION

Each year over $3 billion is spent worldwide in the control of insect pests. Microbial pesticide sales, which today represent less than 1% of that total, are estimated to reach 50% by the year 2000 (Klausner, 1984, re Environmental Protection Agency estimates). Major producers of chemical pesticides are examining microbials as a possible answer to concerns about the dumping of manufactured chemicals into the environment. Concerns about DDT and EDB and the recent tragic events in Bhopal (*Chemical and Engineering News,* 1985a, for review) have accelerated this research.

Microbials are a virtually untapped insecticide resource. Only 14 microorganisms have EPA approval for use as bioinsecticides: 6 bacteria, 4

PROTEIN ENGINEERING
Applications in Science,
Medicine, and Industry

viruses, 3 fungi, and 1 protozoan. The 14 microbials are formulated into only 100 products. Yet over 1000 microorganisms or microbial products are known to be effective against insects. The formulations possible are expected to be many times the number of microbials licensed. For example, 1600 chemicals are presently approved for pest control and these appear in 45,000 formulations (Malone, 1985).

Bioinsecticides, also called biorational insecticides, have a fundamental advantage over chemical pesticides in that the bioinsecticides have a narrower and therefore often "safer" species range. The microbial products can be formulated to target particular insect pests and leave the plants and other animals of an ecosystem unharmed. Of the thousands of known insect pests, only a few hundred are considered harmful or destructive enough to warrant control. Most of these harmful pests have pathogenic organisms associated with them (Paul, 1981).

II. *Bacillus thuringiensis kurstaki*

A. A Commercial Success

The dominant and most successful bioinsecticide today is a mixture of crystals and spores from the lepidopteran pathogen *Bacillus thuringiensis kurstaki* (*B.t.k.*). In this chapter we will discuss the *B.t.k.* product as a model bioinsecticide. The commercial success of the present *B.t.k.* product allows us to focus on the implications of genetically engineering the crystal-toxin genes and the probability of generating new, equally successful recombinant products.

In the Western Hemisphere and Europe, Abbott Laboratories (North Chicago, Illinois), Sandoz (Basel, Switzerland), and Solvay (Brussels, Belgium) are major producers of the *B.t.k.* pesticide. The Abbott Laboratories' product (Dipel) has been found effective on over 200 crops and against more than 55 pest lepidopteran species. *B.t.k.* has become the organism of choice for control of spruce budworm and the gypsy moth.

First the crystal/spore preparations are delivered to the surfaces of the plants to be protected. Then the toxic crystals are ingested by the feeding larvae and the crystal is solubilized in the alkaline larval midgut. The protoxin polypeptide is cleaved and activated by specific proteases to create a very effective gut toxin. The mode of action of the toxin is not fully understood (Lüthy and Ebersold, 1981, for review; Knowles *et al.*, 1984), but the effect on the larvae is clear. At lethal concentrations of the toxin, larval feeding stops within 10 min. Death finally results from dehydration, predation, or, in some species, septicemia resulting from germination of the *B.t.* spores in the larvae.

B. Advantages of *Bacillus thuringiensis* Pesticides

There are many advantages in using *B.t.* pesticides. The toxin is effective against a wide range of lepidopteran larval pests, yet there is no toxic effect on vertebrates (including humans, helpful insects, or the predators of the pests. Recent innovative technologies in fermentation and manufacturing allow *B.t.* to be cost-competitive with chemical pesticides. Although *B.t.* control is not instantaneous, control of the pests is rapid compared to that observed with most viral and fungal pesticides. The products are stable in storage, especially as dried powders. No natural resistance to *B.t.* has been reported during normal field applications. Another advantage of *B.t.* is that it can easily be used in an integrated management program along with chemical pesticides. Finally, because *B.t.* is considered safe, these pesticides have federal approval for use up to the day of harvest.

III. CLONING FOR A BETTER *Bacillus thuringiensis* PRODUCT

A. Why Clone the *Bacillus thuringiensis* Toxin Gene?

The *B.t.* crystal is composed primarily of large polypeptides encoded in some strains by multiple, but homologous, plasmid-borne genes (discussed below). This makes cloning of the toxin genes into *Escherichia coli* or *Bacillus* very attractive. The hope is that the cloned genes can then be manipulated so as to increase the toxin yields or expand the species range.

Several approaches can be pursued to increase the species range. In the course of such research, toxin yields may also increase. One approach is to develop a broad-range bioinsecticide. This can be useful where helpful insects are still left unharmed. In contrast, a narrow-range bioinsecticide can be useful where specific insects are targeted that previously were only marginally susceptible to the crystal toxin. The new toxins can then be formulated with existing preparations to expand the product line manyfold.

The species range can also be increased by transforming the cloned toxin genes into non-*B.t.* microorganisms that can survive for short periods of time in environmental niches difficult to reach by present spraying techniques (Beardsley, 1984; Sterling, 1985). A similar approach places the toxin gene in a nonviable strain of *Pseudomonas,* which then acts as a protective coat for the toxin (*Chemical and Engineering News,* 1985b). Soluble (i.e., noncrystalline) toxins may also be successful if there are parallel advances in manufacturing and delivery technology to ensure persistence of the polypeptides in the field.

Finally, transfer of crystal-toxin genes into strains containing a different pesticidal toxin has been reported for a *B.t. thuringiensis* gene inserted into a *B.t. israelensis* strain (Klier *et al.,* 1983). Both the *B.t.t.* and the *B.t.i.* toxins were active. The hope is that such constructions will show synergistic toxin interactions that provide activity against harmful insects previously unaffected by the two toxins when delivered alone or together in formulations.

B. Success Is Pest Control in the Field

To improve on the *B.t.* product, one must realize that success in manufacturing and delivering the toxin to the field are integrally related to the mode of action of the crystal and the biology of the insect pest. Real success is control of the larval pests in the field. Improvements in the toxin must persist through several levels of research, development, and production.

After fermentation, the crystal/spore/medium preparation is then dried and formulated for field delivery. Delivery in the field depends in part on the plant to be protected and/or the larval pest to be controlled. For example, ground spraying can be successful on smaller farm acreage, whereas forest protection requires large-scale spraying from the air. The carriers that accompany the crystal/spore mixture affect spraying efficiency and persistence in nature and represent a technology as important to the success of the pesticide as are fermentation and manufacturing.

C. Critical Considerations for Industrial Application

Cloning a better *B.t.* product therefore involves several critical considerations. First, the new product's improved toxicity must translate to the field. Unfortunately, it can take years before the extent of this improvement can be measured. This is true even of nonrecombinant products.

The time needed to obtain federal approval of recombinant products may add considerably to the time before a product can be considered successful.

Second, it is economically desirable to integrate and maintain as much as possible the wealth of fermentation, manufacturing, and formulation technology accumulated over the past 20 or more years in the production of the new product. Improvements in yield during fermentation, for example, can easily be lost if difficulties arise in formulation or delivery to the field.

Third, *B.t.* bioinsecticides are safe. The *B.t.k.* crystal is considered nontoxic to humans, pets, livestock, crops, and helpful insects. It is advantageous to have federal approval for use of these pesticides up to the day of harvest or where chemical pesticides are not as acceptable. If experiments are not designed to preserve these properties, genetic manipulation of the crystal peptide may easily eliminate this advantage.

D. An Integrated Research Effort Is Necessary

When these critical considerations are taken into account, the genetic engineering of an improved *B.t.* pesticide no longer appears a straightforward task. The gene is cloned into a foreign organism such as *E. coli,* manipulated, and returned to some desired host for scaleup. The gene now resides on a recombinant vector, not its natural plasmid. After genetic manipulation, the polypeptide produced will differ in some way from the original. It may not even crystallize. Furthermore, multiple toxin genes may be interacting in the present production strains to produce the successful products we know today. Our data suggest that multiple toxin genes interact in the cells of a bacterial culture to produce heterogeneous crystal preparations, and these may be less effective if some genes are not present. Therefore, cloning a single *B.t.* toxin gene is only the beginning for understanding the genetics of *B.t.* crystal formation and toxicity.

The problem should now be placed in perspective. We must admit that we do not adequately understand the genetics of our present strains. We do not adequately understand the biochemistry of the present products or the mode of action of the present toxins. Yet we hope to improve on the product through a molecular biology approach. This is *not* to say that new products are impossible, but rather that the success of a *B.t.* cloning program depends on an integrated research effort at all levels of production.

IV. BIOTOXICITY ASSAYS AS A BARRIER

A. Testing Many Target Species

In any industrial cloning program, the first stage is often cloning of the desired gene into an appropriate host such as *E. coli*. Increasing the expression level and scaleup of the product follow. For many cloning projects these stages bring the cloning effort well along toward a commercial product. For a bioinsecticide this is just the beginning. The bioassay alone can be a formidable barrier. Insect toxicity assays are notoriously sensitive to the assay conditions. Internal standards are essential, and multiple tests are averaged together. Since live organisms are being tested, and, in the case of *B.t.k.*, small larvae are being treated, the tests are very labor-intensive (i.e., see Dulmage *et al.*, 1971, 1976). At the very least, costly breeding colonies of cabbage loopers (*Trichoplusia ni*) must be maintained, since this is the industrywide test organism. When newly targeted pest species are to be tested, the cost increases sharply. For many lepidopteran larvae, for example, the assays are poorly defined, and for many more the larvae cannot be maintained economically in the laboratory. An enzyme-linked immunosorbent assay (ELISA) (Smith and Ulrich, 1983) or a cell culture assay (Johnson and Davidson, 1984) can be helpful, but each of these assays is another step removed from the reality of field testing.

B. Bioassays of Clone Extracts

Fortunately, in B. *thuringiensis* the major crystal-toxin genes have been identified by cloning. At least six laboratories have reported cloning the structural crystal-toxin (delta-endotoxin) gene. The genes have been cloned from three different serovars: *B.t. kurstaki* (Schnepf and Whiteley, 1981; Held *et al.*, 1982; Adang *et al.*, 1985; McLinden *et al.*, 1985), *B.t. thuringiensis* (Klier *et al.*, 1982), and *B.t. sotto* (Shibano *et al.*, 1985). Restriction maps of the genes cloned earlier suggested that these clones were distinctly different from each other, though all of them reportedly produced polypeptides of similar size that immunoreacted with crystal polypeptide antibody and were the size of polypeptides found in the crystal. Each clone was found to be toxic to a lepidopteran—at least qualitatively. Recently, new data have been reported (Wabiko *et al.*, 1985; Schnepf *et al.*, 1985) for the *B.t. thuringiensis* gene which support a greater similarity than first suggested between this gene and the gene of clone ES12, containing the *B.t.k.* gene isolated by Schnepf and Whiteley.

The *B.t. sotto* gene appears nearly identical to the gene of ES12, based on a comparison of the published nucleotide sequences. It is important to note that none of the cloned genes can be compared on the basis of their toxin phenotype. Meaningful quantitative measurements or bioassays with multiple pest species have not been reported for the cloned toxic genes. Unfortunately, the toxicity of the parent *B.t.* strain cannot substitute for clone biotoxicity data if multiple toxin genes are expressing in the parent strain. Our preliminary surveys (unpublished results) suggest that multiple toxin genes in *B.t.* strains is the rule and not the exception.

In the Western Hemisphere, most *B.t.* products for lepidopterans are from *B.t.k.* Therefore the *B.t.k.* cloning data, especially data relating to strain HD-1 are of particular interest from a commercial standpoint. The two cloned *B.t.k.* genes from HD-1 or a derivative strain have been localized on plasmids (Kronstad *et al.*, 1983) or a plasmid and the chromosome (Held *et al.*, 1982). Our sequence analysis suggests that the genes we cloned from the Dipel strain are at least partly homologous to the gene of ES12 isolated by Schnepf and Whiteley. We see no homology between our cloned genes and the second reported *B.t.k.* gene (we thank Drs. Gary Held and Lee Bulla for use of their clone). In our crystal-minus strains only the genes sharing homology to the gene of ES12 are lost, as revealed by Southern hybridization. Genes that are homologous to the gene of ES12 therefore appear to be important, if not essential, for the formation of the crystal, and they probably code for major crystal components.

It is interesting to note that ES12 was identified by immunoreaction with antibody made against the 135,000-dalton polypeptide isolated by sodium dodecyl sulfate–polyacrylamide gel electrophoresis (SDS–PAGE). The second reported *B.t.k.* clone (Held *et al.*, 1982) produced polypeptides that immunoreacted with antibody made against total solubilized crystal. This suggests that the second gene may code for a component of the crystal that is much more immunoreactive in solubilized total crystal preparations, though perhaps a minor component of the crystal. This conclusion may be misleading, however, since a minor crystal component may be a critical component when the total toxic spectrum of a crystal preparation is considered.

In our laboratories, expression levels for the second cloned gene in *E. coli* were very low and toxicity for *T. ni* was not recorded even at the higher concentrations. However, this brings out an important point: bioassays are not sensitive to very low levels of toxin. In testing *B.t.* products, increasing the concentration of the preparation generally brings the killing levels within the range where quantitative LC_{50} determinations can be made. However, *B.t.* preparations are relatively pure crystal/spore preparations compared to *E. coli* clone extracts. Crystal-toxin concentrations in *E. coli* clone extracts are lower than in *B.t.*. Where low-level clone expression means no crystal formation, concentrating an *E. coli* clone extract is not equivalent to raising the concentration of crystals and spores in a *B.t.* extract. Larval bioassays for *B.t.* were developed for and internally standardized to intact crystals which are very stable. The stability of noncrystalline *B.t.* toxins has not been carefully tested, but breakdown products do occur. It is likely that new assay standards and conditions will have to be established before noncrystalline *B.t.* toxin in *E. coli* can be assayed as reliably as *B.t.* crystalline preparations can now be assayed.

Even where an LC_{50} can be calculated for a clone extract, the results are still only qualitative if the extracts are not adjusted for the percentage of toxin protein in the extract. For preparations from similar *B.t.* strains the percentage of toxin is not a major variable since the cultures are fermented until spores and crystals are free in the culture and other cell proteins are largely degraded. In *E. coli* and *Bacillus subtilis* cultures, however, the percentage of toxin protein compared to total protein is much lower than in *B.t.*, even when crystals are made and/or older cultures are assayed. In addition, the expression levels of different toxin genes are not always equal, and where some degradation is observed, only very crude estimates of the percentage of toxin protein are possible.

It is important to keep in mind that toxin gene expression levels in *E. coli* and *B. subtilis* may bear no resemblance to relative expression in *B.t.* Not only is expression vegetative in *E. coli* and *B. subtilis* (discussed below) and not tied to sporulation as it is in *B.t.*, but also in *E. coli* transcription initiates from a fortuitous start site that *B.t.* does not recognize (Wong *et al.*, 1983).

The test organism is also important in interpreting biotoxicity results. For Dipel formulations we find that *Manduca sexta* (tobacco hornworm) is very easily killed. *Trichoplusia ni, Heliothis* sp., and *Spodoptera* sp., in that order, are increasingly difficult to kill. Our data suggest that different cloned genes can have differential toxicity to different hosts. Therefore, if a cloned gene is tested against a less susceptible host, a comparison with other cloned genes is of limited value until many different host species are tested.

V. THE CRYSTAL-TOXIN GENES OF *Bacillus thuringiensis kurstaki*

A. The Three Genes of the Dipel Strain

We have cloned three crystal-toxin genes from the present Dipel production strain (Fig. 1). Two genes (the 6.6 gene, clone HBHE, and the 5.3 gene, clone pC2; gene names are explained below) were detected in a DNA library by immunoreaction with antibody prepared against full-length polypeptide isolated from an SDS–PAGE gel. The remaining gene (gene 4.5 : clone pAN45) was cloned by homology to the first two genes. Each gene has been isolated on pUC vectors in *E. coli* or pBD64 in *B. subtilis*. The clones begin upstream at the *Hin*cII site and extend to an appropriate site downstream including all translation and transcription termination signals. The three cloned genes are homologous over much of their length, though they differ in their restriction patterns in the carboxyl half of each gene.

The three genes can be distinguished in a preparation by Southern hybridization. If a 700-bp *Eco*RI fragment from the conserved amino end of any of the genes is hybridized to total or plasmid DNA digested with the *Hin*dIII restriction enzyme, the three genes can be identified as three different-sized bands on a Southern blot. Toxin genes found in strains derived from the HD-1 strain can be referred to as the 6.6 gene, the 5.3 gene, and the 4.5 gene, since the three *Hin*dIII bands measure 6.6, 5.3, and 4.5 kb in length, respectively. Although these *Hin*dIII fragments do not contain the entire gene, they do serve as useful markers for the genes when analyzing a *B.t.* strain related to the present commercial strains. Each gene does have an internal restriction pattern that is diagnostic; however, Southern hybridization with these restriction cuts is cumbersome. We see no conflicting results when enzymes other than *Hin*dIII are used for *B.t.k.* strain surveys. Similar *Hin*dIII hybridizations were used in an earlier survey (Kronstad *et al.*, 1983). Therefore, we have used the *Hin*dIII patterns in this chapter for consistency with the earlier work.

B. History and Stability of the Dipel Strain

We surveyed Dipel preparations produced from 1975 to 1984. A representative spore/crystal preparation was spread on L-agar plates and isolated colonies were inspected for the presence of the three genes. Without exception, the three-band pattern appeared on the Southern blot for the isolates (Fig. 2A). The survey led to two observations. First, an earlier

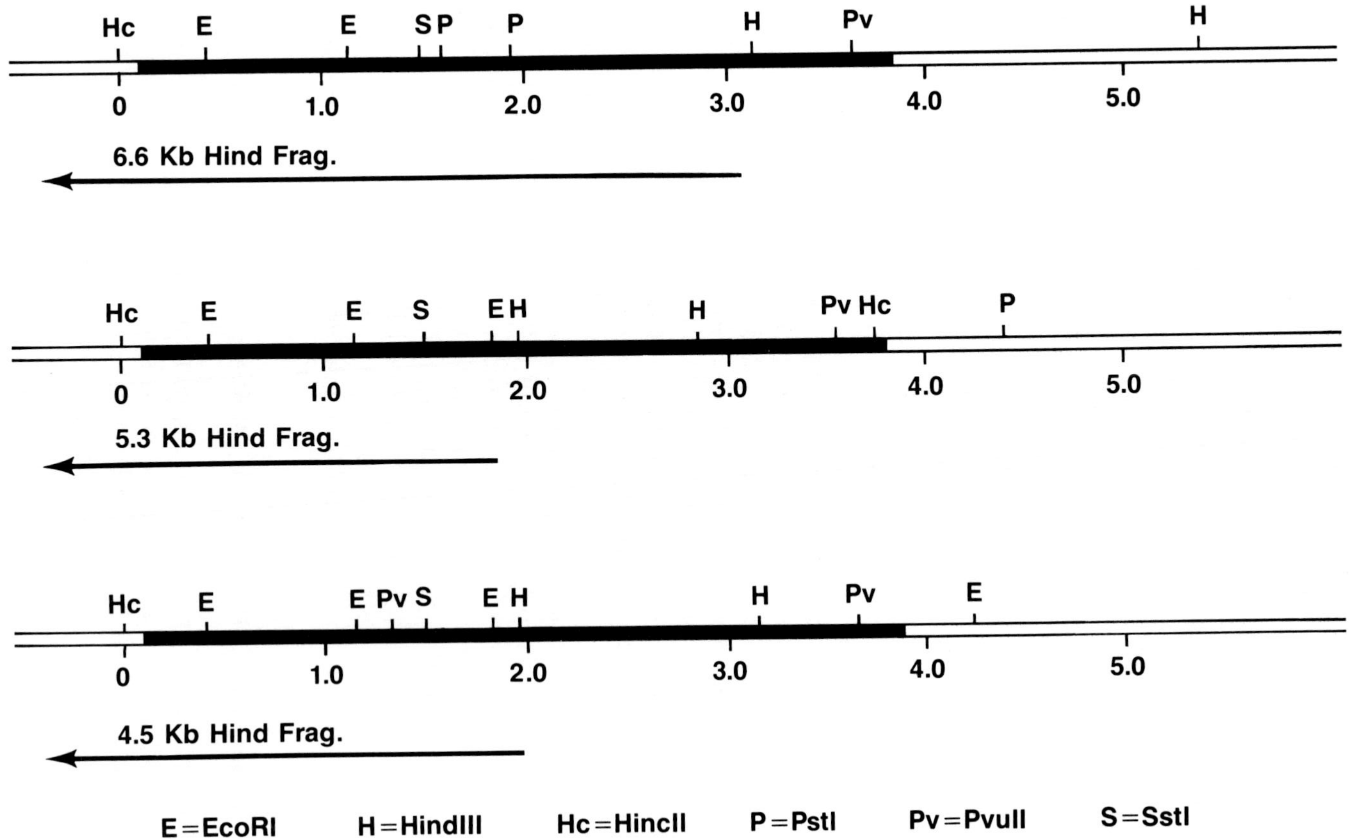
Hc
E
E
S
P
P
H
Pv
H
0
1.0
2.0
3.0
4.0
5.0
6.6 Kb Hind Frag.
Hc
E
E
S
E
H
H
Pv
Hc
P
0
1.0
2.0
3.0
4.0
5.0
5.3 Kb Hind Frag.
Hc
E
E
Pv
S
E
H
H
Pv
E
0
1.0
2.0
3.0
4.0
5.0
4.5 Kb Hind Frag.
E=EcoRI
H=HindIII
Hc=HincII
P=PstI
Pv=PvuII
S=SstI

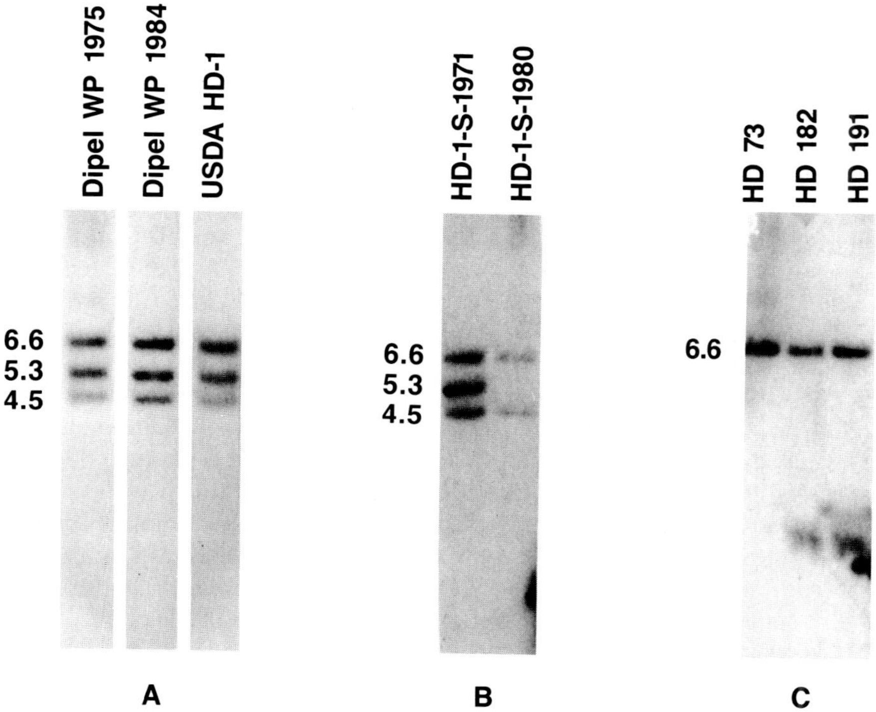

Fig. 2. Southern blot hybridization band patterns for the 700-bp *Eco*RI internal gene probe hybridized to *Hin*dIII cut total *B.t.k.* DNA isolated from strains discussed in the text. Hybridizations were at 65°C with final washes in .3*M* NaCl, .3*M* Na Citrate (pH 7.0).

strain isolated from a Dipel preparation and found to contain only two genes (Kronstad *et al.,* 1983) does not represent the Dipel production strain used throughout this 10-year period, which includes the Dipel strain used today. The HD-1 strain available from the USDA also shows the three-band pattern. Interestingly, a strain we isolated from a vial of *B.t.* powder found at the USDA and dated May 31, 1968, also showed a three-band pattern. This date is after Howard Dulmage reportedly discovered HD-1 and prior to commercialization of the strain (Dulmage, 1970, 1973). Unfortunately we do not know whether this was the discovery prepara-

---

Fig. 1. Restriction site maps of three delta-endotoxin structural genes cloned from the Dipel strain. Each gene has been subcloned into pUC vectors, beginning upstream at the *Hinc*II site and extending downstream to an appropriate site including all transcription termination signals. Shaded areas represent coding regions. The *Hin*dIII fragments discussed in the text extend from inside the genes to sites upstream.

tion, but it does suggest that a three-gene HD-1 strain has existed since before HD-1 was commercialized.

Second, the survey suggested that the industrial fermentation process does not lead to the loss of toxin genes since all survey isolates were from postfermentation dried powders. Only a few isolates were tested for each production year, but all isolates showed the same three-band pattern. We are continuing this survey, using gene-specific probes isolated from the cloned genes. This will allow us to test many more isolates by colony hybridization. At present, however, there is no indication that gene loss occurs in commercial fermentation of *B.t.k.*

This is not to say that toxin genes cannot be lost from a *B.t.k.* strain. Indeed, isolates with either two genes (6.6 and 4.5) or one gene (5.3) can be found after heat treatment in laboratory cultures for a few days at 42°C (data not shown). In fact, all three genes can be lost by this method, producing a crystal-minus strain. The major crystal-toxin genes in the *B.t. kurstaki* serovar (the Dipel strain is an example) are all plasmid-borne (Kronstad *et al.*, 1983, and our laboratory). The heat treatment apparently induces loss of the toxin genes by loss of the large plasmids that the genes are on. The instability of these plasmids under laboratory conditions is well documented (Gonzalez *et al.*, 1981). Fortunately, this instability does not appear to be induced in fermentation broths, which are generally held at 30–32°C.

An important observation should be made here. The *B.t.* strains maintained in laboratories run the risk of plasmid loss or rearrangement. The frequency of plasmid changes is unknown. However, transposonlike qualities have been related to the toxin gene in two studies (Kronstad and Whiteley, 1984; Lereclus *et al.*, 1984). Strains can be identified by (1) specific gene hybridization patterns, (2) flagellar antigen reactions, or (3) biotoxicity to a specified test insect. The first method requires a cloned gene and Southern analysis; the second requires antibody from many different *B.t.* serovars; and the third requires maintaining large colonies of test larvae. All three methods require time and cost, and none of the methods is foolproof. Lyophilized spore preparations are stable, but strains in continual use in a laboratory or transferred between laboratories should be rechecked frequently.

C. Genetics of the Industry-Standard Strains

Two industrywide standards have been used since the production of Dipel began. We analyzed Southern hybridization band patterns of iso-

TABLE I

LC_{50} Data from Bioassay—Third Instar Larvae

Sample	Mortality 3-day	5-day
	Spodoptera exigua	
HD-1-S-1971	45.8	35.5
HD-1-S-1980	78.0 (69.3)[a]	64.8 (57.6)
	Trichoplusia ni	
HD-1-S-1971	28.7	16.0
HD-1-S-1980	34.4 (30.6)	19.6 (17.4)
LC_{50} Ratios[b]		
Spodoptera	3-day 1971/1980 = 45.8/69.3 = 0.66	
	5-day 1971/1980 = 35.5/57.6 = 0.62	
Trichoplusia	3-day 1971/1980 = 28.7/30.6 = 0.94	
	5-day 1971/1980 = 16.0/17.4 = 0.92	

[a] Values in parentheses are corrected for standard potency, using the HD-1-S-1971 LC_{50} values for *T. ni* as the standard. Units are micrograms of preparation per milliliter of medium.

[b] Corrected for standard potency.

lated colonies from each of the standards. The 1971 standard (HD-1-S-1971) shows the typical HD-1 pattern of three bands. However, the most recent standard (HD-1-S-1980) shows only two bands: 6.6 and 4.5 kb (Fig. 2B). Earlier testing at Abbott and elsewhere suggested that the standards may differ in their toxicity to different targeted pests. We therefore submitted the two standards to biotoxicity assays (Table I). Careful analysis was carried out with an appropriate dilution series of the preparations to establish quantitative LC_{50} values for each standard. Each sample was measured in triplicate, using 100 insects per test. The lower the LC_{50} value measured in the laboratory, the more effective the toxin is expected to be in the field. Toxicity levels for *Spodoptera exiqua* (beet army worm) were significantly different between standards. The 1980 standard ("two gene") was more than one-third less active against *S. exiqua* than the 1971 standard ("three gene"). The activity against *T. ni* (cabbage looper) also varied between the two standards, but the differences in activity were much smaller for *T. ni* than for *S. exiqua*. The critical observation from our results is that the amount of change in activity, when the 5.3 gene is absent, is significantly different when different pests are tested.

D. Multiple Genes: Differential Toxicity

We conclude from these data that multiple toxin genes are present in the *B.t.k.* strains and at least some of the genes contribute differentially to the total toxicity spectrum of a given strain. This is supported by the earlier report (Kronstad *et al.,* 1983) that strain HD73 contains only the 6.6 gene (i.e., a single 6.6-kb *Hin*dIII band lights up on the Southern blot autoradiogram). Crystals of this strain are not immunologically like those of HD-1 (Krywienczyk, 1977). In addition, HD73 is much more effective against *Heliothis virescens* (tobacco budworm) than *T. ni,* whereas HD-1 is more effective against *T. ni* than *H. virescens* (Krywienczyk *et al.,* 1973). We have confirmed that our copy of strain HD73 also has only the 6.6-kb *Hin*dIII fragment appearing on an autoradiogram (Fig. 2C). Also, HD162, HD182, HD187, and HD191 show only the presence of the 6.6 gene. These strains are all k73 immunological crystal types (Dulmage *et al.,* 1982). When the polypeptides of HD73 were visualized on an SDS–PAGE gel, only one polypeptide of 135,000 daltons was seen. This corresponds to the molecular weight of the gene product produced by the cloned 6.6 gene. It seems likely that HD73, and probably other k73 crystal-type strains, produce a relatively pure crystalline product from a single structural toxin gene (6.6).

Both the immunological and toxicological phenotypes of the HD73 strain 6.6 gene differ from the combined phenotypes of the HD-1 strain genes, even though the 6.6 gene is present in both strains. This suggests that the 4.5 and/or 5.3 genes affect the toxicological and immunological properties of the Dipel strains in ways that differ from the effect of the 6.6 gene. Since activity against *Helothis* sp. is lower when the 6.6 gene is present with the 4.5 and 5.3 genes, this is an example of an antagonistic interaction between toxin genes for *Heliothis* toxicity. This may not be a direct antagonistic interaction between polypeptides, or for sites on the gut lining, but rather competition for sites in the heterogeneous crystal. In contrast, *T. ni* killing is somewhat greater for the three-gene strain, HD-1, than the single-gene strain, HD73 (Krywienczyk *et al.,* 1978). This illustrates the critical need for biotoxicity assays against many different test organisms.

It should be noted that the HD-1 strain was chosen for commercialization on the basis of preliminary data suggesting that the strain had a greater effective species range. In addition, however, the strain is at least 16 times more effective against a single lepidopteran (*T. ni*) than the *B.t.* strains of the 1960s. Thus both the species range and potency have been increased in *B.t.* products with the discovery of the HD-1 strain (Dulmage, 1970, 1973). We do not know how many, if any, of the HD-1 toxin genes were present in the production strains of the 1960s.

E. Expression of the Cloned Genes

Although the 4.5 gene isolated by Dr. Whiteley's group was not isolated from the Dipel strain, the restriction patterns do suggest that the two 4.5 genes are very similar. Since Whiteley has reported extensively on this gene (Schnepf and Whiteley, 1981, 1985, and see references therein), we will focus here on the remaining two genes.

Qualitatively, products of both the 5.3 and 6.6 genes cloned in *E. coli* or *B. subtilis* are toxic to *T. ni*. Quantitative measurements of the toxicity of the gene product are in progress for both *T. ni* and *S. exiqua*. Preliminary results of the LC_{50} measurements for the 6.6 clone in *E. coli* show much greater toxicity for *T. ni* than the 1960s *B.t.* product strains, though not as great toxicity as the Dipel strain. However, the arguments made earlier against making comparisons of *E. coli* to *B.t.* must apply to these preliminary data, which are not corrected for percentage of toxin protein in the preparations. We know that degradation products appear regularly in the *E. coli* clone preparations. It is not clear if quantitative biotoxicity tests of *E. coli* extracts can ever be made, when corrected for percentage of toxin, especially when estimated from densitometry scans of stained SDS–PAGE gels. Although we have no data to indicate that these toxicity levels could be easily translated to the field, it is clear that the toxin is still very active when produced by foreign hosts.

The clones produce a peptide of either 130,000 daltons (5.3 gene) or 135,000 daltons (4.5 and 6.6 genes) that can be detected by Western blotting (Fig. 3). A degradation product of approximately 60,000 daltons often appears in *E. coli* preparations. Isolates of the Dipel strain or the USDA HD-1 strain show peptides of both 135,000 and 130,000 daltons being produced. This suggests that at least two of the known genes express polypeptides in an HD-1 culture.

In *E. coli,* expression of the clones occurs early and throughout vegetative growth. In *B. subtilis,* expression is also vegetative. The cloned gene continues to be expressed at high levels vegetatively in a sporulation-minus mutant (SpoHI) (Shivakumar *et al.,* 1986). This is surprising since both *B. subtilis* and *B.t.* are gram-positive sporulating species and in *B.t.* the crystal-toxin is produced only at 3–4 hr into sporulation (Andrews *et al.,* 1981). We suspect that the vegetative polymerase of *B. subtilis* recognizes the transcription sequence normally recognized only during sporulation by the *B.t.* polymerases. Or perhaps, as in *E. coli* (Wong *et al.,* 1983), *B. subtilis* recognizes a fortuitous vegetative sequence that *B.t.* ignores. Since two transcription start sites are reportedly recognized in *B.t.* with two distinct recognition sequences, it could be that the toxin gene is also being expressed during sporulation in wild-type *B. subtilis*. The vegeta-

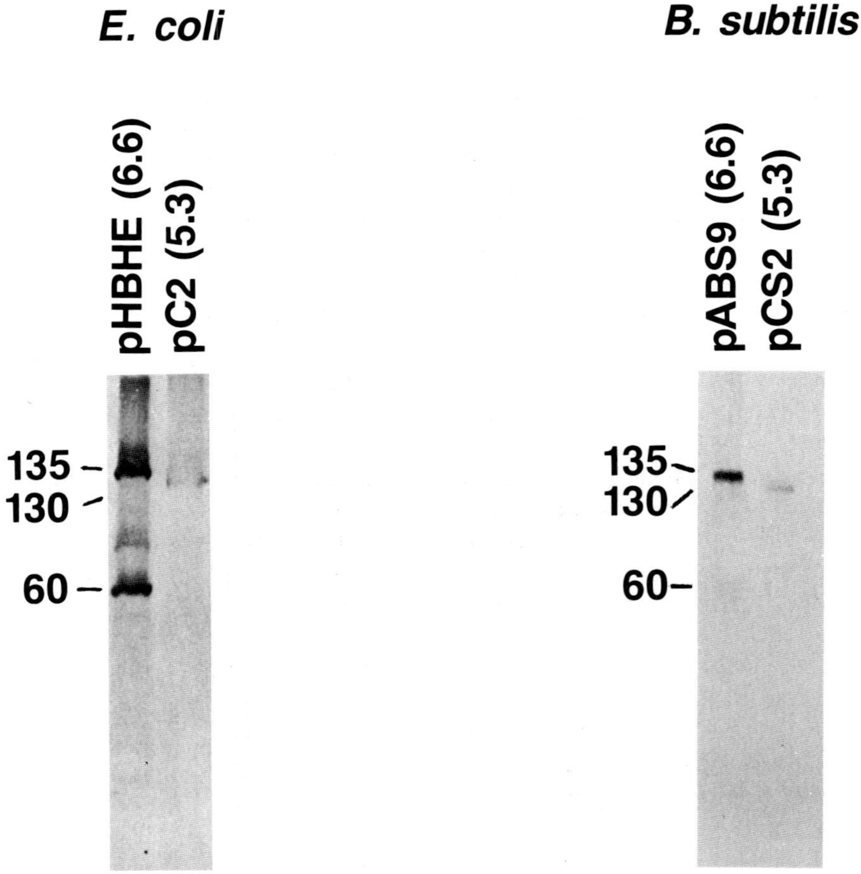

Fig. 3. Western blot analysis of two genes cloned from the Dipel strain. Immunoreactivity was detected by using rabbit antibody prepared against full-sized toxin polypeptide isolated from an SDS–PAGE gel. The 60-kilodalton band is believed to be a breakdown product. Clones pABS9 and pCS2 contain insert DNA from pHBHE and pC2, respectively, transferred from a pUC vector to pBD64 and transformed into *B. subtilis*.

tive expression may be so extensive that sporulation stage expression from another start site is being masked.

Electron micrographs of *E. coli* containing the 6.6 toxin gene reveal angular or amorphous granules with a crystalline lattice similar to the *B.t.* crystal lattice (Fig. 4A). Granules without the crystalline lattice are seen for the 5.3 gene (Fig. 4B). In *B. subtilis,* the crystalline lattice is observed for both the 6.6 and 5.3 genes and now a typical *B.t.* -like bipyramidal crystal is seen occupying a large portion of the *B. subtilis* cell (Figs 4C and

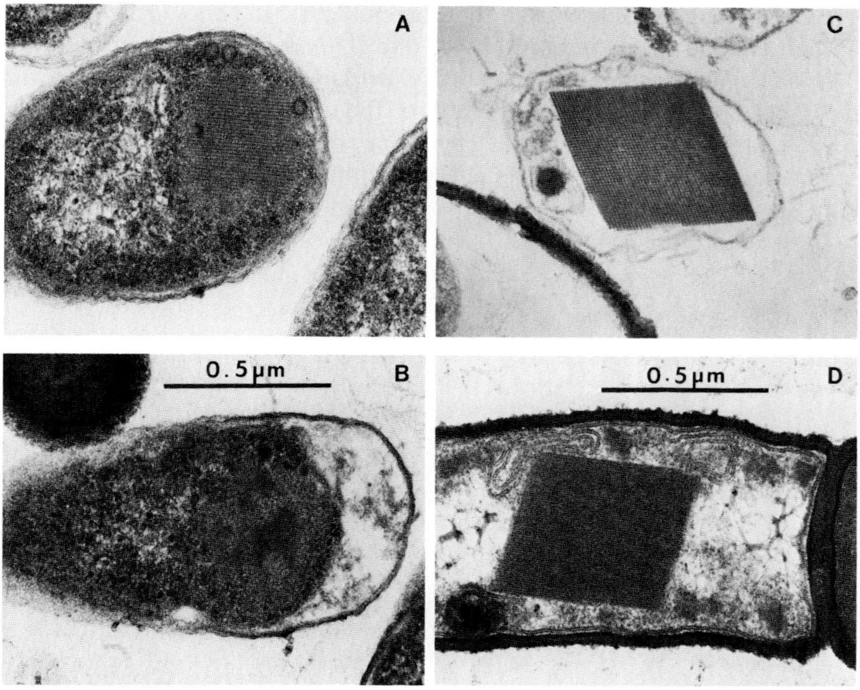

Fig. 4. Electron micrographs of two delta-endotoxin genes of the Dipel strain expressing in *E. coli* (A, 6.6 gene; B, 5.3 gene) and in *B. subtilis* (C, 6.6 gene; D, 5.3 gene).

4D). In both *E. coli* and *B. subtilis* crystal formation appears to be more prevalent in cells in the late log phase and stationary cells, suggesting that the polypeptide is accumulated over time.

VI. CONCLUSION

We find that a successful commercial *B.t.k.* strain (the Dipel strain) contains multiple toxin genes that we believe may interact to produce the toxic crystalline product. Therefore it is inappropriate to describe the *B.t.k.* crystal as though it were always encoded by a single toxin gene or encoded by multiple genes with similar contributions to crystal composition, immunology, or toxicity. This is clearly not true for the successful commercial strains. The plasmid-borne genes of the Dipel strain are all capable of expression and should not be considered "pseudogenes." In addition, at least some of the genes appear to contribute to the strain's total toxic range by differential toxicity to specific lepidopteran pests.

Finally, we have discovered that the cloned toxin genes are expressing at high enough levels in both *E. coli* and *B. subtilis* to form visible granules or crystals. This high level of expression allows us to study in a meaningful way, in *E. coli* and *Bacillus,* the individual components of the *B.t.k.* crystal as separate homogeneous crystalline forms amenable to the present quantitative biotoxicity assays. Multiple gene constructs are helping us to understand additive, synergistic, or even antagonistic effects of toxin gene interaction.

ACKNOWLEDGMENTS

We thank Charlotte Toyra for typing and editing the manuscript, Sharon Kister for assisting with DNA sequencing, and Robert Smith for helpful comments.

REFERENCES

Adang, M. J., Staver, M. J., Rocheleau, T. A., Leighton, J., Barker, R. F., and Thompson, D. V. (1985). *Gene* **36,** 289–300.

Andrews, R. E., Bechtel, D. B., Campbell, B. S., Davidson, L. I., and Bulla, L. A. (1981). *In* "Sporulation and Germination" (H. S. Levinson, A. L. Sonenshein, and D. J. Tipper, eds.), pp. 174–177. Am. Soc. Microbiol., Washington, D. C.

Beardsley, T. (1984). *Nature (London)* **312,** 686.

Chemical and Engineering News (1985a). Feb. 11, pp. 1–78.

Chemical and Engineering News (1985b). June 17, p. 29.

Dulmage, H. T. (1970). *J. Invertebr. Pathol.* **15,** 232–239.

Dulmage, H. T. (1973). *Bull. Entomol. Soc. Am.* **19,** 200–202.

Dulmage, H. T., Boening, O. P., Rehnborg, C. S., and Hansen, G. D. (1971). *J. Invertebr. Pathol.* **18,** 240–245.

Dulmage, H. T., Martinez, A. J., and Pena, T. (1976). *U.S. Agric. Res. Serv., ARS, Tech. Bull.* **1528.**

Dulmage, H. T., Beegle, C. C., deBarjac, H., Reich, D., Donaldson, G., Krywienczyk, J. (1982). *U.S. Agric. Res. Serv., South. Reg. T Rep., ARS-S* **ARM-S-30.**

Gonzalez, J. M., Dulmage, H. T., and Carlton, B. C. (1981). *Plasmid* **5,** 351–365.

Held, G. A., Bulla, L. A., Ferrari, E., Hoch, J., Aronson, A., and Minnich, S. A. (1982). *Proc. Natl. Acad. Sci. U.S.A.* **79,** 6065–6069.

Johnson, D. E., and Davidson, L. I. (1984). *In Vitro* **20,** 66–70.

Klausner, A. (1984). *Bio/Technology* **2,** 408–419.

Klier, A., Fargette, F., Ribier, J., and Rapoport, G. (1982). *EMBO J.* **1,** 791–799.

Klier, A., Bourgouin, C., and Rapoport, G. (1983). *Mol. Gen. Genet.* **191,** 257–262.

Knowles, B. H., Thomas, W. E., and Ellar, D. J. (1984). *FEBS Lett.* **168,** 197–202.

Kronstad, J. W., and Whiteley, H. R. (1984). *J. Bacteriol.* **160,** 95–102.

Kronstad, J. W., Schnepf, H. E., and Whiteley, H. R. (1983). *J. Bacteriol.* **154,** 419–428.

Krywienczyk, J. (1977). *For. Tech. Rep.* [*Can. For. Serv.*] **IP-X-12.**

Krywienczyk, J., Dulmage, H. T., and Fast, P. G. (1978). *J. Invertebr. Pathol.* **31,** 372–375.

Lereclus, D., Ribier, J., Klier, A., Menou, G., and Lecadet, M.-M. (1984) *EMBO J.* **3,** 2561–2567.

Lüthy, P., and Ebersold, H. R. (1981). *In* "Pathogenesis of Invertebrate Microbial Diseases" (E. Davidson, ed.), Chapter 9. Allanheld-Osmun, Montrose, New Jersey.
McLinden, J. H., Sabourin, J. R., Clark, B. D., Gensler, D. R., Workman, W. E., and Dean, D. H. (1985). *Appl. Environ. Microbiol.* **50,** 623–628.
Malone, J. (1985). Remarks before The 1985 Washington Conference on Biotechnology, Arlington, Virginia.
Paul, J. K., ed (1981). "Genetic Engineering Applications for Industry," pp. 21–22. Noyes Data Corp., Park Ridge, New Jersey.
Schnepf, H. E., and Whiteley, H. R. (1981). *Proc. Natl. Acad. Sci. U.S.A.* **78,** 2893–2897.
Schnepf, H. E., and Whiteley, H. R. (1985). *J. Biol. Chem.* **260,** 6273–6280.
Schnepf, H. E., Wong, H. C., and Whiteley, H. R. (1985). *J. Biol. Chem.* **260,** 6264–6272.
Shibano, Y., Yamagata, A., Nakamura, N., Iizuka, T., Sugiasaki, H., and Takanami, M. (1985). *Gene* **34,** 243–251.
Shivakumar, A. G., Gundling, G. T., Benson, T. A., Casuto, D., Miller, M. F., and Spear, B. B. (1986). *J. Bacteriol.* **166,** 194–204.
Smith, R. A., and Ulrich, J. T. (1983). *Appl. Environ. Microbiol.* **45,** 586–590.
Sterling, J. (1985). *Gen. Eng. News* **5,** 1.
Wabiko, H., Held, G. A., and Bulla, L. A. (1985). *App. Environ. Microbiol.* **49,** 706–708.
Wong, H. C., Schnepf, H. E., and Whiteley, H. R. (1983). *J. Biol. Chem.* **258,** 1960–1967.

Index

T

V

X